Springer Series in
OPTICAL SCIENCES 83

founded by H.K.V. Lotsch

Springer
Berlin
Heidelberg
New York
Barcelona
Hong Kong
London
Milan
Paris
Singapore
Tokyo

Springer Series in

OPTICAL SCIENCES

The Springer Series in Optical Sciences, under the leadership of Editor-in-Chief *William T. Rhodes*, Georgia Institute of Technology, USA, and Georgia Tech Lorraine, France, provides an expanding selection of research monographs in all major areas of optics: lasers and quantum optics, ultrafast phenomena, optical spectroscopy techniques, optoelectronics, information optics, applied laser technology, industrial applications, and other topics of contemporary interest.
With this broad coverage of topics, the series is of use to all research scientists and engineers who need up-to-date reference books.

The editors encourage prospective authors to correspond with them in advance of submitting a manuscript. Submission of manuscripts should be made to the Editor-in-Chief or one of the Editors. See also http://www.springer.de/phys/books/optical_science/os.htm

Our book tries to stay focused on the nonimaging lens; but we enjoy excursions even if they are supplementary to the pure goal of understanding the nonimaging Fresnel lens. Of course, the subtitle *Design and Performance of Solar Concentrators* sets the stage and determines the proper ingredients of the story of the nonimaging lens.

We would like to empower you to fully understand the nonimaging lens and the principles of solar concentration, not just by quoting references or undisclosed methodology, but by consciously following up the properties of the device. We aim to communicate the foundations of nonimaging optics and the framework of concentrators. Our ultimate pleasure would be to hear from you that this book excited your spirits and enabled you to design your own nonimaging system – preferably a solar concentrator, naturally.

We would like to acknowledge the following friends and colleagues, whose inspirations, comments, or support this work depended upon: Atsushi Akisawa, Kenji Araki, Jeff Gordon, Serge Habraken, Tamotsu Hiramatsu, Bipin Indurkhya, Takao Kashiwagi, Ken-ichiro Komai, Yanqiu Li, David Mills, Yohei Mizuta, Juan Carlos Miñano, Isao Oshida, Jose Giner Planas, Ari Rabl, Susana Sainz-Trapaga, Roland Winston, and many others.

Tokyo, Paris
May 2001

Ralf Leutz
Akio Suzuki

Preface

Nonimaging Fresnel Lenses: Design and Performance of Solar Concentrators; what are we talking about?

It is easy to forget that you, dear reader, may not be one of those who work in*exactly* the same field as we do: nonimaging optics for the concentration of sunlight.

You may be a researcher in some optical science interested in the core subject of this book: the world's first practical design of a nonimaging Fresnel lens concentrator. You may not be too excited about the collection of solar energy, but you would want a full description of the optical performance of the lens. Which you will get, mostly in terms of nonimaging optics, complete with test results, and set against the competition of imaging Fresnel lenses and mirror-based imaging and nonimaging concentrators.

If you are a solar energy professional, you are likely to be interested in reading why nonimaging optics and solar energy collection go together so well. They do so, because the concentration of solar energy does not demand imaging qualities, but instead requires flexible designs of highly uniform flux concentrators coping with solar disk size, solar spectrum, and tracking errors. Nonimaging optics has been developed to perfection since its discovery in 1965, in dealing with solar power conversion. Much of this experience is useful in nonimaging optical design in other fields where the markets already are more rewarding than in solar power generation, such as optoelectronics.

Depending on your speciality, you will find some sections to exactly fit your needs. You will probably wish we had done more on the topic. If, for example, you were to design a test bed for a photovoltaic multijunction device under concentration, you possibly bought this book for the contents of Chap. 9. In a less specialized way, if you always wanted to know how to design imaging Fresnel lenses, this information is assembled here, too.

This book is a research monograph. No matter how hard we tried to make it comprehensive, it would always be a personal account of work related to our own research. We did not attempt to write a solar energy handbook nor a textbook on nonimaging optics, not because that would be repetitive or redundant (not really!), but because our approach is the exploitation of a merger of two most exciting topics.

For Noriko, Nanami, and Moeka

For Ariya

Dr. Ralf Leutz
Tokyo University
of Agriculture and Technology
BASE
2-24-16 Naka-cho, Koganei-shi
Tokyo 184-8588
Japan
E-mail: ralfsun@yahoo.com

Dr. Akio Suzuki
UNESCO
Natural Sciences Sector
1, rue Miollis
75732 Paris Cedex 15
France
E-mail: a.suzuki@unesco.org

ISSN 0342-4111

ISBN 3-540-41841-5 Springer-Verlag Berlin Heidelberg New York

Library of Congress Cataloging-in-Publication Data applied for.

Die Deutsche Bibliothek – CIP-Einheitsaufnahme
Leutz, Ralf:
Nonimaging fresnel lenses : design and performance of solar concentrators ;
with 44 tables / Ralf Leutz ; Akio Suzuki. - Berlin ; Heidelberg ; New York ; Barcelona ; Hong Kong ; London ; Milan ; Paris ; Singapore ; Tokyo :
Springer 2001
(Springer series in optical sciences ; 83)
(Physics and astronomy online library)
ISBN 3-540-41841-5

Springer-Verlag Berlin Heidelberg New York
a member of BertelsmannSpringer Science+Business Media GmbH

http://www.springer.de

Data prepared by the authors using a Springer TEX macropackage
Fresnel font by Tom 7; fonts.tom7.com
Cover concept by eStudio Calamar Steinen using a background picture from The Optics Project. Courtesy of John T. Foley, Professor, Department of Physics and Astronomy, Mississippi State University, USA.
Cover production: *design & production* GmbH, Heidelberg

Printed on acid-free paper SPIN 10793655 56/3141/di 5 4 3 2 1 0

Ralf Leutz   Akio Suzuki

Nonimaging Fresnel Lenses

Design and Performance of Solar Concentrators

With 139 Figures and 44 Tables

Contents

Executive Summary

Nonimaging Fresnel Lenses: Design and Performance of Solar Concentrators aims at giving a full and clear account of the optics, the design, the performance and possible applications of the world's first nonimaging Fresnel lens solar concentrator.

Chapter 1, *Lenses and Mirrors for Solar Energy*, introduces us to the diversity of solar concentrators by illuminating the metaphorical relation between the use of mirror concentrators for solar thermal systems, on the one hand, and the application of lens concentrators for photovoltaic power conversion, on the other hand. Solar concentrators are classified according to the optical principles they utilize. The parabolic mirror introduces the principles of imaging concentrators. In contrast, a nonimaging lens/mirror combination is discussed.

A general view of the principles and consequences of nonimaging optics is given in Chap. 2, *Nonimaging Optics*. The compound parabolic concentrator (CPC) is investigated in detail. Concepts and methods are presented as a theoretical background for the design of nonimaging concentrators.

The optical fundamentals of Fresnel lenses are presented in Chap. 3, *Fresnel Lens Optics*. Prism optics includes refraction, internal reflection, deviation, and dispersion, all dependent on the wavelengths (colors) of the incident light and on the refractive index of the prism material. It is shown that the refractive power of a prism is smallest in a minimum deviation prism, offering the highest performance in coping with the spectral width of sunlight.

The history of Fresnel lenses is the subject of Chap. 4, *Earlier Fresnel Lenses*. References are given to the evolution of imaging and nonimaging lens designs. Some analytical solutions for imaging Fresnel lenses show their shortcomings, and introduce us to the application of geometrical optics.

Chapter 5, *Nonimaging Fresnel Lens Design*, is a core topic of the present book. We present the numerical solution of the optimum-shaped nonimaging lens. The method of computation (Newton's method) for various constraints is applied to the design of a variety of lenses of different shapes, showing the flexibility of the simulation and the nature of the design process.

The optical performance of the novel nonimaging lens concentrator is assessed in Chap. 6, *Lens Evaluation*, by means of ray tracing. The chapter deals with the definition of concentration ratios, which are designed to make solar concentrators comparable. We also discuss the practical limits of concentration with refractive concentrators.

Chapter 7, *Optimization of Stationary Concentrators*, is an excursion into the optimization of nontracking nonimaging concentrators, which do not follow the position of the sun. A simple, yet universal radiation model is developed, and the optimum characteristics of stationary concentrators are derived. The mathematical method is coordinate transformation.

Two lens designs, one stationary, the other tracking in one axis, are prototyped in Chap. 8, *Prototype Design, Manufacturing and Testing.* Practical considerations and the redesign of the lens for manufacturing form a major part of this chapter. It is concluded with preliminary tests of the nonimaging lens under sunlight and moonlight, where the three-dimensional angles of incidence of the light prove to affect the performance of two-dimensional linear Fresnel lenses.

Photovoltaic systems, in particular the ones using multijunction devices, are sensitive to local changes in flux uniformity and to local misrepresentations of the solar design spectrum caused by the optics of the concentrator. The flux density and the color behavior of the nonimaging lens are likely to affect photovoltaic system performance, underlining the significance of the assessments in Chap. 9, *Concentrated Sunlight and Photovoltaic Conversion.* In the course of the deliberations, solar disk size and solar spectra are evaluated in terms of lens performance and photovoltaic cell efficiency. A brief introduction of the cost of photovoltaic systems of different concentration ratios follows.

Solar Thermal Concentrator Systems (Chap. 10) require a thermodynamic approach. We focus on the evaluation of solar-assisted sorption air-conditioning in Australasian climates. An exergy analysis of a solar thermal concentrator describes the usefulness of this approach without necessarily endorsing it.

In the concluding Chap. 11, *Solar Concentration in Space*, the discussion focuses on lenses and mirrors for solar concentration in space. With the first solar concentrator supplying power to a satellite launched only in 1998, we describe the experience in this advanced field. Since Fresnel lenses and parabolic mirrors are being contemplated for concentration in space, we resume the open discussion initiated in the first chapter of this book by asking: *Lenses or Mirrors?*

1 Lenses and Mirrors for Solar Energy

1.1 Photovoltaic or Thermal Concentration?

When the design simulation of the nonimaging Fresnel lens solar concentrator was completed, we thought of the lens as being a direct competitor to the compound parabolic concentrator (CPC), which is a mirror-based solar thermal concentrator. We still think this to be true, but we are also investigating the role of the Fresnel lens in photovoltaic (pv) applications. Two questions arise: first, why does 'refractive lens' sound like 'pv', and 'reflective mirror' like 'solar thermal'? And second, assuming the distinction to be merely historical, why are we considering and testing the nonimaging lens for use in the photovoltaic conversion of sunlight, and not only for solar thermal applications, as originally intended?

Boes and Luque [18] try to explain why lenses have been used almost exclusively in photovoltaics, and mirrors in solar thermal systems. They point out that Fresnel lenses offer more flexibility in optical design, thus allowing for uniform flux on the absorber, which is one of the conditions for efficiency in photovoltaic cells. Furthermore, Fresnel lenses are said to be less prone to manufacturing errors, since the errors at the front and back faces of the prism are indeed partially self-correcting (p. 103), while an angular error in the mirror's slope leads to about twice this error in the reflected beam. This is true for flat Fresnel lenses, where the front faces of the prisms blend into a horizontal surface, and also for shaped lenses, in particular nonimaging lenses.

On the other hand, imaging Fresnel lenses are still very prone to movements of the focal point due to nonparaxial incidence, especially when compared to nonimaging mirrors, which have been available longer than nonimaging lenses. Ideal nonimaging concentrators produce uniform radiation on flat absorbers, the main characteristic of 'ideal' being the condition that the first aperture of the concentrator be filled completely by uniform radiation, or radiation from a Lambertian source. Only then will the second aperture (the absorber) receive uniform flux. The sun itself may satisfy the Lambertian approximation, although its brightness is not uniform and its wavelength-dependent brightness changes significantly from its center to its outer areas. Since nonimaging concentrators are designed according to one or two pairs of acceptance half-angles, the concentrator accepts light other than the almost paraxial rays of the sun (acceptance half-angle $\theta_s = \pm 0.275°$, equalling the half-angle of the solar disk), and the concentrated flux is not uniform. Secondary concentrators can be used to make the flux on the absorber more uniform, but the price to be paid usually is rejection of at least some rays, and additional reflection losses.

Imaging Fresnel lenses may be designed aspherically, with corrections in each prism for uniform flux, but both focal foreshortening and longitudinal focal movements require high-precision tracking. Prisms split white light into its color components. Refraction indices are wavelength-dependent, and a truly uniform flux will possibly remain an illusion, although we will see that the object of this work, the optimum-shaped nonimaging Fresnel lens, mixes colors at the absorber.

Photovoltaic cells require homogeneous flux and white light for optimum performance. The 'hot spot' problem is created by mirrors as well as by lenses and is of concern not only for photovoltaic energy conversion but also for solar thermal applications.

Discussions of illumination topics can be found in [18, 49, 182]. Although the authors' approaches come from different directions according to the field they are most familiar with, no clear technical link between 'lens' and 'pv' or 'mirror' and 'thermal' could be established.

Historical aspects are apt to throw more light onto these metaphorical connections. The ability to concentrate has been known for both lenses and mirrors for two millenia: according to legend, Archimedes instructed Greek soldiers to use their shields to concentrate the sun's rays in order to burn the sails and superstructure of the invading Roman fleet off Syracuse in 212 B.C. The geometry and power of 'burning mirrors' was known to the Chinese, too [68].

The first lenses in history were probably glass balloons filled with water. They were used by Roman doctors to burn out wounds. Mouchot [113] cites Plinius (23–79 A.D.) with respect to these earliest solar concentrators, and reports further that thin lenses with a solid body of glass appear in Italy at the end of the 13th century.

Fresnel lenses made of glass were used soon after their practical discovery by Augustin Jean Fresnel in 1822 as collimators in lighthouses. The reason for their success was that they were considerably lighter than singlets and absorbed less radiation than their oil-covered mirror predecessors. Even today, lighthouse lenses are manufactured from glass to withstand the high temperatures present. Parabolic mirrors, on the other hand, have been used for large-scale solar thermal applications since the beginning of the 20th century: in 1913, a $35\,\mathrm{kW_{mech}}$ collector field consisting of a $1233\,\mathrm{m}^2$ area of parabolic troughs was installed in Egypt for irrigation, before World War I destroyed further efforts in solar thermal power generation [46].

The first idea for the collection of solar energy for industry and recreation comes from Leonardo da Vinci, who in 1515 proposed a parabolic mirror four miles across. In 1866, Mouchot designed and ran a solar-powered steam engine [68, 113]. Solar air and solar water heating for housing application was tested and documented for four houses in the United States by the end of World War II [86] and the thermodynamic properties of solar energy collection were well researched.

The photovoltaic effect was discovered already in 1839 by Edmond Becquerel [47]. More important, however, was the coincidence that the world saw both the invention of practical Fresnel lenses due to the availability of acrylic plastic *and* the development of efficient silicon solar cells for applications in space in the early 1950s. Polymethylmethacrylate (PMMA) is a lightweight, clear, and stable polymer with optical characteristics close to those of glass, and superior utilizability for the manufacturing of Fresnel lenses. The properties of PMMA were reported during World War II by Johnson (see [135]). Since the late 1940s experimental Fresnel lenses were built mainly for optical applications for electrotechnology, such as sensors [107].

From the beginning, Fresnel lenses and photovoltaics were the domain of companies and large research institutions. The link between both fields may have been electrotechnology, where experiences in pv and optical sensors are overlapping. Confusingly, the nonimaging concentrator CPC, a mirror, had been invented in 1965 for the reflection of Čerenkov radiation onto a sensor, and it took more than a decade for it to become the state-of-the-art of solar thermal energy collection. The CPC found other applications in astronomy (in combination with a lens [54]) and laser technology.

The advantages of nonimaging concentrators were realized by the solar thermal community. Some research work was carried out in the development of nonimaging Fresnel lenses [33, 76, 90], with the first two works in the list aiming at solar thermal applications, but earlier failures with imaging lenses [51, 117] led lenses to fall into thermal oblivion. Modern solar thermal markets are established, and concepts have been developed that do not include Fresnel lenses. Research institutions and corporations are organized with strict separation of solar thermodynamic and solar electrotechnical departments, reinforcing the status quo.

Similarly, mirrors never really found their way into the photovoltaic community and market. The imaging Fresnel lens of O'Neill [126] is probably the best-known commercially introduced concentrator technology for photovoltaics. It has never been published in the predominantly thermal *Solar Energy Journal.* There are a few Fresnel lenses which have been specifically designed for photovoltaic applications [36, 60, 110]. These lenses were all developed in private companies, and presented at the *IEEE Photovoltaic Specialists Conferences.* We found only one early paper [92] proposing mirrors for photovoltaic concentration, for an application in outer space. This paradigm has only recently been broken with the EUCLIDES project [153], which uses parabolic trough concentrators and bifacial cells.

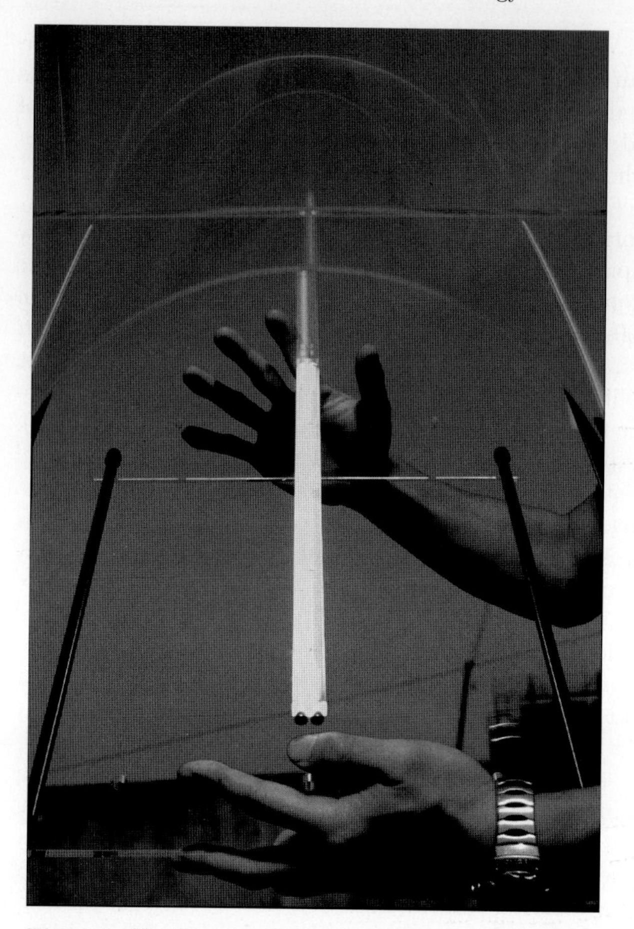

Fig. 1.1. The first prototype of the nonimaging Fresnel lens under the sun of Tokyo, May 1999. Acceptance half-angles $\theta = 2°$, $\psi = 12°$

Having said all this, and having found only historical reasons for a distinction between photovoltaics and solar thermal developments, why follow the same path, and design and test a novel nonimaging Fresnel lens for photovoltaics?

The answer is that the novel nonimaging Fresnel lens has been designed with a thermal application in mind. Thermal requirements differ from those in photovoltaics. Medium temperatures can be achieved by reducing conductive and convective heat losses, tracking is more problematic due to transport of the working fluid, and 'hot spots' pose less problems. Absorber design is of some difficulty as its shape is often not flat; heat pipes or fluid operations have to be installed.

Testing the performance of the collector is relatively simple in photovoltaics. Although concentrator cells must be applied, output is easily measured, whereas solar thermal application testing requires larger collector arrays. The decision to produce a nonimaging Fresnel lens prototype with acceptance half-angle pairs of $\theta = \pm 2°$ (cross-sectional) and $\psi = \pm 12°$ (perpendicular) was based on the potential of easier recognition of optical errors when absorber and angles are chosen to be smaller, and the geometrical concentration ratio is selected to be higher, here 19.1. This lens can be seen under the sun in Tokyo, Japan, in Fig. 1.1.

A second concentrator, for a stationary solar collector, has been built. With acceptance half-angles of $\theta = \pm 27°$ and $\psi = \pm 45°$ this concentrator features a geometrical concentration ratio of 1.7, but does not require any tracking at all. This stationary concentrator is the result of an acceptance half-angle optimization by means of a radiation model. The low concentration model is the optimum compromise of concentration ratio and amount of intercepted yearly radiation.

The nonimaging Fresnel lenses described in this book are the result of a historical development that offers the best of reflective and refractive technologies for any application in the collection and concentration of solar radiation.

1.2 Classification of Solar Concentrators

The wide field of optics provides the laws that govern solar concentrator design. Consequently, there are numerous concentrator designs and design ideas. These can be traced to an optical principle, and a solar concentrator classification according to the optical principles of reflection, refraction, dispersion, and fluorescence shall be attempted. The following treatment is by no means exhaustive, nor are the classification principles authoritative, or even commonly agreed upon. The optical principles are:

- reflection, for mirror concentrators;
- refraction, for concentrator geometries based on lenses, or Fresnel lenses;
- dispersion, for concentrators based on the dispersive power of prisms or holograms;
- fluorescence (luminescence), for concentrators of global radiation by means of fluorescent dyes embedded in a flat plate glass or plastic.

Solar concentrators are often based on a combination of these optical principles: there are concentrators with mirror primaries and lens secondaries, and vice versa. Systems comprising a number of successive mirrors have been proposed, both as Cassegrain and compound parabolic geometries.

Cassegrain optics is usually employed to reduce the constructive length of a concentrator by folding the optical path of the ray bundle. In the conventional geometry, as in a telescope, a primary parabolic mirror reflects incident rays onto a hyperbolic reflector closer to the source; the secondary hyperbolic mirror reflects the rays in turn through a hole in the primary mirror, where an image can be observed or energy can be collected. Cassegrain magnification being high, as desired in astronomy, implies low concentration ratios, which may be increased with a tertiary concentrator. Tertiary concentration and folding of the path means additional reflection losses.

Two classes of solar concentrators can be distinguished, based on their optical design and imaging properties:

- imaging concentrators, which generally are not ideal;
- nonimaging devices, which can be ideal concentrators.

'Ideal' means that a concentrator geometry's design may approach the thermodynamic limit of concentration, where the absorber will equal or even surpass the sun in its brightness, and will have the sun's temperature of 5777 K. Imaging optics usually do not reach ideal levels of concentration, whereas nonimaging concentrators can be ideal. Design procedures for imaging and nonimaging geometries are entirely different. Nonimaging concentrators, while only discovered in 1965, are well suited for the collection of solar energy, because the goal is not the reproduction of an accurate image of the sun, but instead the collection of energy.

Table 1.1 gives an overview of the optical principles used in the design of some solar concentrators, and their classes. Variations of mirrors and lenses can be constructed according to imaging or nonimaging design procedures, and as two-dimensional (2D) concentrators focusing on a line, or as three-dimensional (3D) concentrators of rotational symmetry, focusing on a point. Of course, nonimaging concentrators produce a focal area of finite size, rather than the in theory infinitely small focus of imaging concentrators.

Example: Parabolic Trough

The parabolic trough reflector is the most common solar concentrator. It is used in solar thermal power generation as a line-focusing parabolic trough and as a point-focusing paraboloidal dish. The reflector's shape (Fig. 1.2) is defined by the focal length f. The horizontal axis x cuts along the cross-section of the concentrator, the dependent axis z denotes the height of the mirror above the vertex,

$$z = \left(\frac{x^2}{4f}\right) . \tag{1.1}$$

The parabolic trough is not a thermodynamically ideal concentrator. The geometric concentration ratio C_{g}, defined as the ratio between collector entry aperture width D and receiver width d, can be calculated as

$$C_{\mathrm{g}} = \frac{D}{d} = \frac{\sin \psi_{\mathrm{r}}}{\sin \sigma_{\mathrm{D}}} \tag{1.2}$$

Table 1.1. Optical principles and classes of solar concentrators

Principle	Imaging class	Nonimaging class
Reflection	parabolic trough, paraboloidal dish [177]; Fresnel reflectors [108]; spherical reflector with tracking receiver [45]; Cassegrain optics [147]; heliostat [74]	compound parabolic concentrator (CPC) [182]; hyperboloid (flow line concentrator) [182]; derived and other shapes; higher order concentrators
Refraction	Fresnel lens	nonimaging Fresnel lens; compound elliptic rod lens [191]
Dispersion, diffraction[a]	prism, hologram [16]; diffraction grating	
Fluorescence (luminescence)	–	fluorescent planar concentrators [187]

[a] The limit of geometrical optics, i.e. that structures must be large in comparison to wavelengths, enables the distinction of classes; not applicable in the case of diffraction.

for line-focusing parabolic troughs with a cylindrical receiver. The size of the image of the degraded sun on the receiver is [177] $\sin\sigma_{\mathrm{D}} = \sigma_{\mathrm{S}} + \sigma_{\mathrm{B}}$, where the half-angle of the sun is $\sigma_{\mathrm{S}} = 4.65$ mrad and the half-angle of the combined beam spread errors, $\sigma_{\mathrm{B}} = 10.0$ mrad, amounts to approximately twice the size of the Gaussian image of the sun (there is a difference in reflection errors between the longitudinal and perpendicular directions of incidence [143]).

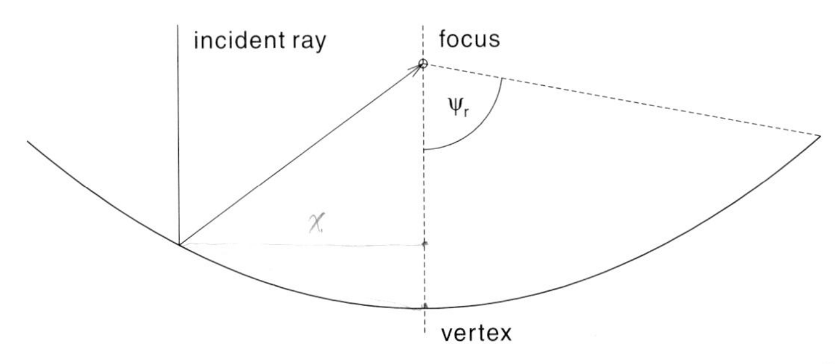

Fig. 1.2. Parabolic concentrator with rim angle or aperture angle $\psi_{\mathrm{r}} = 80°$ and focal length of 30% of rim width

The angle of maximum concentration (not of optimum operation) is $\psi = 90°$, replacing the rim angle ψ_r in (1.2). One finds for the maximum concentration ratio of the parabolic trough [177]:

$$C_g = \frac{1}{\sin \sigma_D} \approx 70 , \tag{1.3}$$

which is roughly one third of the ideal value of the geometrical concentration ratio for line-focusing nonimaging concentrators. The image errors expressed by $\sin \sigma_D$ serve as a parameter of the concentration ratio of the imaging concentrator. We will see that this is not the case for nonimaging designs.

Example: Fluorescent Planar Concentrator

Another example from Table 1.1 is the fluorescent collector [160, 187], shown in Fig. 1.3, which shall be described here for its ability to collect global solar radiation. Most other concentrators collect only direct radiation, rejecting incoming diffuse radiation as long as it does not have the same direction as the beam radiation.

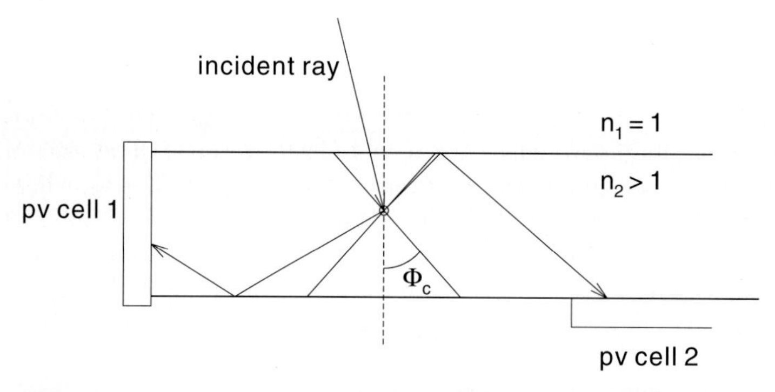

Fig. 1.3. Fluorescent concentrator, with refractive index $n_2 = 1.49$ and critical angle for total internal reflection $\Phi_c = 42.16°$. Positioning of photovoltaic cells to collect emitted photons is shown. Cross-sectional view

Fluorescence is the absorption of electromagnetic radiation and its subsequent emission at a longer wavelength (Stokes-shift). Light at shorter, often ultraviolet, wavelengths stimulates the dye molecule embedded in a plastic or glass sheet; the isotropically emitted light peaks at a longer wavelength.

The following list gives a brief introduction to the factors limiting the optical efficiency η_{opt} of the fluorescent concentrator. The first two and the last point in the list are part of geometrical optics, whereas the remaining efficiencies are uncommon design parameters of solar concentrators. The factors are

- R: Fresnel reflection coefficient at the surface.

- η_{trap}: fraction of light trapped within the boundaries of the plate of refraction index $n_2 \approx 1.5$ because of total internal reflection for angles greater than the critical angle Φ_{c}. From [152] this is given by

$$\eta_{\mathrm{trap}} = \sqrt{\frac{1-(n_2/n_1)^2}{1-(1/n_1)^2}} \,. \tag{1.4}$$

  The rays leave the material either at the end of the plate, where photovoltaic cells can be mounted, or at a place where the refractive index boundary is interrupted, e.g. by water drops or by the solar cell 2 in Fig. 1.3.
- η_{abs}: absorption efficiency due to the absorption spectrum of the dye.
- η_{qua}: quantum efficiency of the dye.
- η_{stokes}: energy loss due to the Stokes-shift.
- η_{dye}: light conduction efficiency in the plate, restricted by reabsorption by the dye as a function of η_{qua}.
- η_{mat}: 'matrix efficiency', loss caused by scattering and absorption in the matrix.
- η_{TIR}: efficiency of total internal reflection (TIR).

Characteristical efficiencies of the fluorescent solar collector are given in Table 1.2 [187].

Reported concentration ratios [187] are 1–4% at the sides of a square plate of 0.16 m^2 area, and can be increased by increased collector size. These optical concentration ratios must be evaluated in comparison with other stationary, nontracking concentrators. Also, the conversion efficiency of photovoltaic cells optimized for one energy gap is potentially high.

Table 1.2. Fluorescent solar collector: expected losses and efficiencies for two stages of input solar spectrum[a]

Quantity	UV–550 nm	550–800 nm
$1-R$	0.98	0.99
η_{trap}	0.74	0.74
η_{abs}	0.29	0.35
η_{qua}	1.0	0.75
η_{stokes}	0.75	0.8
η_{dye}	0.8	0.8
$\eta_{\mathrm{mat}} + \eta_{\mathrm{TIR}}$	0.95	0.95
η_{opt}[b]	0.12	0.12

[a] Source: [187].
[b] Approximate, as some coefficients of the product may depend on the same variables.

There are scattering and reabsorption processes connected to the dye. The energy of the Stokes-shift is freed as heat in the collector material, which has to be kept at ambient temperature not only to prevent material damage and increase the longevity of the dye, but also to prevent back reactions of fluorescent photons with thermal photons resulting in photons of short wavelength. Thermodynamically, the concentration of fluorescent collectors is limited to a colour flux of radiance temperature corresponding to the sun's temperature of 5777 K.

The efficiencies of fluorescent concentrators have so far been limited by their characteristic of absorbing mostly ultraviolet radiation. To overcome this problem of low η_{abs}, stacked fluorescent materials can be designed and new dyes may be found. The trapping efficiency is comparatively high, given that global radiation is collected. This is an advantage for the installation of solar concentrators in areas with high diffuse radiation fractions.

Functions of Nonimaging Concentrators

Form follows function, and the concentrator's optical principle (its form) serves a purpose of application (its function). A high geometrical concentration ratio and low optical errors related to beam spread or average number of reflections, and the associated losses, will often be conflicting design concepts. Concentration ratio and tracking requirements are usually positively correlated. The assembly of reflector or refractor, on the one hand, and receiver, on the other, can define the concentration ratio that is practically obtainable. Table 1.3 gives an overview of the functionality of nonimaging concentrator concepts for attaining low, medium, and high concentration ratios.

The functional analysis in Table 1.3 identifies critical factors for each concentrator type to determine its position relative to the other nonimaging concentrators in the matrix. The compound parabolic concentrator is

Table 1.3. Suitability of nonimaging concentrator concepts for different concentration ratios C and tracking requirements

	$C \approx 2$ stationary	$C \approx 20$ 1-axis azimuthal tracking	$C \approx 200,\ 2000$ 2-axis polar tracking
Compound parabolic concentrator CPC	++	+	–
Nonimaging Fresnel lens FLC	+	++	+
Lens/mirror integrated concentrators RX [102], RXI [104]	–	+	++

well suited for stationary and thermal solar applications. Its mirrors can be truncated without much penalty in efficiency. The CPC can be designed with large acceptance half-angles, allowing sunlight from most positions of the sun in the hemisphere to enter the concentrator aperture. Being an ideal concentrator, the rays will be reflected onto the absorber. The mirror geometry of the CPC can be manufactured from metal, and can be used in evacuated tube-type solar collectors for hot water or process steam generation.

When designed with smaller acceptance half-angles and higher concentration ratio, the mirrors of the CPC become increasingly large, and the CPC deep. The average number of reflections for each ray increases. The angle of reflection increases as well, and losses rise. If the reflectance of the mirror is not exceptionally high, and the nonimaging concept not a necessity (as in gold-plated detectors carried by satellites), imaging solutions like the parabolic trough are to be preferred for higher concentration ratios.

The nonimaging Fresnel lens, when compared to the CPC, shows slightly more optical losses when designed for stationary concentration, but is essentially immune to the CPC's practical limit of concentration due to the increased number and angle of reflections. For high and very high concentration ratios, dispersion becomes the dominating problem, and aberrations similar to those in imaging Fresnel lenses occur. While the nonimaging Fresnel lens is superior to imaging lenses in color mixing and flux concentration, other concentrator concepts are as well suited for high-concentration applications.

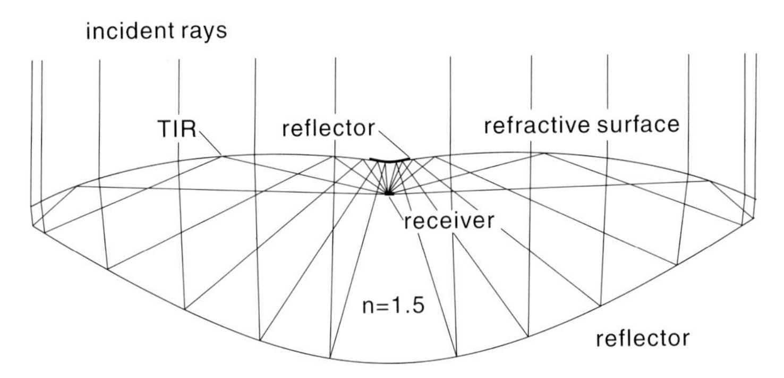

Fig. 1.4. Ideal nonimaging lens/mirror concentrator (RXI) of acceptance half-angle $\theta = \pm 1.0°$. Refraction, reflection, and total internal reflection (TIR). Ray tracing for incident rays entering with $\theta_{\text{in}} = 1.0°$ from the right. (Personal communication from J.C. Miñano, Instituto de Energía Solar, Ciudad Universitaria, Madrid, Spain)

The integrated lens/mirror concentrator RXI (Fig. 1.4) directs incident light, first refracted at the surface of a transparent material and then reflected at the mirrored back of this lens, onto a receiver embedded in the system. For large acceptance half-angles and low concentration ratios, the absorber becomes increasingly large and effectively shades the mirror, and performance is limited. For high concentration ratios, the RXI concept is unsurpassed in its aspect ratio, which is defined as the ratio of the aperture diameter to the concentrator height. The optical characteristics of the RXI of acceptance half-angles $\theta = \pm 0.5°$, $\theta = \pm 1.0°$, and $\theta = \pm 5.0°$ as a result of three-dimensional ray tracing are given in Table 1.4. The design procedure for the RXI concentrator [104] is a brilliant example of a nonimaging design, and a very good example of how to write about it.

Table 1.4. Geometric characteristics of the nonimaging integrated lens/mirror concentrator RXI[a] with refractive index $n = 1.5$. The system has rotational symmetry

Acceptance half-angle $\pm\theta$ deg	0.5	1.0	5.0
Geometrical concentration ratio C_g, -	29546	7387	296
Total transmission[b], %	96.9	97.3	86.9
Aspect ratio (thickness/entry aperture diameter), -	0.278	0.279	0.332
Receiver half-diameter[c], -	1.0	1.0	1.0
Dielectric thickness, -	95.6	47.9	11.4
Entry aperture diameter, -	343.8	171.9	34.4
Diameter of the front metallic reflector, -	23.2	11.2	10.8

[a] Source: [104].
[b] Total transmission is defined as the ratio of power reaching the receiver to the power impinging the entry aperture within the angular cone of the acceptance half-angle. Reflection and absorption losses are not considered. Shadowing losses due to the front reflector are taken into account.
[c] Nonimaging design reference length. All following lengths are expressed as multiples of the absorber half-width (for this concentrator of rotational symmetry, this is the absorber radius).

Imaging concentrators are generally intended for high concentration ratios; 'imaging' by definition corresponds to a geometrical concentration ratio of infinity. In practice, the geometrical concentration ratio will be defined by optical losses due to system design and manufacturing, as well as by the size, brightness distribution, and spectral irradiance of the sun.

For the remainder of this book, the geometrical concentration ratio will be distinguished from the optical concentration ratio, which is defined as the product of geometrical concentration and optical efficiency for solar rays incident at a particular angle.

2 Nonimaging Optics

2.1 Nonimaging Concentration

The nonimaging optical system, by definition, does not produce an image of the light source. Instead, it is designed to concentrate radiation at a density as high as theoretically possible. Nonimaging optics has been used for detecting Čerenkov radiation in a fission reactor in the 1960s. Čerenkov radiation is a weak signal and has a limited angle of emission; these characteristics demand the use of the nonimaging concentrator, as will be seen in the course of this section. A detector for Čerenkov radiation with a nonimaging concentrator was installed at the Fermi Laboratory of the University of Chicago, and at the Argonne National Laboratory in the USA by Hinterberger and Winston [182]. Earlier work on this ideal concentrator was undertaken by Baranov in the former USSR [27, 182].

The nonimaging concentrator can concentrate low-density solar radiation without a sun-tracking mechanism. It may be designed as a stationary solar concentrator. This characteristic attracts solar researchers and many studies have been performed since the 1970s, led by the team of the University of Chicago. Most earlier works focused on the design and theoretical fundamentals of stationary concentrators. Their maximum concentration ratio has a theoretical maximum of roughly four in terms of radiation flux density. Following these low-concentration studies, researchers' concerns have shifted towards the design and realization of collectors that highly concentrate solar radiation. In principle, the nonimaging optical concentrator is ideal, and can achieve a maximum concentration ratio of $n^2 \times 43\,400$ with a transparent material of refractive index n filling the concentrator. The refractive index of air is $n = 1$.

Nonimaging optics has evolved from geometrical optics. In contrast to imaging devices, nonimaging systems do not necessarily create an image of the object (the source) in the focal plane. The goal of nonimaging optics is not photographic accuracy, but the collection of rays incident at a first (entry) aperture of the optical system. W. T. Welford, writing about the "connections and transitions between imaging and nonimaging optics"states that:

> The conventional optical designer starts with a specification stating, *inter alia*, the required aperture and field of view and the aberration

> tolerances to be met over the field of view; there is often a tolerance of distortion but this is never framed in a way indicating that a certain concentration ratio is required. For a nonimaging concentrator design, on the other hand, we start with a quite different specification: there could be for example a specified entry angle and aperture, and a concentration ratio, which specifies the exit aperture; the designer is not in the least interested in where the input rays emerge as long as they *do* emerge, and in place of a rather detailed imaging tolerance specification there is only the efficiency at the specified geometrical concentration ratio. (*his emphasis*) [180]

One of the advantages of imaging designs is that they offer the possibility for a variable field of view (FOV, or field stop, [148]). The second, or exit aperture, of the imaging system is of variable size (e.g. in a camera), whereas nonimaging systems are usually designed with fixed second apertures. In fact, the size of the exit aperture may be one of the initial constants used in the design of the nonimaging system.

Nonimaging concentrators have certain properties that are best explained by the concepts of geometrical concentration ratio and edge ray principle. Both are related and are tools of geometrical optics. It then will become clear what the entry angle is and how the concentration ratio of a nonimaging concentrator is calculated. The efficiency of the collector mentioned by Winston in [180] is influenced by the properties of the nonimaging concentrator. While rays in an imaging system generally pass through forming a wavefront, more or less neatly arranged on the way from object to image, individual rays within the nonimaging concentrator may give a disoriented impression when being reflected or refracted any number of times before exiting the second aperture.

Definition of Concentration Ratios

Solar concentrators are often characterized by the geometric concentration ratio C and the optical efficiency η. An imaginary concentrator is depicted in Fig. 2.1. We suppose that the radiative energy Φ_1 (in Watt) passes through the first aperture of area S_1 (in m^2); the radiative energy Φ_2 is emitted from the second aperture of area S_2. The concentrator's geometric concentration ratio and optical efficiency are given by

$$C = \frac{S_1}{S_2}\,, \tag{2.1}$$

$$\eta = \frac{\Phi_2}{\Phi_1}\,, \tag{2.2}$$

respectively. These are the general definitions of the geometric concentration ratio and the optical efficiency of a concentrator.

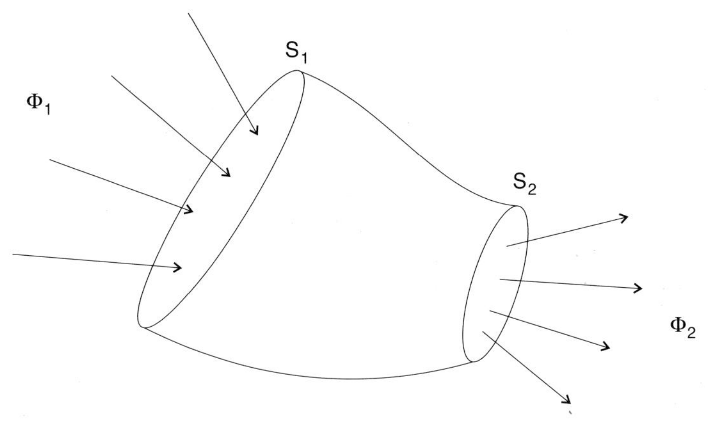

Fig. 2.1. The nonimaging concentrator

When using such concentrators, the energy density on the surface of the radiation receiver is of prime concern in order to gain energy from the solar equipment, such as solar photovoltaic panels or solar heat collectors. The radiation receiver is usually placed at the exit aperture S_2 of the concentrator. The concentrator's performance can be accounted for by the density ratio between the entry and exit apertures. This is called the optical concentration ratio or the optical effectivity, which is defined as

$$\eta_C = \frac{(\Phi_2/S_2)}{(\Phi_1/S_1)} = \eta\, C\,. \tag{2.3}$$

If the concentrator is a perfect concentrator in terms of optical losses, i.e. $\eta = 1$, the optical concentration ratio and the geometric concentration ratio are identical: $\eta_C = C$.

The concept of geometrical concentration ratio is essential to understanding nonimaging concentrators. A concentrating optical system may be pictured as in Fig. 2.2, with entry and exit apertures. The power accepted by the optical system is determined by the radius of the entry aperture a and the acceptance half-angle θ, which is the semiangle of the beams accepted. The refractive indices of the material before the entry aperture and of the transparent material filling the concentrator are marked n and n', respectively. Consider a two-dimensional (2D) concentrator, for example a line-focusing solar collector, where the geometrical concentration ratio C is defined as the ratio of the entry aperture width a to the exit aperture width a':

$$C = \frac{a}{a'}\,. \tag{2.4}$$

Setting a starting point $P\,(y, z)$ of a ray with direction cosines (M, N) anywhere in front of the entry aperture, the movement of P along the y axis, which is the axis of concentration, can be written as $\mathrm{d}y$. The changes of

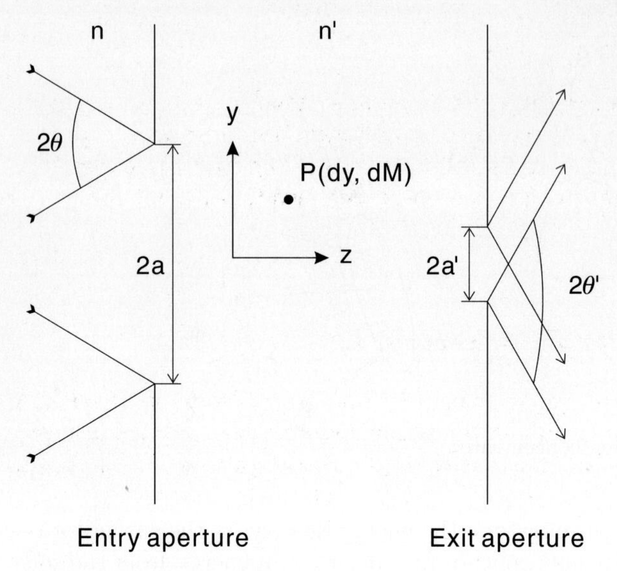

Fig. 2.2. The concept of étendue and the theoretical maximum concentration ratio of a two-dimensional optical system. (Following [182])

direction of the ray relative to the same axis may be written as dM. The influence of the refractive index n is considered, and the 'generalized étendue' (or Lagrange invariant) for entry and exit apertures becomes for reasons of symmetry

$$n \, \mathrm{d}y \, \mathrm{d}M = n' \, \mathrm{d}y' \, \mathrm{d}M' \ . \tag{2.5}$$

The étendue, or 'throughput', has also been called the optical equivalent of the mechanical volume [12]. Integration of (2.5) over y and M yields

$$4na \, \sin\theta = 4n'a' \, \sin\theta' \ . \tag{2.6}$$

Using (2.4) results in the concentration ratio of the ideal linear concentrator:

$$C = \frac{a}{a'} = \frac{n' \, \sin\theta'}{n \, \sin\theta} \ . \tag{2.7}$$

Maximum concentration is achieved when θ' reaches its limit of $\pi/2$. Furthermore, the refractive index of the materials involved is often $n = 1.0$ for air or vacuum. This is usually the case in solar energy applications. The refractive index n here is the refractive index of the transparent material in between the concentrator and absorber. For the nonimaging Fresnel lens (and possibly other refractive concentrators) there is an additional influence of the refractive index of the lens itself. This is discussed in Sect. 6.5. The maximum linear concentration is

$$C_{\mathrm{2D,\,max}} = \frac{n}{\sin\theta} \,. \tag{2.8}$$

The considerations for (2.5–2.8) are equally true for a collector of rotational symmetry, i.e. a 3D concentrator. Concentration in that case happens along both y axis and x axis, the third axis of the coordinate system in Fig. 2.2, and yields a maximum theoretical concentration ratio in air of

$$C_{\mathrm{3D,\,max}} = \frac{n}{\sin^2\theta} \,. \tag{2.9}$$

Thermodynamic Limit of Concentration

The thermodynamic limit of solar concentration can also be deduced from the constellation of the earth and the sun in space (following [31]). The geometrical concentration is defined as the ratio of the aperture surface to the absorber surface:

$$C = \frac{A_{\mathrm{a}}}{A_{\mathrm{abs}}} \,. \tag{2.10}$$

Assuming that the sun is a blackbody radiator of temperature $T_{\mathrm{s}} = 5777\,\mathrm{K}$, the amount of radiation incident on the earth on the first aperture of the concentrator is

$$q_{\mathrm{s}\to\mathrm{a}} = \sigma A_{\mathrm{a}} T_{\mathrm{s}}^4 \left(\frac{r_{\mathrm{s}}}{l}\right)^2 , \tag{2.11}$$

where r_{s} is the radius of the sun, l the distance sun-to-earth, and $\sigma = 5.67 \cdot 10^{-8}\ \mathrm{Wm^{-2}K^{-4}}$ is the Stefan–Boltzmann constant. The size of the solar disk observed by the collector can be expressed in angular terms as

$$q_{\mathrm{s}\to\mathrm{a}} = \sigma A_{\mathrm{a}} T_{\mathrm{s}}^4 \sin^2\theta_{\mathrm{s}} \,, \tag{2.12}$$

where $\theta_{\mathrm{s}} = 0.275°$ is the half-angle of the sun. Imagine the ideal black absorber to emit radiation at a temperature T_{abs}. This radiation is thought to be completely received at the sun, and so

$$q_{\mathrm{abs}\to\mathrm{s}} = \sigma A_{\mathrm{abs}} T_{\mathrm{abs}}^4 \,. \tag{2.13}$$

The thermodynamic maximum of concentration can only be reached if $T_{\mathrm{abs}} = T_{\mathrm{s}}$ and $q_{\mathrm{s}\to\mathrm{a}} = q_{\mathrm{a}\to\mathrm{s}}$. From (2.12) and (2.13) follows that the maximum concentration of the solar image on the absorber of a three-dimensional concentrator is

$$C_{\mathrm{3D,\,max}} = \frac{1}{\sin^2\theta_{\mathrm{s}}} \approx 43\,400 \,. \tag{2.14}$$

For linear concentrators

$$C_{\mathrm{2D,\,max}} = \frac{1}{\sin\theta_{\mathrm{s}}} \approx 208 \,. \tag{2.15}$$

The thermodynamic approach yields the same equations for determining the ideal concentration ratio of solar concentrators as the geometrical approach described alongside the edge ray principle. Additionally, the thermodynamic way allows for calculation of the maximum temperature of an absorber under the sun, once the real concentration ratio of the concentrator is known:

$$T_{\text{abs, max}} = T_{\text{s}} \sqrt[4]{\frac{C}{C_{\text{max}}}} \,. \tag{2.16}$$

It is assumed that there are no heat losses from the absorber. The implications of real and ideal concentration ratios are discussed in Sect. 6.5 in connection with the nonideal performance of the nonimaging Fresnel lens.

Although now the upper theoretical limit of the concentration ratio for nonimaging systems is known, nothing follows from here for the shape of the actual collector to be designed, except that the tool for the designer of a nonimaging collector is the edge ray principle [182]. An application of the edge ray principle for the design of the nonimaging Fresnel lens solar concentrator is pictured in Fig. 2.3.

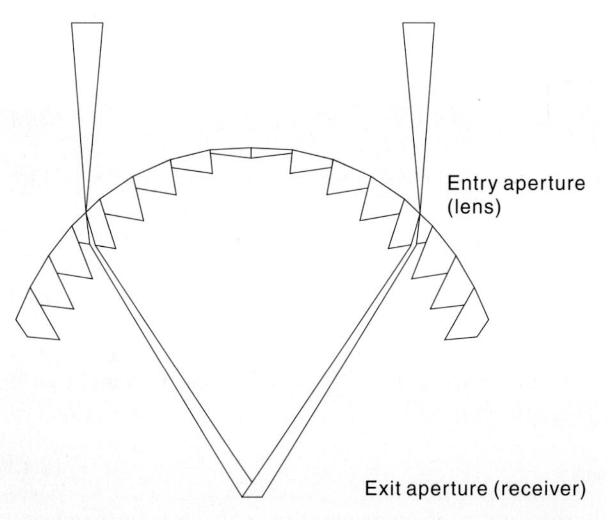

Fig. 2.3. The edge ray principle for yellow light applied to the nonimaging Fresnel lens of acceptance half-angles $\theta = \pm 2°$ in the cross-sectional plane (plane of paper) and $\psi = \pm 12°$ in the plane perpendicular to it. The prisms are assumed to be small, the extreme edge rays at each prism are the extreme edge rays at the exit aperture, or absorber

The width of each prism of the lens in Fig. 2.3 is assumed to be small in comparison to the width of the absorber. The extreme edge rays entering the first surface of each of the prisms are refracted twice, and reach the absorber at its corners, thus being extreme edge rays at the system's second aperture.

Interestingly, the left edge ray reaches the absorber at its leftmost corner, and the edge ray entering the system from the right hits the rightmost reach of the absorber. In the case of the compound parabolic concentrator (CPC) with flat absorber, the rightmost edge ray reaches the absorber level at its left fringe, and vice versa for the left edge ray.

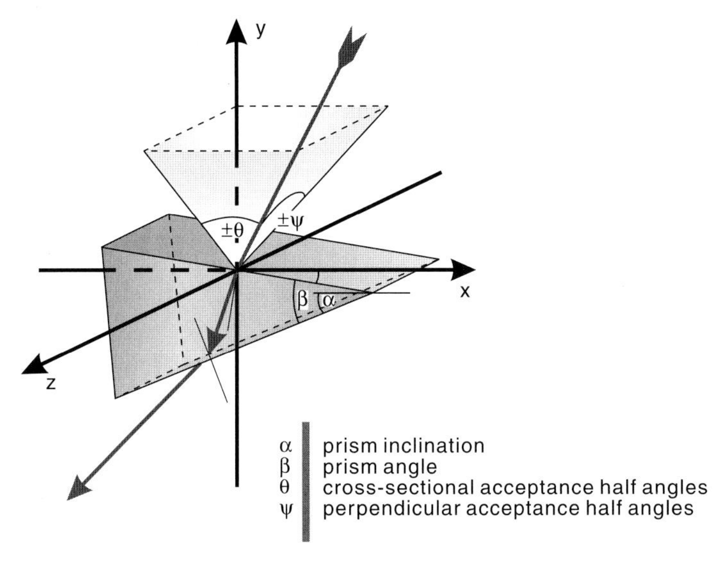

Fig. 2.4. Three-dimensional prism design process. The incident ray entering the prism is described by a cross-sectional and a perpendicular incidence angle. The twice-refracted ray leaves the prism after total reflection at the prism's back has been checked

Figure 2.4 is a schematic of the edge ray entering the prism's front surface described by the acceptance half-angles θ and ψ. The ray is refracted towards the back of the prism, where it is refracted at the boundary between the prism and air under the constraints of the total reflection. Finally, the ray leaves the prism. The x axis in Fig. 2.4 coincides with the cross-sectional cut of the Fresnel lens to be designed. In Sect. 5.2, the design of the nonimaging Fresnel lens will be described in detail, and the central role of the edge ray principle will become clear.

The successful application of the edge ray principle in the design of a nonimaging system does not ensure that this system is ideal. As will be shown during the discussion of the ideal and nonideal properties of the nonimaging Fresnel lens in Sect. 6.5, some of the incident rays at the Fresnel lens will miss the absorber (if $\psi > 0$), although incident within the borders of the acceptance half-angles, which are the extreme edge rays. In the case of the 3D CPC, some skew rays will be reflected back out through the first aperture. Only for the linear CPC, have ideal designs been accomplished [182].

It is important to note that ideal concentrators generally do not provide homogeneous radiation flux on the second aperture. They do so only when the first aperture is completely filled with uniform radiation, i.e. radiation of Lambertian quality. The sun is a Lambertian source only to a rough approximation, and fills the entry aperture of a concentrator only when the acceptance angle of the collector and the solid angle of the solar disk are of equal size.

Grilikhes [49] points out that simple concentrators, such as flat absorbers with flat side mirrors (boosters, see also Sect. 9.5 on homogenizers), provide uniform radiation even for orientation errors of the system. These simple concentrators may be classified as nonimaging approximations, as the location of the mirrors and their orientation may be designed according to the edge ray principle.

2.2 Generalized Ideal Concentration

To simplify the discussion in this section, any concentrator is assumed to be optically ideal: there is no optical loss. Then the concentration ratio of the optically ideal concentrator is represented by C.

As opposed to the imaging optical system, the nonimaging system is not constrained by the geometrically strict relation between real and virtual images. For the study of the geometry of nonimaging optical systems, the concepts of radiance (in $\mathrm{W/m^2sr}$) and that of the Lambertian radiator are important. All radiators or absorber surfaces to be dealt with here are regarded as being Lambertian and having a blackbody surface.

The necessary and sufficient condition for achieving maximum ideal concentration of radiative energy is that the emitter and the receiver are in thermodynamical equilibrium with each other.

A blackbody sphere of temperature T emits energy with an energy density of $G_\mathrm{s} = 4\pi R^2 \sigma T^4$ from its surface, where R is the radius of the sphere and σ is the Stefan–Boltzmann constant ($\sigma = 5.67 \cdot 10^{-8}\,\mathrm{Wm^{-2}K^{-4}}$). The point P in Fig. 2.5 is set apart from the center of the spherical emitter by the distance D, and the radiation density is diluted to $G_\mathrm{D} = G_\mathrm{s}\,(R/D)^2$ (in $\mathrm{W/m^2}$).

An ideal concentrator facing the blackbody at P at distance D from the center of the sphere must have the concentration ratio

$$C = \frac{G_\mathrm{s}}{G_\mathrm{D}} = \left(\frac{D}{R}\right)^2 , \tag{2.17}$$

so that the blackbody absorber of the concentrator is in the same thermodynamical condition as the radiation source (see Sect. 2.1). This thermodynamical condition implies that the energy density on the absorber surface is perfectly uniform and, on any point on that surface, radiation strikes equally from all angles. The absorber of the ideal concentrator is a Lambertian radiator.

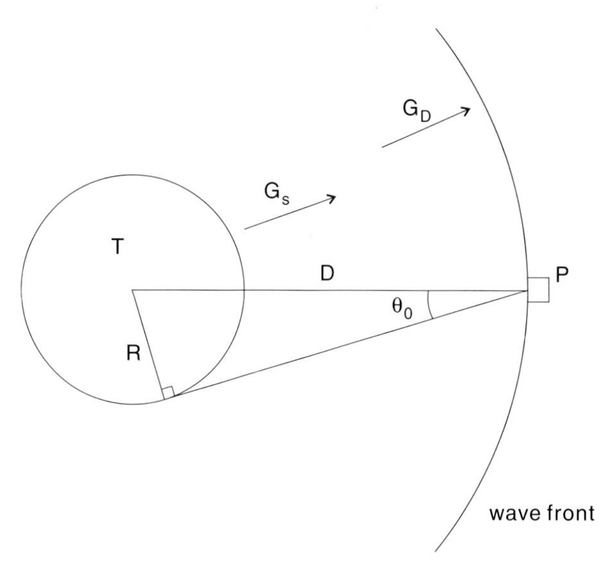

Fig. 2.5. Blackbody radiation at temperature T

In Fig. 2.5, a minute concentrator at P is shown schematically with axial symmetry about the axis of revolution D of the system. The spherical emitter shall represent the sun. The angle subtended by the sun at the concentrator is represented by $\theta_0 = \theta_s$, which can be calculated from the relation

$$\sin\theta_s = \frac{R}{D} \,. \tag{2.18}$$

Substitution of this relation into (2.17) gives the ideal concentration ratio of the minute concentrator as

$$C = \frac{1}{\sin^2\theta_s} \,. \tag{2.19}$$

Let the point P be on earth and the spherical emitter be the sun. The radius of the sun is $R = 6.96 \cdot 10^8$ m; the average distance between earth and sun is $D = 1.496 \cdot 10^{11}$ m. The maximum ideal concentration ratio becomes $C = 43\,400$. Here the angle subtended by the sun at P is $\varepsilon_0 = \theta_s = 0.275° \approx 1/4°$. The moon has a similar angular radius and can be used as a light source instead of the sun (see Sect. 8.6).

For further study on the ideal concentrator, it is worthwhile to investigate the paraboloidal solar concentrator (often called a parabolic dish) for comparison. This concentrator is popular, e.g. as the optical system for solar furnaces for obtaining very high concentration ratios.

For the sake of brevity, the brightness distribution of the solar disk is assumed to be uniform. In Fig. 2.6, the solar radiation subtended with half-angle ε_0 at P on the paraboloid is reflected towards the focal point F. The

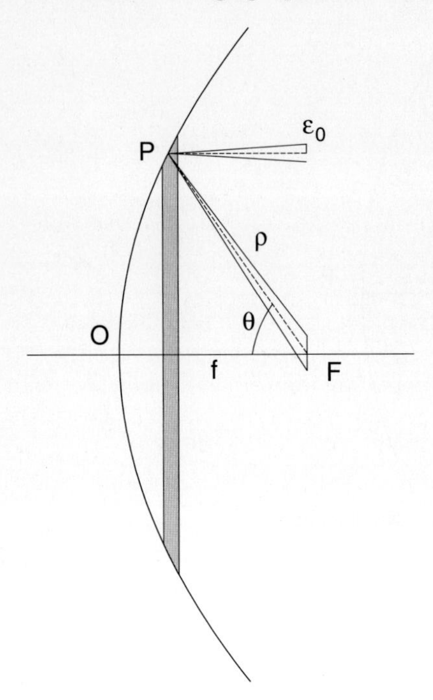

Fig. 2.6. Geometric view of highly concentrating paraboloidal concentrator

image of the sun on the plane vertical to the center axis becomes oval. A ring element of paraboloid represented by P (grey part in Fig. 2.6) illuminates a circle having a radius of $(\rho\varepsilon_0/\cos\theta_s)$, where ρ is the distance between the focal point F and P, and θ is the angle between $\overline{PF}$ and the optical axis of the paraboloid. The distance ρ is geometrically a function of θ, which is described by

$$\rho(\theta) = \frac{2f}{1+\cos\theta}\,, \tag{2.20}$$

where f is the focal length of the concentrator. Then, the area inside the illuminated circle

$$S = \pi\left(\frac{\varepsilon_0\rho(\theta)}{\cos\theta}\right)^2 \tag{2.21}$$

is also a function of θ. Therefore, the illuminated area of the paraboloid with a rim angle ψ_r can be expressed as

$$S = \pi\left(\frac{\varepsilon_0\rho(\psi_r)}{\cos\psi_r}\right)^2\,. \tag{2.22}$$

By having G_0 (in W/m^2) represent the direct component of the solar radiation, the concentration ratio for the illuminated area at the focus becomes (using radians)

$$C = \frac{1}{G_0} \frac{G_0 \cdot \pi \left(\rho\left(\psi_\mathrm{r}\right) \sin \psi_\mathrm{r}\right)^2}{\pi \left(\varepsilon_0 \, \rho(\psi_\mathrm{r}) / \cos \psi_\mathrm{r}\right)^2} = \left(\frac{\sin 2\psi_\mathrm{r}}{2\varepsilon_0}\right)^2 . \tag{2.23}$$

This concentration is maximized for $\psi_\mathrm{r} = \pi/4$, which results in $C =$ 10 900. This value is approximately one quarter of the theoretical maximum described in (2.19). In reality, concentration by the paraboloid yields a radiation intensity distribution around the exact focal point. The maximum concentration at the focal point is given by 43 400 at $\psi_\mathrm{r} = \pi/2$. Note that this concentration is identical with the ideal limit.

Equation (2.19) was derived assuming that both the radiation emitter and the concentrator are surrounded by a material of refractive index $n = 1$. Taking into account the refractive index, as shown in Fig. 2.7, two cases can be distinguished: (a) the blackbody emitter is located inside the refractive medium and the concentrator is situated in a vacuum; and (b) the concentrator is set inside the refractive medium while the emitter remains in the vacuum.

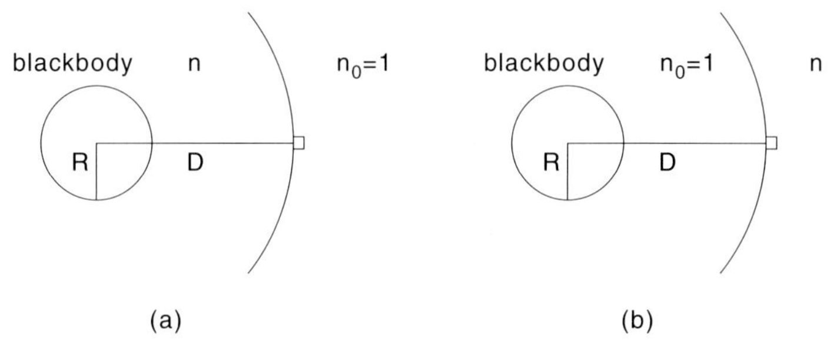

Fig. 2.7. Influence of refractive media on concentration: (**a**) blackbody emitter inside refractive medium, concentrator in air or vacuum, (**b**) concentrator inside refractive medium, emitter in air or vacuum

The speed of light in the refractive medium becomes c/n, where c represents the speed of light in vacuum and n is the refractive index of the medium. Then the distance D in Fig. 2.5 can be read as (D/n) for case (a) and as (nD) for case (b). For both cases, (2.19) can be rewritten as

$$C = \frac{G_\mathrm{s}}{G_\mathrm{D}} = \begin{cases} \left(\frac{D/n}{R}\right)^2 & \text{case (a) ,} \\ \left(\frac{nD}{R}\right)^2 & \text{case (b) .} \end{cases} \tag{2.24}$$

The subtended angle of the spherical emitter in the vacuum at D, θ_s, can be used to simplify this equation to

$$C = \begin{cases} \left(\dfrac{1}{n \sin \theta_s}\right)^2 & \text{case (a)} , \\ \left(\dfrac{n}{\sin \theta_s}\right)^2 & \text{case (b)} . \end{cases} \tag{2.25}$$

In general, the refractive index of the medium surrounding the emitter n_B and that of the medium surrounding the concentrator n_C unify the above two equations, and so we obtain

$$C = C_{3D} = \left(\frac{n_C}{n_B \sin \theta_s}\right)^2 . \tag{2.26}$$

This means that a solar concentrator made from a refractive material is able to obtain a higher concentration ratio than a simple concentrator consisting of reflectors. For example, if a solar concentrator is made with a transparent material of $n_C = 1.5$, it can reach a concentration ratio of 97 700 under ideal concentrating conditions. Note that this higher concentration does not mean a higher temperature than the solar radiance temperature; rather, both the emitter and receiver must emit at identical radiance temperatures, under the second law of thermodynamics.

As for the 2D concentrator's concentration ratio, the following equation can be derived from a discussion similar to (2.26):

$$C_{2D} = \frac{n_C}{n_B \sin \theta_s} = \sqrt{C_{3D}} . \tag{2.27}$$

2.3 Lagrange Invariant

An optical path connecting the real and virtual images while passing through an optical system conserves the product of location, angle of incidence, and refractive index of the space. This value is known as the Lagrange invariant [21].

Figure 2.8 shows the spaces 1 and 2. The optical system is sandwiched between the two spaces. A monochromatic beam L_1 in space 1 passes through the optical system, becoming L_2 in space 2. Here the direction cosine of L_1 at (x_1, y_1, z_1) is described by (L_1, M_1, N_1). Similarly, the direction cosine of L_2 at (x_2, y_2, z_2) can be expressed as (L_2, M_2, N_2).

Photometry theory states that the energy δF (in Watt) emitted from an element of surface $\mathrm{d}S$ (in m^2) in the wavelength spectrum enclosed by λ and $\lambda + \mathrm{d}\lambda$ (in m) per solid angle $\mathrm{d}\Omega$ (in sterad) is

$$\delta F(\lambda) = B(\lambda) \cos \theta \, \mathrm{d}S \, \mathrm{d}\Omega \, \mathrm{d}\lambda , \tag{2.28}$$

where $B(\lambda)$ is the photometrical brightness in $\mathrm{Wm^{-3}sr^{-1}}$, which is a function of the location of $\mathrm{d}S$ and the radiation direction. The angle θ is the emitting

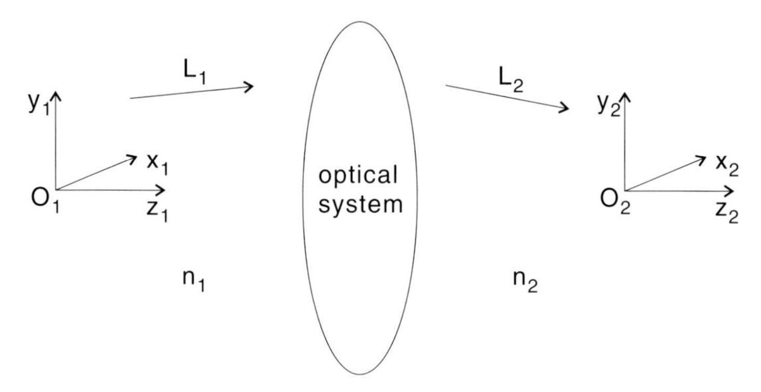

Fig. 2.8. Two spaces separated by an optical system

angle in radians between the normal vector of the surface and the radiation. The surface is supposed to be perfectly black so that Planck's law of radiation gives

$$B(\lambda)\,\mathrm{d}\lambda = 2\,n^2\,\frac{h\,c^2}{\lambda^5}\frac{\mathrm{d}\lambda}{\exp(hc/\lambda kT)-1} = n^2 f(T,\lambda)\,\mathrm{d}\lambda\ , \tag{2.29}$$

where n is the refractive index of the radiation space. In the equilibrium condition, the surface temperature is a constant, and thus $f(T,\lambda) = \text{constant}$. Integrating (2.29) over the total wavelength region yields a function of only the temperature and the refractive index:

$$\int_0^\infty B(\lambda)\mathrm{d}\lambda = n^2 F(T) \qquad (\mathrm{Wm^{-2}sr^{-1}})\ . \tag{2.30}$$

In Fig. 2.8, the radiation energy of L_1 can be transmitted through the optical system without any losses, since the optical system is assumed to be an ideal one. This means that the quantity expressed in (2.28) is conserved in both spaces 1 and 2. From (2.28) and (2.29), this conservation can be expressed as

$$\begin{aligned} n_1^2\,F(T)\,\cos\theta_1\,\mathrm{d}S_1\,\mathrm{d}\Omega_1 &= n_2^2\,F(T)\,\cos\theta_2\,\mathrm{d}S_2\,\mathrm{d}\Omega_2\ , \\ n_1^2\,\cos\theta_1\,\mathrm{d}S_1\,\mathrm{d}\Omega_1 &= n_2^2\,\cos\theta_2\,\mathrm{d}S_2\,\mathrm{d}\Omega_2\ . \end{aligned} \tag{2.31}$$

This equation can also be expressed, using the direction cosine (L, M, N) and the relations $\cos\theta = N\,\mathrm{d}x\,\mathrm{d}y$ and $\mathrm{d}\Omega = \mathrm{d}L\,\mathrm{d}M/N$, as

$$n_1^2\,\mathrm{d}x_1\,\mathrm{d}y_1\,\mathrm{d}L_1\,\mathrm{d}M_1 = n_2^2\,\mathrm{d}x_2\,\mathrm{d}y_2\,\mathrm{d}L_2\,\mathrm{d}M_2\ . \tag{2.32}$$

This is the general expression of the Lagrange invariant in nonimaging optics. It represents an analogy of Liouville's theorem of the conservation of phase space volume [188].

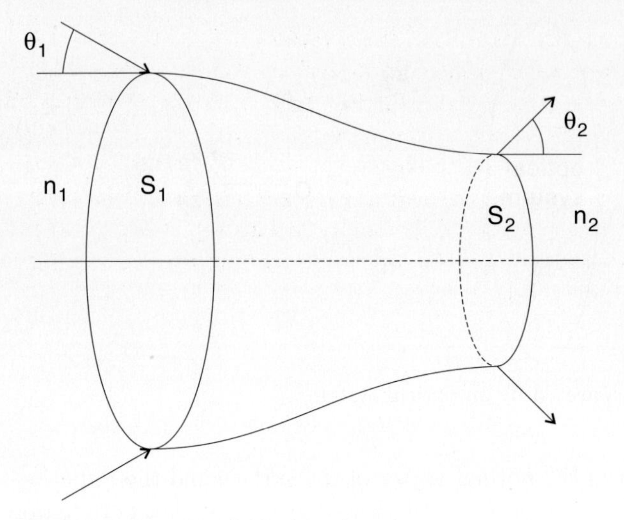

Fig. 2.9. Rotational ideal nonimaging concentrator

Let us take the example of the rotational 3D concentrator depicted in Fig. 2.9 to deduce the concentration ratio from the Lagrange invariant. This concentrator receives radiation at the acceptance half-angle θ_1 at the aperture 1 and emits the radiation with the acceptance half-angle θ_2. Applying (2.31) to this concentrator results in

$$n_1^2\, S_1 \left(\int_{\theta=0}^{\theta=\theta_1} \mathrm{d}(\cos\theta) \right)^2 = n_2^2\, S_2 \left(\int_{\theta=0}^{\theta=\theta_2} \mathrm{d}(\cos\theta) \right)^2 , \tag{2.33}$$

where S_1 and S_2 are the aperture areas of entry and exit. Therefore, the concentration becomes

$$C = \frac{S_1}{S_2} = \frac{n_2^2 \sin^2\theta_2}{n_1^2 \sin^2\theta_1} . \tag{2.34}$$

This concentration becomes a maximum if $\theta_2 = \pi/2$. This is a condition of the ideal concentrator. The result agrees with (2.26) and we understand that the emitted radiation should be uniformly dispersed within the hemispherical region. This conclusion is a necessary condition for the ideal concentrator.

2.4 Nonimaging Mirrors

As seen from (2.26), (2.27) and (2.34), the nonimaging optical system features a certain angular range of incident rays. This angle is called the acceptance angle and, for the ideal nonimaging system, all rays incident within its range will definitely arrive at the exit aperture of the concentrator. On the other hand, rays incident at angles larger than the acceptance angle are forced out of the entry aperture after several internal reflections. We study this unique property for the two-dimensional compound parabolic concentrator (CPC) which is known as an ideal nonimaging concentrator.

The cross-section of the 2D CPC is shown in Fig. 2.10. This concentrator has an axis-symmetric geometry. The incidence angle θ_0 at the entry aperture, which is the utmost angle of incidence, is called the edge ray. The acceptance angle of the CPC is defined by $2\theta_0$. The incidence angle from $-\theta_0$ to $+\theta_0$ can be concentrated onto the exit aperture.

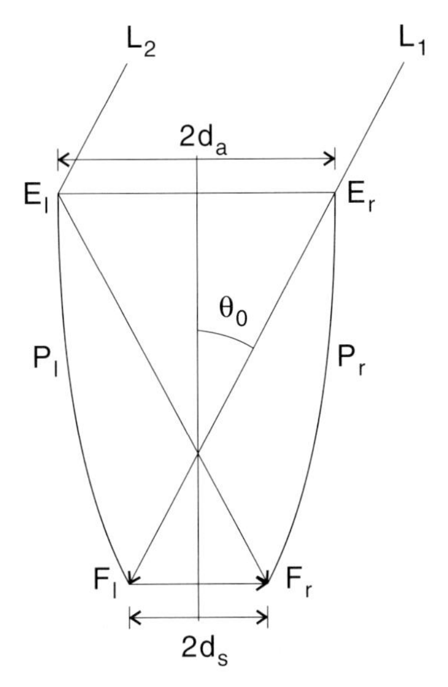

Fig. 2.10. Cross-sectional view of the two-dimensional compound parabolic concentrator (CPC)

Two parabolic curves, P_l and P_r, are inclined by $+\theta_0$ and $-\theta_0$, respectively, from their vertical positions. The focal point of P_l for the edge rays is F_r and that of P_r is F_l. In Fig. 2.10, an edge ray L_2 is depicted striking at F_r after one reflection at E_l. If this concentrator is designed to have the largest possible aperture, the tangent line of the curve P_l at E_l becomes vertical or parallel to the y axis. Thus, $\overline{E_\mathrm{r}F_\mathrm{l}}$, which is a part of L_1, and $\overline{E_\mathrm{l}F_\mathrm{r}}$, which is a part of L_2 are symmetrical to each other about the axis. Assuming an arbitrary position of the wavefront for all edge rays between L_1 and L_2 before arriving at the entry aperture, the optical path lengths of all rays from the wavefront to the focal point are the same. The optical path length is defined as the geometrical length multiplied by the refractive index of the medium.

The geometrical concentration ratio of this concentrator is expressed by $C = (d_a/d_s)$. Applying the parabolic line equation (2.20) to the CPC of Fig. 2.10, we can obtain the following two equations:

$$2\,d_s = \frac{2f}{1+\cos(\pi/2-\theta_0)}\,, \tag{2.35a}$$

$$\frac{d_a+d_s}{\sin\theta_0} = \frac{2f}{1+\cos(\pi-2\theta_0)}\,, \tag{2.35b}$$

where f represents the focal length of the parabola. These equations result in $C = 1/\sin\theta_0$, which is identical to (2.27), provided that $n_B = n_C = 1$. If this CPC is filled with an optical material having the refractive index n, the incidence angle θ_0 can be replaced by $\theta_0' = \sin^{-1}(\sin\theta_0/n)$, which is given by Snell's law of refraction (3.2). This yields $C = n/\sin\theta_0$; the rule that all edge rays have the same optical length is still valid.

The CPC is designed so that the bottom end of one parabolic element becomes the focal point of the opposite one; all edge rays at angles of incidence θ_0, and reflected by one parabola, fall on the terminal point of the opposite parabola. The foci are identical to the edges of a flat absorber, or the exit aperture. All edge rays impinge on the point after one reflection. All rays having smaller angles of incidence will irradiate the absorber region. Thus, all rays within $\pm\theta_0$ will be collected on the absorber. This is called the edge ray principle [182].

A ray which has a slightly larger angle of incidence is rejected from the entry aperture after several reflections between the two parabolic reflectors. In other words, the absorber part cannot be seen when looking through the entry aperture at such angles. No ray with a larger angle of incidence than θ_0 can reach the exit aperture.

This CPC is an ideal concentrator only when the two edge rays, L_1 and L_2 in Fig. 2.10, are parallel. If the size of the CPC is much smaller than or very distant from the radiation emitter, the CPC will work as an ideal concentrator to a good approximation. If this approximation is not valid, the curve P_l of Fig. 2.10 cannot be represented by a parabola because the edge ray L_1 is no longer parallel to L_2.

Tailored Method

In Fig. 2.10, assume an arbitrary point P on the reflector P_r. Defining the angle $\angle PF_lF_r = \theta$ and the length $\overline{F_lP} = \rho$, a differential equation can be derived from their geometric relation:

$$\frac{1}{\rho}\frac{\mathrm{d}\rho}{\mathrm{d}\theta} = \tan\alpha\,, \tag{2.36}$$

where α is the angle of reflection of a ray incident from the focal point F_1. The reflected ray on the point P is identical with the edge ray of the concentrator except for its direction. Equation (2.36) is the differential form of the parabolic curve. It can be applied to the design of nonimaging reflective concentrators in general. This is called the Tailored method [65].

Cone-Type Concentrators

The CPC is an ideal concentrator in theory, but manufacturing the exact reflector curve with a specular surface is not a simple task, since the reflector curvature becomes larger as the reflector gets closer to the entrance aperture. If we were able to form the nonimaging concentrator only with flat mirrors, the concentrator shape could be fabricated easily and precisely. It is actually possible to design a quasi-nonimaging concentrator consisting of only flat mirrors.

As an example, we study a V-shaped concentrator in two dimensions (2D), not a cone-shaped one in 3D. It is assumed that the reflectivity of the mirror is unity.

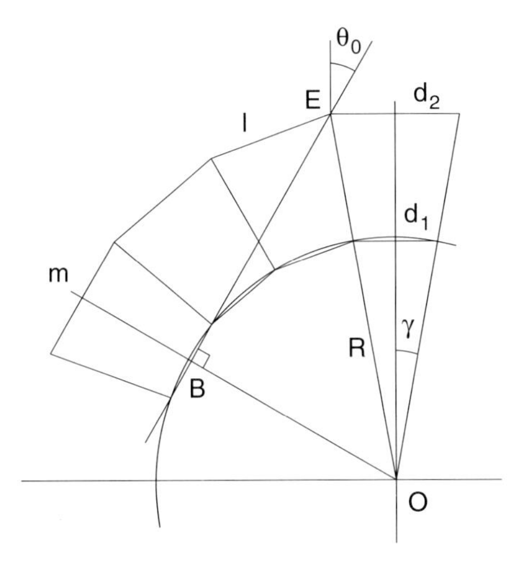

Fig. 2.11. Conceptual view of V-grooved nonimaging concentrator

Figure 2.11 shows the design concept for such a V-shaped concentrator. The concentrator has an entrance aperture of width $2d_2$, an exit aperture of $2d_1$, and a height h. The images of the reflector are all located on a circle of radius R. The apex angle of the V-shaped concentrator is represented by 2γ. We imagine a ray striking at point E at the left edge of the entrance aperture with an angle of incidence θ_0. The ray arrives at the center of the exit aperture of the concentrator's mth image with an angle of incidence of $\pi/2$. The geometric relation of $\triangle EBO$ gives

$$R = \frac{1}{\cos\gamma} = \frac{h\sin(\theta_0 + \gamma)}{\cos\gamma - \sin(\theta_0 + \gamma)} , \tag{2.37}$$

where $\gamma = (\pi/2 - \theta_0)/2m$. The concentration ratio $C_V (= d_2/d_1)$ becomes

$$C_V = \frac{\cos\gamma}{\sin(\theta_0 + \gamma)} . \tag{2.38}$$

Equation (2.38) converges to the ideal value of 2D concentration, $1/\sin\theta_0$, as $\gamma \to 0$. The smaller γ becomes, the larger the concentration becomes. In order to achieve smaller angles γ, the V-shaped concentrator must have small acceptance angles and/or be constructed to yield a larger number of images m. In the case of more images m, the V-concentrator gets taller and the reflection losses increase. Thus, the concentrator increases its concentration ratio while sacrificing its optical efficiency in a trade-off. From a practical point of view, a reasonable number of images m would be two, resulting in a concentration ratio of 1.37 for $\gamma = 15°$ with $\theta_0 = 30°$. If the concentrator were made of an optical material with a refractive index of 1.5, the concentration ratio would improve to 1.58 for $\theta_0 = 30°$ and $m = 2$ ($\gamma = 17.6°$). There are various design methods for this sort of V-concentrator, but all designs commonly incur optical losses due to a number of reflections for higher concentration.

Super-High Concentration with Nonimaging Concentrators

As previously described, the solar concentration of an ideal nonimaging concentrator on earth is 43 400 n^2, where n is the refractive index of the concentrator material. While the 3D CPC is not exactly an ideal concentrator, its geometric concentration is that of the ideal one; the optical efficiency is not unity only due to some rejected skew rays. The height of the concentrator for the subtended angle of the sun, $\theta_s = 0.275°$, is 43 400 times the radius of the circular absorber. This is not a realistic geometry for constructing a concentrator, which would be 434 m high with an absorber radius of 1 cm.

In order to design a more practical concentrator, the combination of a paraboloidal concentrator as primary and a nonimaging as secondary has been studied in [65, 182]. If the secondary concentrator is an ideal nonimaging one, this combined system has the concentration ratio

$$C = \left(n \frac{\cos\psi_r}{\theta_s} \right)^2 , \tag{2.39}$$

where ψ_r is the rim angle of the primary dish mirror. The same rim angle is also the acceptance half-angle of the secondary nonimaging concentrator. The effect of the shadow of the secondary concentrator falling onto the primary is not considered. The smaller the rim angle, the larger the collector's concentration becomes. For example, conditions of $n = 1.5$ and $\psi_r = 10°$ yield a concentration ratio of 94 700.

In an experiment under similar conditions carried out at the University of Chicago at a concentration ratio of 102 000, a higher energy density than that on the surface of the sun was observed. Note that, even though the energy density was higher than on the surface of the sun, the receiver's temperature cannot exceed the sun's temperature [65].

3 Fresnel Lens Optics

3.1 Reflection and Refraction

Studying Fresnel lenses means understanding geometrical optics. As Fresnel lenses are essentially chains of prisms, the optics of the prism play a vital part in designing Fresnel lenses. Geometrical optics is well understood, and this knowledge is readily available in handbooks [21, 66, 157]. However, classical geometrical optics deals with refractive imaging devices.

Imaging Fresnel lenses can be designed with the lens maker's formula. The nonimaging Fresnel lens with convex curvature requires a novel approach. Finite-size prisms do not allow for an analytical solution of the relative positions of prism, ray, and absorber. The design solution is a numerical simulation. Thus, geometrical optics has to be enhanced by adding the principles of nonimaging design.

Reflection and refraction are two basic principles in geometrical optics. A ray will be reflected and/or refracted when incident on a surface of a medium of different (higher or lower) refractive index. For the two cases of reflection in a plane (2D) and in space (3D):

$$\text{angle of incidence} = \text{angle of reflection} , \tag{3.1}$$

where both angles are measured from a common line perpendicular to the boundary surface. This line is called the normal. A schematic for reflection and refraction is given in Fig. 3.1a,b.

Refraction is characterized by Snell's law, which was essentially stated in 1621 by the Dutch astronomer and mathematician Willebrord Snell (1591–1626). In France, the law is called Descartes' law, since he used the ratio of the sines first. Referring to Fig. 3.1b for the definitions of the symbols:

$$\frac{\sin \Phi}{\sin \Phi'} = \frac{n'}{n} . \tag{3.2}$$

The refractive index n of a material is defined as the ratio of the speed of light in vacuum, $c = 2.997925 \cdot 10^8$ m/s, to the speed of light in the medium, v:

$$n = \frac{c}{v} . \tag{3.3}$$

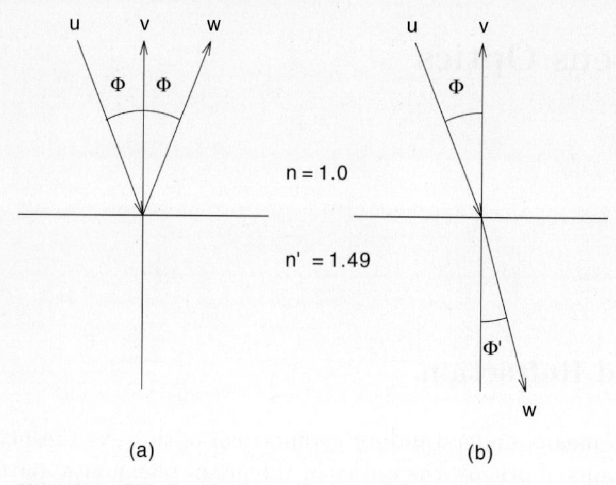

Fig. 3.1. Schematic for reflection (**a**) and refraction (**b**) of incident ray on a boundary surface (two-dimensional view)

When only one refractive index n_{D} is given, this refers to yellow light at the wavelength $\lambda_{\mathrm{D}} = 589.2\,\mathrm{nm}$.

A ray entering a material of higher refractive index will be refracted towards the normal; a ray leaving a substance of higher refractive index into a material of lower refractive index will be refracted away from the normal.

Refractive indices for some transparent substances are given in Table 3.1. The relationship between the refractive index and the density of the material is 'irrational', as there is no apparent (e.g. linear) connection. The refractive index is a function of the wavelength λ and depends strongly on a natural frequency unique to each material. This frequency is thought to interfere with the movement of the ray within the material in much the same way as resonance affects sound. For a discussion see [66].

This resonance is also responsible for selective reflection and absorption over a certain bands of wavelength. No transparent material is known to be unaffected by selective absorption, although the effects on overall transmittance are small.

Reflection and refraction (the cases in Fig. 3.1a,b) can be calculated for the two-dimensional case. In Fig. 3.1a a ray $\boldsymbol{u}$ is reflected as vector $\boldsymbol{w}$ at a surface described by its normal $\boldsymbol{v}$. As (3.1) holds true, the angle of incidence equals the angle of reflection Φ. Since all angles are located in the same plane,

$$\boldsymbol{w} = a\boldsymbol{u} + b\boldsymbol{v}\ . \tag{3.4}$$

Scalar multiplication with $\boldsymbol{u}$ yields the expression

$$\boldsymbol{w} \bullet \boldsymbol{u} = a\,(\boldsymbol{u} \bullet \boldsymbol{u}) + b\,(\boldsymbol{u} \bullet \boldsymbol{v})\ . \tag{3.5}$$

Scalar products have the form

Table 3.1. Indices of refraction n_{D} (deviation of yellow light at $\lambda_D = 589.2\,\mathrm{nm}$)[a]. Dispersion in the visible region is defined as the difference $n_{\mathrm{F}} - n_{\mathrm{C}}$ between the refractive indices for the blue line at wavelength $\lambda_{\mathrm{F}} = 486.1\,\mathrm{nm}$ and the red line at $\lambda_{\mathrm{C}} = 656.3\,\mathrm{nm}$

Material	n_{D}	n_{F}	n_{C}	$n_{\mathrm{F}} - n_{\mathrm{C}}$
Vacuum	1.000	1.000	1.000	0.000
Air	1.000	1.000	1.000	0.000
Water	1.333	–	–	–
Methylmethacrylate	1.493	1.485	1.495	0.009
Borosilicate crown 1 glass	1.500	1.498	1.505	0.008
Diamond	–	2.410	2.435	0.025

[a] Source: [66].

$$\boldsymbol{x} \bullet \boldsymbol{y} = |\boldsymbol{a}||\boldsymbol{b}| \cos\varphi, \qquad 0 \leq \varphi \leq \pi . \tag{3.6}$$

When the vectors are unit vectors ($|\boldsymbol{x}| = |\boldsymbol{y}| = 1$), the scalar product equals the cosine of the angle enclosed by the vectors. From Fig. 3.1a

$$\cos\Phi = (-\boldsymbol{u} \bullet \boldsymbol{v}) . \tag{3.7}$$

From (3.7) and the definition that $\boldsymbol{u} \bullet \boldsymbol{u} = 1$, the unit vectors in (3.5) reduce to enclosed angles and can be expressed as

$$a - \cos\Phi\, b = -\cos 2\Phi . \tag{3.8}$$

This system of linear equations for two unknown variables is completed by a second equation; for reasons of symmetry (3.5) is written as

$$\boldsymbol{w} \bullet \boldsymbol{v} = a\,(\boldsymbol{u} \bullet \boldsymbol{v}) + b\,(\boldsymbol{v} \bullet \boldsymbol{v}) . \tag{3.9}$$

A scalar multiplication with $\boldsymbol{v}$ and the application of the same rules as above yields

$$-\cos\Phi\, a + b = \cos\Phi . \tag{3.10}$$

Cramer's rule for linear (n, n) equation systems offers a convenient solution for this problem. Cramer's rule can be applied because the system is linearly independent due to the independence of the vectors leading to its formulation [25]. Cramer states that

$$x_{\mathrm{i}} = \frac{D_{\mathrm{i}}}{D} , \tag{3.11}$$

where D is the determinant of a matrix composed of the factors preceding the x_{i} in the system of linear equations. For the case of the reflected vector

$$D = \begin{vmatrix} 1 & -\cos\Phi \\ -\cos\Phi & 1 \end{vmatrix} = 1 - \cos^2\Phi = \sin^2\Phi\,, \tag{3.12a}$$

$$D_1 = \begin{vmatrix} -\cos 2\Phi & -\cos\Phi \\ \cos\Phi & 1 \end{vmatrix} = -\cos 2\Phi + \cos^2\Phi\,. \tag{3.12b}$$

By substituting

$$\cos 2\Phi = \cos^2\Phi - \sin^2\Phi \tag{3.13}$$

in (3.12b), a can be calculated to be

$$a = \frac{D_{\mathrm{i}}}{D} = \frac{\sin^2\Phi}{\sin^2\Phi} = 1\,. \tag{3.14}$$

From (3.10)

$$b = -2\cos\Phi\,. \tag{3.15}$$

If the normal $\boldsymbol{v}$ in Fig. 3.1 is defined as pointing in the opposite direction, the sign of b changes.

In a very similar procedure, the constants a and b are described for the case of refraction (Fig. 3.1b). Contrary to the reflection case, the angles Φ and Φ' are not equal, but obey Snell's law (3.2), namely

$$n\sin\Phi = n'\sin\Phi'\,. \tag{3.16}$$

From (3.4), and (3.7), after applying the scalar multiplications of vector $\boldsymbol{w}$ with $\boldsymbol{u}$ and $\boldsymbol{v}$, respectively, it follows

$$\boldsymbol{w} \bullet \boldsymbol{u} = a - \cos\Phi\, b = \cos(\Phi - \Phi')\,, \tag{3.17a}$$

$$\boldsymbol{w} \bullet \boldsymbol{v} = -\cos\Phi\, a + b = -\cos\Phi'\,. \tag{3.17b}$$

Cramer's rule can be applied in a similar manner:

$$D = \begin{vmatrix} 1 & -\cos\Phi \\ -\cos\Phi & 1 \end{vmatrix} = 1 - \cos^2\Phi = \sin^2\Phi\,, \tag{3.18a}$$

$$D_1 = \begin{vmatrix} \cos(\Phi - \Phi') & -\cos\Phi \\ -\cos\Phi & 1 \end{vmatrix} = \cos(\Phi - \Phi') - \cos\Phi\,\cos\Phi'\,. \tag{3.18b}$$

Using the theorem

$$\cos(\Phi - \Phi') = \cos\Phi\,\cos\Phi' + \sin\Phi\,\sin\Phi'\,, \tag{3.19}$$

it follows that the expression for the constant a (with $n_{\mathrm{air}} = 1.0$ and (3.16)) is given by

$$\begin{aligned} a &= \frac{D_1}{D} = \frac{\sin\Phi\,\sin\Phi'}{\sin^2\Phi}\,, \\ &= \frac{\sin\Phi'}{\sin\Phi} = \frac{1}{n'}\,. \end{aligned} \tag{3.20}$$

From (3.17b) the expression for the constant b is given by

$$b = -\cos\Phi' + \frac{1}{n'}\cos\Phi \,. \tag{3.21}$$

The constants a and b permit the calculation of the reflected or refracted vector $\boldsymbol{w}$ from the vectors $\boldsymbol{u}$ and $\boldsymbol{v}$ in its components:

$$\boldsymbol{w}.u = a\ \boldsymbol{v}.u + b\ \boldsymbol{u}.u \,, \tag{3.22a}$$

$$\boldsymbol{w}.v = a\ \boldsymbol{v}.v + b\ \boldsymbol{u}.v \,, \tag{3.22b}$$

$$\boldsymbol{w}.w = a\ \boldsymbol{v}.w + b\ \boldsymbol{u}.w \,. \tag{3.22c}$$

Calculation of reflection and refraction using vector algebra offers the advantage of simplicity over the conventional method using lengths and angles. This becomes obvious for the three-dimensional case (compare [143] and [21]).

Table 3.2. Factors for calculating reflection and refraction at the prism. $n' = 1.49$, refractive index of PMMA

Case	Factor a	Factor b
Refraction at the prism's first surface	$1/n'$	$-\cos\Phi_1' + (1/n')\cos\Phi_1$
Refraction at the prism's second surface	n'	$-\cos\Phi_2' + n'\cos\Phi_2$
Reflection at the prism's second surface	1	$-2\cos\Phi_2$

The constants a and b have been summarized in Table 3.2 for convenience. Three cases are significant for this evaluation: (i) refraction at the first surface of the prism, (ii) refraction at the second surface of the prism, and (iii) total reflection on the second surface of the prism. The refractive index changes from case (i) to cases (ii, iii) from n_{acrylic} to n_{air}. For the nomenclature used refer to Fig. 5.9.

3.2 Total Internal Reflection

Total internal reflection plays a vital role for the design and evaluation of optical elements. Imagine a ray passing from a medium like air into a medium of higher refractive index. The angle of refraction will always be smaller than the angle of incidence: the ray is refracted towards the normal. As all incident rays will be refracted to an angle of refraction smaller than the angle of incidence, there must exist a range of angles of refraction that refracted rays never reach. In Fig. 3.2a this range is located between the horizontal and the refracted ray 2′.

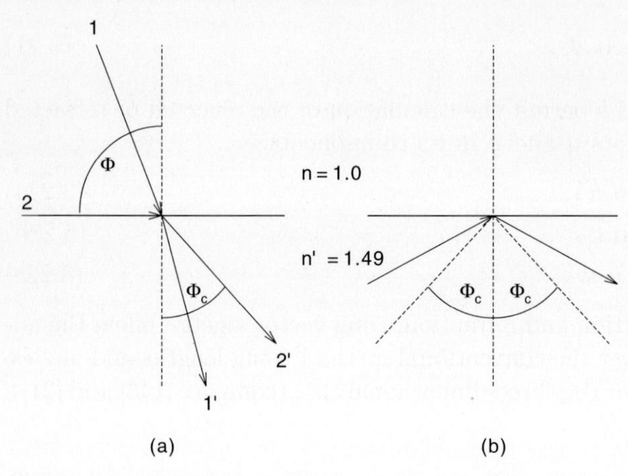

Fig. 3.2. Total reflection: (**a**) the critical angle Φ_c as result of the refracted ray 2′ incident at 90°, (**b**) total reflection beyond the critical angle

A ray entering a medium with higher refractive index at an angle of incidence $\Phi = 90°$ defines the critical angle, or maximum angle of refraction. This critical angle can be calculated by substituting $\Phi = 90°$ in Snell's law:

$$\frac{\sin \Phi_c}{\sin \Phi} = \frac{n}{n'} ,$$

$$\sin \Phi_c = \frac{n}{n'} . \tag{3.23}$$

For an acrylic prism ($n' = 1.49$) surrounded by air, $\sin \Phi_c = 0.671$ and $\Phi_c = 42.16°$.

The principle of reversibility states that a reflected or refracted ray, if reversed in its direction, will retrace its original path. When applying the principle of reversibility to the calculation of the critical angle, the significance of total reflection becomes clear: a ray that intends to leave the medium of higher refractive index is subject to total reflection when trying to do so at an angle greater than Φ_c. Figure 3.2b illustrates total reflection for angles of incidence greater than the critical angle.

There is little energy lost in the process of total reflection. Total reflection facilitates transport of light in optical fibers with small reflection losses. Also, prisms can be designed that reflect rays in the direction parallel to their direction of incidence (such as the 'triple mirror' prism).

Total reflection is a three-dimensional process. The critical angle forms a cone around the normal to the boundary surface. Accordingly, the angles of incidence that still allow for refraction (i.e. allow the ray to pass the boundary) form a similar cone in the material with higher refractive index.

3.3 Deviation

The two surfaces of a prism form the prism angle β. Due to this angle the deviation produced by the refraction at the first surface is not annulled by the refraction at the second surface as in the case of refraction at the plane-parallel plate, but is further increased. Increasing deviation increases both the chances for total reflection and chromatic dispersion.

For the two-dimensional prism (Fig. 3.3), the total deviation is calculated as the sum of the deviations produced at each surface:

$$\delta = \phi_1 - \phi_1' + \phi_2 - \phi_2' = \phi_1 + \phi_2 - \beta \,. \tag{3.24}$$

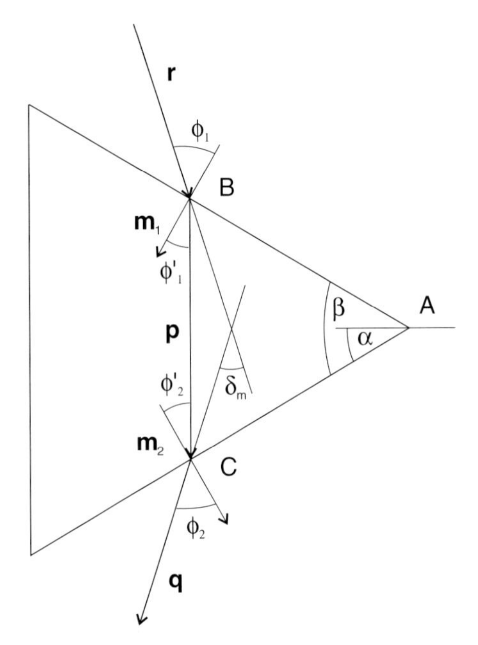

Fig. 3.3. Minimum deviation prism. α is the prism inclination, β is the prism angle, δ_{m} the angle of (minimum) deviation, $\boldsymbol{r}$, $\boldsymbol{p}$, and $\boldsymbol{q}$ are the incident ray, the ray after refraction at the first surface, and after refraction on the second surface (internal refraction), respectively. $\boldsymbol{m}_{\mathrm{i}}$ are surface normals, ϕ_{i} are angles of incidence and refraction

Equation (3.24) does not hold true for the three-dimensional prism since the angles to be added no longer lie in one plane. A convenient way to calculate the total deviation is the scalar product of the vectors $\boldsymbol{r}$ and $\boldsymbol{q}$:

$$\boldsymbol{r} \bullet \boldsymbol{q} = |\boldsymbol{r}||\boldsymbol{q}| \cos\delta, \qquad 0 \leq \delta \leq \pi \,. \tag{3.25}$$

The two vectors form a plane. Given that the vectors have previously been normalized (i.e., $\boldsymbol{r}$ and $\boldsymbol{q}$ are unit vectors), it follows that

$$\delta = \arccos(\boldsymbol{r} \bullet \boldsymbol{q}) \ . \tag{3.26}$$

When there is no perpendicular component of the incident ray (the z–component in Fig. 2.4), then the results of (3.24) and (3.25) are identical. For most applications there will be additional deviations resulting from incidence from a direction off the cross-sectional plane.

The dependency of the minimum deviation δ_{m} on the angle of incidence at the first surface ϕ_1 has been visualized in Fig. 3.4. Perpendicular incidence is added, from a solely cross-sectional case ($\psi = 0$) to incidence incorporating a component of 45° in the perpendicular direction.

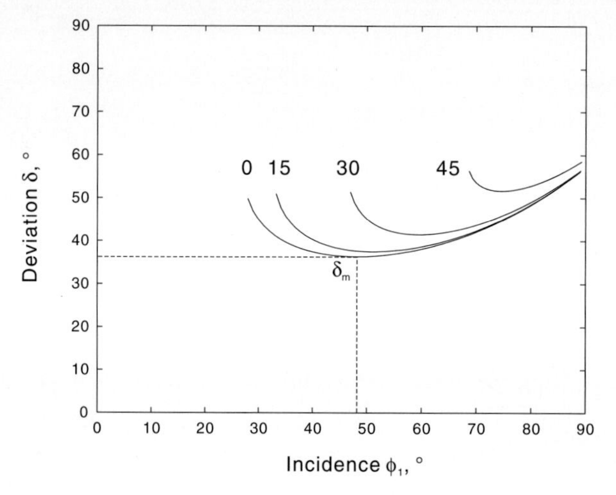

Fig. 3.4. Deviation for a prism of $\alpha = 60°$. δ_{m} is the angle of (minimum) deviation, ϕ_1 is the angle of incidence on the first surface. 3D cases for 0, 15, 30, and 45° perpendicular incidence are shown

That minimum deviation exists can be proved from (3.24) [21]. For the two-dimensional case follows from the general equation for the angle of deviation δ (3.24), expressed by the outer angles of incidence and refraction

$$\delta + \beta = \phi_1 + \phi_2 \ , \tag{3.27}$$

and for the inner angles of refraction and incidence

$$\beta = \phi_1' + \phi_2' \ . \tag{3.28}$$

Applying Snell's law of refraction results in

$$\sin \phi_1 = n' \sin \phi_1' \ , \tag{3.29a}$$

$$\sin \phi_2 = n' \sin \phi_2' \ . \tag{3.29b}$$

where n' is the refractive index of the prism material. The prism will be surrounded by air. The deviation, dependent on the angle of incidence ϕ_1, will have an extremum when its first derivative equals zero:

$$\frac{\mathrm{d}\delta}{\mathrm{d}\phi_1} = 0 \,. \tag{3.30}$$

Due to the relation (3.27), deriving ϕ_2 for ϕ_1 can give a maximum value of -1:

$$\left(\frac{d\phi_2}{d\phi_1}\right)_{\text{extr}} = -1 \,. \tag{3.31}$$

From (3.28), (3.29a), and (3.29b) follows that

$$\begin{aligned} \frac{\mathrm{d}\phi_1'}{\mathrm{d}\phi_2} &= -\frac{\mathrm{d}\phi_2'}{\mathrm{d}\phi_1} \,, \\ \cos\phi_1 &= n' \cos\phi_1' \frac{\mathrm{d}\phi_1'}{\mathrm{d}\phi_1} \,, \\ \cos\phi_2 \frac{\mathrm{d}\phi_2}{\mathrm{d}\phi_1} &= n' \cos\phi_2 \frac{\mathrm{d}\phi_2'}{\mathrm{d}\phi_1} \,, \end{aligned} \tag{3.32}$$

resulting in

$$\frac{\mathrm{d}\phi_2}{\mathrm{d}\phi_1} = -\frac{\cos\phi_1 \cos\phi_2'}{\cos\phi_1' \cos\phi_2} \,, \tag{3.33}$$

which, following (3.31), for an extremum becomes

$$\frac{\cos\phi_1 \cos\phi_2'}{\cos\phi_1' \cos\phi_2} = 1 \,. \tag{3.34}$$

To illuminate the prerequisites for the minimum deviation, (3.29a) and (3.29b) are substituted into (3.34) to yield

$$\frac{1 - \sin^2\phi_1}{n'^2 - \sin^2\phi 1} = \frac{1 - \sin^2\phi_2}{n'^2 - \sin^2\phi_2} \,. \tag{3.35}$$

The above equation is satisfied by

$$\phi_1 = \phi_2 \,, \tag{3.36a}$$

$$\phi_1' = \phi_2' \,, \tag{3.36b}$$

denoting the conditions for an extremum of the deviation. To determine the nature of the extremum, the second derivative of the deviation $\mathrm{d}^2\delta/\mathrm{d}\phi_1^2$ must be evaluated. Equations (3.27) and (3.33) give

$$\begin{aligned} \frac{\mathrm{d}^2\delta}{\mathrm{d}\phi_1^2} &= \frac{\mathrm{d}^2\phi_2}{\mathrm{d}\phi_1^2} \,, \\ &= \frac{\mathrm{d}\phi_2}{\mathrm{d}\phi_1} \frac{\mathrm{d}}{\mathrm{d}\phi_1} \left(\log\left(-\frac{\mathrm{d}\phi_2}{\mathrm{d}\phi_1} \right) \right) \,, \\ &= \frac{\mathrm{d}\phi_2}{\mathrm{d}\phi_1} \left(-\tan\phi_1 - \tan\phi_2' \frac{\mathrm{d}\phi_2'}{\mathrm{d}\phi_1} + \tan\phi_1' \frac{\mathrm{d}\phi_1'}{\mathrm{d}\phi_1} + \tan\phi_2 \frac{\mathrm{d}\phi_2}{\mathrm{d}\phi_1} \right) \,. \end{aligned} \tag{3.37}$$

For the extreme case ($\phi_1 = \phi_2$, $\phi_1' = \phi_2'$), and using (3.31), (3.32), and (3.29a), (3.29b), the following equation is obtained:

$$\left(\frac{\mathrm{d}^2\delta}{\mathrm{d}\phi_1^2}\right)_{\mathrm{extr}} = 2\tan\phi_1 - 2\tan\phi_1' \frac{\cos\phi_1}{n'\cos\phi_1'} ,$$

$$= 2\tan\phi_1 \left(1 - \frac{\tan^2\phi_1'}{\tan^2\phi_1}\right) . \tag{3.38}$$

The refractive index of the prism material $n' > 1$, and thus $\phi_1 > \phi_1'$. Since $0 < \phi_1 < \pi/2$, $\tan\phi_1 > 0$. The last term in (3.38) does not become negative, and so the overall equation obeys the relation

$$\frac{\mathrm{d}^2\delta}{\mathrm{d}\phi_1^2} > 0 . \tag{3.39}$$

The deviation reaches a minimum when the rays pass through the prism in a symmetrical way, i.e. (3.36a) and (3.36b) hold true. Every prism has one, and only one angle of incidence, for which $\phi_1 = \phi_2$, that results in minimum deviation.

Minimum deviation is also understood as a consequence of the principle of the reversibility of light. If reversed in its direction, the ray will follow the same passage through the prism as on its initial pass. There will be only one angle of incidence resulting in minimum deviation.

Although the preceding proof has been conducted for a two-dimensional prism (for the cross-sectional plane) only, the principle of the reversibility of light is equally true for the three-dimensional case: there can be only one angle of incidence giving minimum deviation. This relation between equal angles of incidence and the reversibility of light can be shown for the three-dimensional case. A prism with a prism angle $\alpha = 19°$ and inclined by $-6°$ is evaluated with a ray-tracing algorithm. The rays are incident at 45° in the perpendicular plane while the incidence in the cross-sectional is simulated over the range $0 \leq \phi_{\mathrm{c-s}} \leq \pi/2$. The incident rays are defined in the same coordinates as the prism inclination.

The angles of incidence ϕ_1 and ϕ_2 (reversibility!) are calculated for the 3D case with the help of the scalar product between the surface normals, and the rays. The angles are not signed, as they form cones around the surface normals. In Fig. 3.5, the deviation for the angles of incidence ϕ_1 and ϕ_2 is developed. The angles always form pairs, as required by the principle of the reversibility of light. For the case of minimum deviation, again $\phi_1 = \phi_2$.

3.4 Refractive Indices

The speed of light in a medium varies with color. Since the refractive index n of a material is defined as ratio of the speed of light in vacuum to the speed

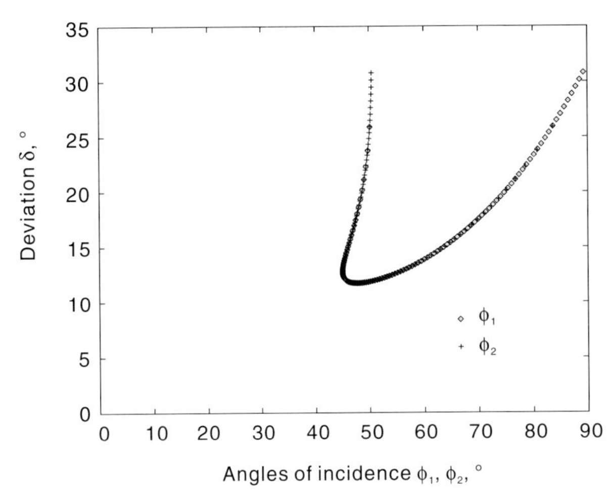

Fig. 3.5. Minimum deviation for a prism for three-dimensional incidence (45° in the perpendicular plane, $0 \leq \phi_{\mathrm{c-s}} \leq \pi/2$ in the cross-sectional plane). $\alpha = -6°$ is the prism inclination, $\beta = 19°$ is the prism angle. The angles of incidence ϕ_1 and ϕ_2 demonstrate the principle of the reversibility of light in its application to the phenomena of minimum deviation in the 3D case

of light in the material, the refractive index is a function of the color of light, i.e. its wavelength:

$$n = n(\lambda) \ . \tag{3.40}$$

Dispersion occurs due to the color-dependent refraction. Shorter wavelengths (ultraviolet, blue) are refracted further away from the surface normal than longer wavelengths (red, infrared). Table 3.3 lists some of the wavelengths commonly used to describe the dispersion of an optical material. In most cases, three wavelengths are used. These are the spectral lines for helium at 587.6 nm (yellow light), and hydrogen at 486.1 nm (blue) and 656.3 nm (red light), respectively [157]. If only one refractive index is given, usually $n_{\mathrm{D}} = 589.2$ nm is used [66].

Borosilicate crown glass (e.g. BK7) is an optical glass containing boric oxide, along with silica and other ingredients. Its index of refraction and its Abbe number are both relatively low. BK7 is optically very homogeneous and offers transmission rates up to 95% (for a thickness of 10 mm) over a bandwidth of $350 < \lambda < 2700$ nm. This is desirable for solar applications, since the solar spectrum covers approximately a range of 250–8000 nm, with about 93% of the energy transmitted in the band covered by the borosilicate glass. Furthermore, borosilicate glass is produced in large quantities at moderate cost, which is another advantage for its application in solar collectors.

Figure 3.6 compares the solar spectral irradiance with the transmittance of BK7 and polymethylmethacrylate (PMMA), and their respective refractive indices.

Table 3.3. Dispersion: standard wavelengths and refractive index n_{BK7} of BK7 glass[a]

Wavelength, nm	Designation	Spectral source	n_{BK7}
656.27	C	Hydrogen	1.51432
589.29	D	Sodium (doublet)	1.51673
587.56	d	Helium	1.51680
486.13	F	Hydrogen	1.52238
365.01	i	Mercury	1.53627

[a] Sources: [155, 157].

Low cost is an additional advantage of polymethylmethacrylate. The transmittance of PMMA almost reaches that of BK7 glass over the solar spectrum. Data for a sample of $d = 3.2\,\mathrm{mm}$ [43] is reproduced in Fig. 3.6(middle). A sample of $d = 1.0\,\mathrm{mm}$ has been measured for comparison. Further measurements of another sample of general purpose polymethylmethacrylate show that reflection at the surface, and not absorption within the material, is the leading cause for transmission losses. Reflection losses account for less than 10% of transmission if the angle of incidence is kept below 55° [61].

The refractive index can be plotted as a function of the wavelength for a material. The common approach and the industrial standard since Schott abandoned its Schott dispersion formula [157] is called the Sellmeier formula. This formula, found in 1871, is not entirely empirical but has a physical basis in describing the dispersion of uncoupled molecules (they are assumed to respond with resonance to the passing light waves, and in turn alter the velocity of the light):

$$n = \sqrt{1 + \sum_{\mathrm{j}=1}^{3} \frac{a_{\mathrm{j}} \lambda^2}{\lambda^2 - b_{\mathrm{j}}}}\,, \tag{3.41}$$

where the wavelength λ is in μm.

Six constants (Table 3.4) are needed to calculate the Sellmeier formula for three combinations of refractive indices and wavelengths. The graph can then be mean-square fitted and plotted. Another option is to use the electronic Schott glass catalog [155] that contains data for various glasses. Data can be saved and plotted, or calculated using the Sellmeier formula, as in Fig. 3.6(bottom).

The refractive index of PMMA is sightly smaller than the one presented for BK7 glass. The wavelength-dependent refractive index may be calculated with the Cauchy formula or with the empirical Hartmann formula. Both are of limited accuracy, around ±0.005 in the visible range of light for the Cauchy

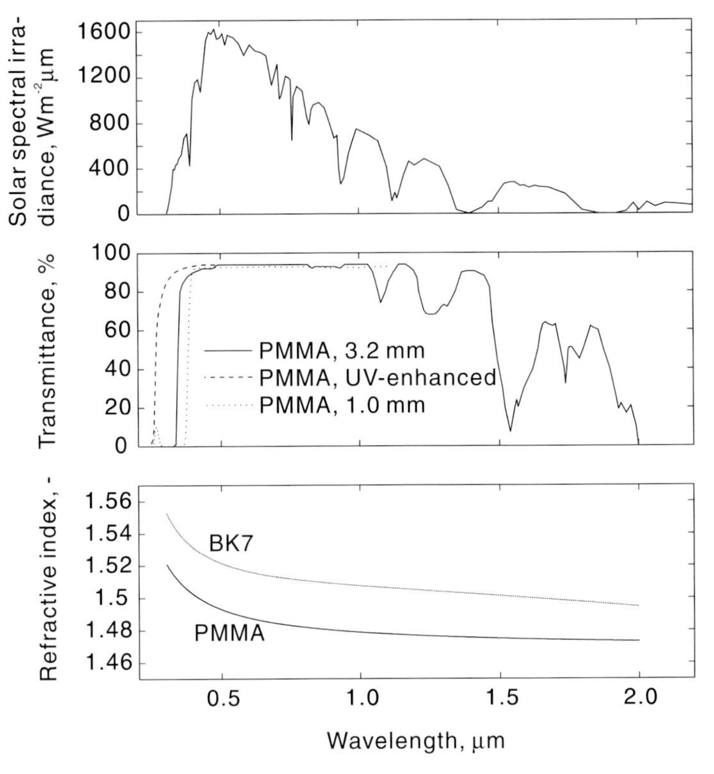

Fig. 3.6. Top: Solar spectral irradiance. **Middle**: Transmittance of general purpose acrylic and acrylic with enhanced transmittance for ultraviolet rays (straight and broken lines [43]); dotted line is for authors' measurements of 1.0 mm sample. **Bottom**: Refractive indices of BK7 glass [155] calculated with the Sellmeier formula, and polymethylmethacrylate PMMA [135] calculated with the Hartmann formula. All data plotted as a function of wavelength

formula [157]. Equation (3.42) is a formula for PMMA, based on Cauchy [122] with λ in Å:

$$n = 1.4779 + \frac{5.0496 \cdot 10^5}{\lambda^2} - \frac{6.9486 \cdot 10^{11}}{\lambda^4} \,. \tag{3.42}$$

The Hartmann formula is explained for PMMA by Oshida [135] with the constants for acrylic with λ in Å.

$$n = n_0 + \frac{C}{\lambda - \lambda_0} = 1.4681 + \frac{93.42}{\lambda - 1,235} \,. \tag{3.43}$$

Data obtained with Hartmann's formula has been plotted in Fig. 3.6. For a discussion of the transmittance and the refractive indices for PMMA in the infrared see [30].

The refractive index is temperature dependent. An expression for the D-line is (t is the temperature in ℃)

Table 3.4. Constants of dispersion formula (Sellmeier formula) for BK7 glass[a]

a_1	1.03961212
a_2	$2.31792344 \cdot 10^{-1}$
a_3	1.01046945
b_1	$6.00069867 \cdot 10^{-3}$
b_2	$2.00179144 \cdot 10^{-2}$
b_3	$1.03560653 \cdot 10^{2}$

[a] Source: [155].

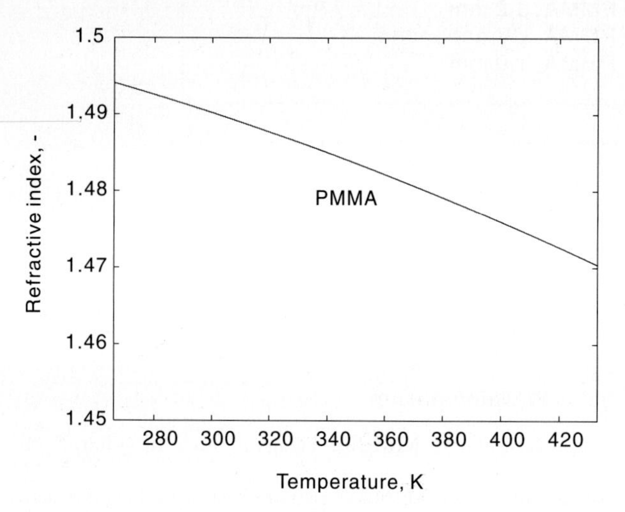

Fig. 3.7. Temperature dependence of refractive index of PMMA at wavelength of yellow light (D-line at 589.29 nm), between glass point and melting point of PMMA at 266 K and 433 K, respectively

$$n_{\mathrm{D}} = 1.4933 - 1.1 \cdot 10^{-4} t - 2.1 \cdot 10^{-7} t^2 \,. \tag{3.44}$$

The transition temperatures of PMMA are relatively low at 266 K for glass and 399 K for rubber, while the range for melting is 433–473 K [138]. Temperature-induced changes in the refractive index are relatively small, and have been plotted for the D-line using (3.44) in Fig. 3.7. The effects on temperature for Fresnel lenses in actual applications can be neglected, unless very high accuracy in imaging devices is required.

Even smaller than the effects of changes in ambient temperature are the effects of changes in relative humidity on the refractive index. Kouchiwa [72] examined those changes for an injection-molded PMMA singlet lens after having exposed it for a period of 30 days to changes from a defined standard condition of 20℃ and 65% relative humidity. He found an empirical equation including both the effects of temperature and relative humidity, and added a third part to his equation (3.44):

$$n_D = n_0 - 1.25 \cdot 10^{-4} \Delta t - 2.1 \cdot 10^{-7} \Delta t^2 - 1.1 \cdot 10^{-3} \Delta rh \,, \qquad (3.45)$$

where Δt, and Δrh are the changes in temperature and relative humidity from the standard condition. Equations (3.44) and (3.45) are nice examples for inaccuracies in secondary sources. Both may be used due to the linear dependencies and the adjustment of n_0. Kouchiwa's original formula (3.45) has been plotted for two temperatures 20℃ and 40℃ over the range of relative humidity in Fig. 3.8.

Refractive materials can also be described with one value only. The Abbe number, or V_d-value, contains the mentioned three refractive indices n_F, n_d, and n_C for blue, yellow, and red light, respectively:

$$V_\mathrm{d} = \frac{n_\mathrm{d} - 1}{n_\mathrm{F} - n_\mathrm{C}} \,. \qquad (3.46)$$

It is then possible to compare various types of glasses and other refractive materials visually. In Fig. 3.9 data from Schott's 'glass file' is plotted. The overview has been enlarged by the addition of some important plastics, namely polymethylmethacrylate (PMMA), polystyrene (PS), polycarbonate (PC), and styrene acrylonitrile (SA). The organics are compared in Table 3.5 with the BK7 glass.

From Fig. 3.9 and Table 3.5 it is apparent that the optical properties of some plastics, particularly those of polymethylmethacrylate do not deviate much from some glasses. Their indices of refraction over the band of wavelengths are similar. From an optical point of view, either material may be used

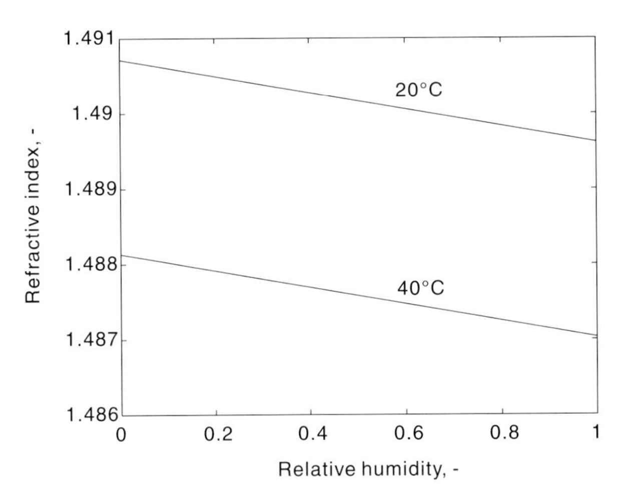

Fig. 3.8. Dependence of refractive index of PMMA on relative humidity and temperature

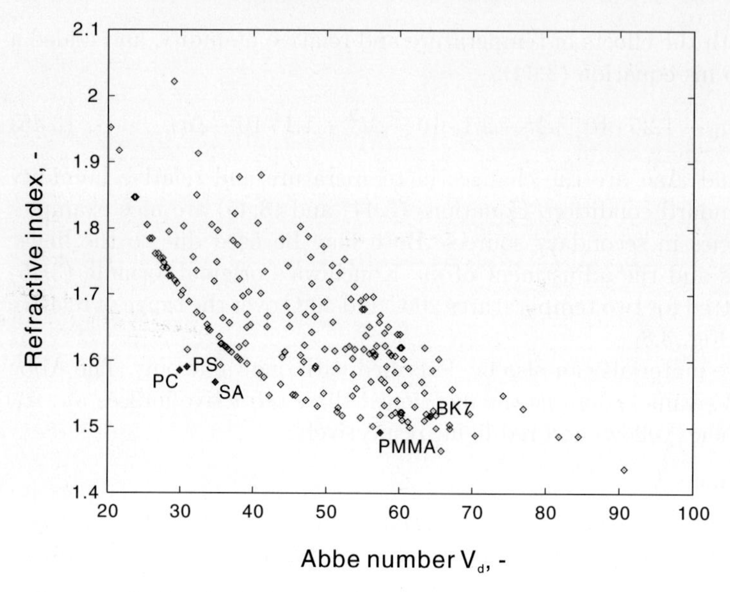

Fig. 3.9. Refractive indices and Abbe numbers of Schott Glaswerke glasses [155], and some plastics (see text) important for optical applications [157]

Table 3.5. Optical properties of some plastics and BK7 glass[a]. Refractive index n_d, Abbe number V_d, and temperature coefficient dn/dT

Material	n_d	V_d	dn/dT per °K
Polymethylmethacrylate (PMMA)	1.4918	57.4	−0.000105
Polystyrene (PS)	1.5905	30.9	−0.000140
Polycarbonate (PC)	1.5855	29.9	−0.000107
Styrene acrylonitrile (SA)	1.5674	34.8	−0.000110
BK7 glass (BK7)	1.5168	64.2	−0.000170

[a] Source: [157].

for solar energy applications. The differences are the production processes, costs, and material longevities under ultraviolet radiation.

3.5 Minimum Dispersion

Solar light is polychromatic light. The dispersion of a prism is of interest for solar concentration. It can be shown that minimum dispersion occurs at minimum deviation. The minimum deviation prism is to be preferred for solar applications so as to keep the absorber as small as possible. A proof for the congruence of minimum dispersion and minimum deviation is given by Born and Wolf in [21].

Imagine a pencil of light with a uniform wavefront BB' incident on a prism (Fig. 3.10). The refracted ray bundle (line CC') will no longer have a uniform wavefront, but will be a function of the wavelength λ. It has been shown before that the refractive index is a function of the wavelength of the passing light. According to (3.40) the deviation δ depends on the wavelength λ:

$$\delta = \delta(\lambda) \ . \tag{3.47}$$

The dependence of the deviation on the wavelength can be expressed as

$$\frac{\mathrm{d}\delta}{\mathrm{d}\lambda} = \frac{\mathrm{d}\delta}{\mathrm{d}n}\frac{\mathrm{d}n}{\mathrm{d}\lambda} \ . \tag{3.48}$$

In (3.48) the deviation can be evaluated in two factors. The first factor on the right-hand side, $\mathrm{d}\delta/\mathrm{d}n$, depends completely on the geometry of the ray and the prisms. The second factor, $\mathrm{d}n/\mathrm{d}\lambda$, characterizes the dispersive power of the prism material (as in Fig. 3.6). Keeping the angle of incidence $\phi_1 =$ constant, and using the equations for the angle of deviation expressed in (3.27) and (3.28), we obtain

$$\begin{aligned} \frac{\mathrm{d}\delta}{\mathrm{d}n} &= \frac{\mathrm{d}\phi_2}{\mathrm{d}n} \ , \\ \frac{\mathrm{d}\phi_1'}{\mathrm{d}n} &= -\frac{\mathrm{d}\phi_2'}{\mathrm{d}n} \ . \end{aligned} \tag{3.49}$$

Since ϕ_1 is still constant, Snell's law of refraction applied in (3.29a) and (3.29b) yields the differentiations

$$\sin\phi_1' + n\cos\phi_1' \frac{\mathrm{d}\phi_1'}{\mathrm{d}n} = 0 \ ,$$

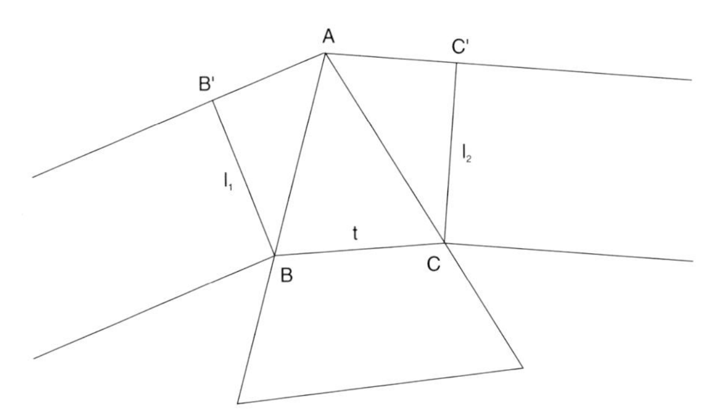

Fig. 3.10. Dispersion by a prism. Conventions used in the proof by Born and Wolf [21]

$$\cos\phi_2 \frac{\mathrm{d}\phi_2}{\mathrm{d}n} = \sin\phi_2' + n\cos\phi_2' \frac{\mathrm{d}\phi_2'}{\mathrm{d}n} . \tag{3.50}$$

Combining the previous four equations yields

$$\frac{\mathrm{d}\delta}{\mathrm{d}n} = \frac{\sin(\phi_1' + \phi_2')}{\cos\phi_2 \, \cos\phi_1'} = \frac{\sin\beta}{\cos\phi_2 \, \cos\phi_1'} . \tag{3.51}$$

For the triangle ABC (see Figs. 3.3 and 3.10) one gets

$$AC = \frac{\cos\phi_1'}{\sin\alpha} t . \tag{3.52}$$

For the triangle ACC':

$$AC = l_2 \frac{1}{\cos\phi_2} . \tag{3.53}$$

The deviation depending on the wavelength reaches a minimum when $l_1 = l_2$. The principle of the reversibility of light requires this symmetry. The same argument from the principle of the reversibility of light also requires minimum deviation.

Minimum deviation means minimum dispersion. Every wavelength has its own angle of minimum deviation, dependent on its refractive index. The refractive indices of common materials vary smoothly over the wavelengths of interest for solar energy applications. Choosing an average wavelength and refractive index for the design of the prisms will minimize the beam spread of white light due to dispersion. In solar energy applications this allows for an absorber of minimum size to be installed.

Using (3.48), (3.51) with (3.52) and (3.53) results in

$$\frac{\mathrm{d}\delta}{\mathrm{d}\lambda} = \frac{t}{l_2} \frac{\mathrm{d}n}{\mathrm{d}\lambda} . \tag{3.54}$$

The angular dispersion (i.e. the angle by which the emergent wave front (l_2) is rotated when the wavelength is changed by $\Delta\lambda$) is

$$\Delta\delta = \frac{t}{l_2} \frac{\mathrm{d}n}{\mathrm{d}\lambda} \Delta\lambda . \tag{3.55}$$

If the pencil of light completely fills the prism, then $t = b$, where b is the base of the prism. The design of the base of a prism is thus defined by the angle of incidence, and the refractive index of the prism material. Or, the design of the shape of a prism should depend on its refractive power.

Earlier Fresnel Lenses

4.1 History of Fresnel Lenses

Refraction occurs when a ray passes through surfaces separating materials with different indices of refraction. The contour of these surfaces define the focusing properties of the optical element. The bulk of material between the refracting surfaces has little effect on the optics of the lens.

Probably centuries ago, it was realized that the lens needs only one shaped surface to define its focal abilities. This type of lens is called plano-convex. In 1748, Georges de Buffon cut away the inside material of this plano-convex lens (if he was only theoretically doing so, with the first actual echelon lens to be constructed by Abbè Rochon some thirty years later, remains unclear). Buffon started from the flat side, and left stepped rings following the original spherical shape of the lens (Fig. 4.1a). Buffon's work was followed by that of Condorcet and Sir D. Brewster, both of whom worked with built-up lenses made of stepped annuli [43].

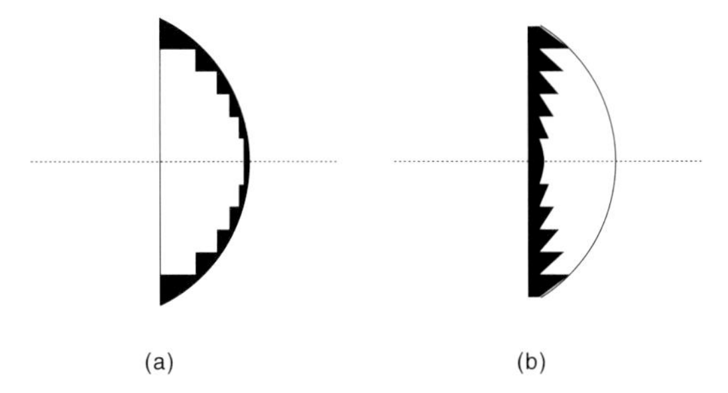

Fig. 4.1. Buffon's stepped surface lens of 1748 (**a**), and Fresnel's lens of 1822 (**b**). Light enters the three-dimensional lenses from the left. The dark parts are material

These lenses were to be used in lighthouses, first along the French Normandy and Brittany coasts. Previously, lenses had been too heavy for use on a lighthouse turntable, and too bulky to be easily manufactured. Instead of lenses, most lighthouses were equipped with metal parabolic mirrors, which, oiled against corrosion, absorbed as much as half of the light during reflection.

If a lens was used, the rays from comparatively weak light sources were partially absorbed in the lens material. Lighthouses were powered by oil lamps in those days. The height of the shipping age in the 18th century fostered the development of what would be known as the Fresnel lens.

The first lighthouse to find entry into the history books, one of the Seven Wonders of the World, was erected in 283 B.C. near Alexandria; it carried a fire until the 1st century. Stone or wooden lighthouses with wood or coal fires were known on the coasts of the North and Baltic Seas from the 13th century on.

Brewster mentioned stepped lenses in an encyclopedia article in 1811, but he was not to be the first one to actually make and use those lenses, as Aničin et al. report [5]. Their article on Fresnel's achievements was triggered by the accidental find of the original 'mèmoire' read before the members of the French Academy of Sciences on 29 July 1822 in Paris:

MEMOIRE
SUR
UN NOUVEAU SYSTEME D'ÉCLAIRAGE
DES PHARES;
PAR M. A. FRESNEL,
INGÉNIEUR AU CORPS ROYAL DES PONTS ET CHAUSSÉES,
ANCIEN ÉLÈVE
DE L'ÉCOLE POLYTECHNIQUE;
LU À L'ACADÉMIE DES SCIENCES, LE 29 JUILLET 1822.

The Fresnel lens was officially invented in 1822, when Fresnel presented his lighthouse turntable to the French Academy of Science and installed (probably another set) at a lighthouse at Cordouan, on the river Gironde. Augustin Jean Fresnel, who was born on 10 May 1788 in Broglie (Eure), and died on 14 July 1827 in Ville d'Avray near Paris, is credited with numerous scientific achievements, among them contributions on diffraction solving the dispute between the corpuscular and wave theories of light, interference, polarization and the transverse nature of light, reflection and refraction, total internal reflection, and last but not least, lighthouse lenses.

There are sixteen lenses used in the novel turntable, pictured in Fig. 4.2. An approximate enlargement of Fresnel's lens is given in Fig. 4.1b. Each set of eight lenses forms an octagon. The rays formed by the larger lenses are named R, while the ones formed by the smaller lenses are called r. The larger set distributes almost two thirds of the light emitted by the central oil lamp L. These lenses are $0.76 \times 0.76\,\mathrm{m}^2$, with a focal distance of 0.92 m. Up to a distance of 33 km, the light reached observers within 6.5°. The smaller set of lenses collects light emitted by the lamp in the upward direction. The mirrors M reflect the concentrated light. The beams from the weaker set precede the light from the main set of lenses by 7°, pictured as the angle between r and R in the top view in Fig. 4.2. This shift doubles the effective duration of light emitted by the lighthouse during rotation.

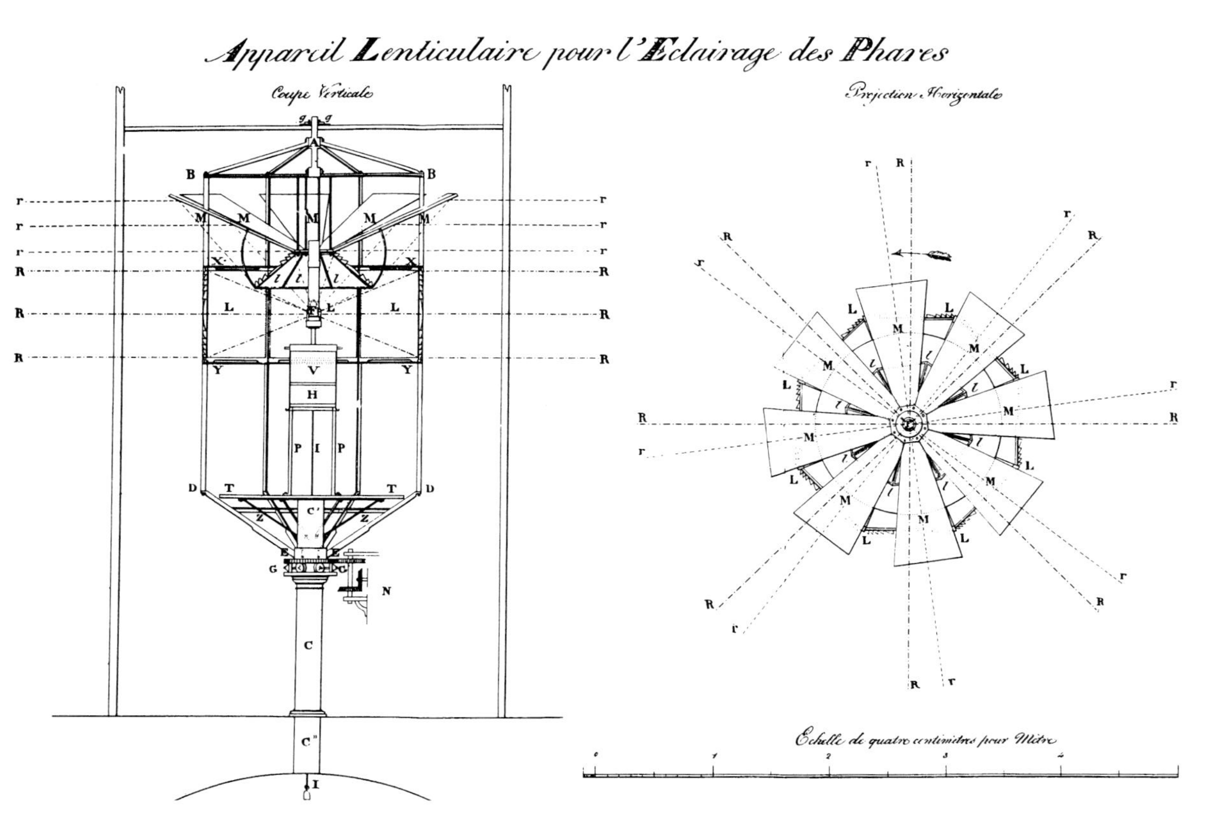

Fig. 4.2. Fresnel's turntable for a lighthouse. Side and top views, with the two sets of lenses forming two octagons. The rays formed by the larger lenses are named *R*, while the ones formed by the smaller lenses (with the mirrors *M*) are called *r*. Original drawing of 1822, reproduced from [5] with kind permission of the publisher and B. Aničin

Fresnel's lenses were most certainly the first aspherical lenses known. Aspherical lenses are conicoids, i.e. rotated conic surfaces. Their mathematical description is more difficult than that of spherical surfaces, but their strength lies in the ability of the aspherical surface to correct spherical aberrations, of which the longitudinal ones are the most important. Longitudinal spherical aberration occurs since the focal lengths for rays refracted from the outer parts of the surface are shorter than those for rays coming from the inner parts of the refracting surface. Spherical aberrations result in image blur. For a complete discussion, refer to [66] or [157].

The mirrors in early lighthouses had been of parabolic shape. As spherical mirrors exhibit spherical aberrations similar to those of spherical lenses, it is very likely that Fresnel had aspherical corrections for his lenses in mind. Aničin et al. [5] try to show that only the limitation of the production process led to the approximation of aspherical steps by spherical ones. In fact, modern Fresnel lenses are free of spherical aberration since each step is designed separately (as opposed to designing one spherical surface for the whole lens) to focus on the focal point. From what is known, Fresnel's lens had aspherical steps, and a spherical center part, and are thus corrected for longitudinal spherical aberrations.

Cited, and translated in [5], Fresnel states his discovery. His wording still serves as the most accurate definition of Fresnel lenses:

> Mais si l'on divise celle-ci en anneaux concentriques, et qu'on ôte à la petite lentille du centre et aux anneaux qui l'entourent toute la partie inutile de leur épaisseur, en leur laissant seulement assez pour qu'ils puissent être solidement unis par leurs bords les plus minces, on conçoit qu'on peut également obtenir le parallélisme des rayons emergens partis du foyer, ou, ce qui revient au même, la reunion au foyer des rayons incidens parallèles a l'axe de la lentille, en donnant a la surface de chaque anneau la courbure et l'inclinaison convenables.
>
> *Yet if we divide (the lens) in concentric annuli and remove from the small lens in the center and from the annuli which encircle it all the unnecessary part of their thickness, leaving barely enough to make possible strong joints along their thinnest borders, one conceives that the parallelity of the rays emerging from the focus could be equally obtained, or, which is tantamount, the reunion in the focus of the rays incident parallel to the axis of the lens (could be obtained) by giving to the surface of each annulus the convenient curvature and inclination.*

Of course, Fresnel's lighthouse lenses are not designed for the concentration of solar rays. They are imaging devices with outward facing grooves and tips. On the other hand, his lenses could well be used for solar concentration. The rays R and r represent a beam of nearly parallel light, directed towards ships nearing the coast where the lighthouse is erected. Assuming the lamp to be a point source, a lens creating a beam of parallel light is called a collimator. Due to the principle of reversibility of light and the allowed approximation of the sun's rays as being parallel, light coming from the outside and passing Fresnel's lighthouse lenses will be refracted and focused towards L. Such a lens acts as a collector.

Notwithstanding the fact that tips facing outwards are of inferior performance when compared to lenses with grooves facing inward, where the prisms are less steep, Fresnel's lenses could have been used as solar concentrators. One might turn Fresnel's lens around, having the grooves face inwards. This, however, does not equalize the performance difference, but makes it worse instead. The larger groove angles foreshorten the focus, and may lead to total internal reflection when the grooves of the aspheric lens, which have been corrected for outward use, are turned inward.

Glass used to be the preferred material for Fresnel lenses until the 1950s, Fresnel lenses were manufactured with similar technologies from the times of Fresnel's first models. Quality lenses were molded, and the grooves were ground and polished. To save production cost, cheaper lenses were cast by pressing glass into molds. The casting would not be further treated, which resulted in the bad optical quality of the lens. The high surface tension of the glass is responsible for rounded grooves and tips, a problem occurring during solidification.

Modern plastics, new molding techniques, and computer-controlled diamond turning machines have improved the quality of Fresnel lenses, and have opened new horizons for the design of Fresnel lenses for numerous applications. Fresnel lenses can be pressure-molded, injection-molded, cut, or extruded from a variety of plastics. Production costs for large outputs are low. Typical applications (some according to [43]) for Fresnel lenses are:

- Lenses (3D) for overhead projectors: condensers for finite conjugates;
- Imaging lenses (2D) for copy machines: collector, focusing a collimated (parallel) beam of light;
- Field lenses, camera focusing screens: condensers where the grooves face the shorter conjugate, which for reasons outlined above is usually avoided;
- Magnifiers or reading lenses creating virtual images;
- Infrared applications, holography, communications (lenses for data transfer), process monitoring, etc.;
- Lighthouses, if not replaced by electronic positioning systems;
- Solar energy applications: collectors of solar rays for thermal and photovoltaic use, special designs for daylighting, etc.

Since a novel Fresnel lens designed for solar applications is the main topic of this book, the wide variety of lenses used for the concentration of solar rays deserves special attention.

The first attempts to use Fresnel lenses for the collection of solar energy occurred at the time when suitable plastics such as polymethylmethacrylate,

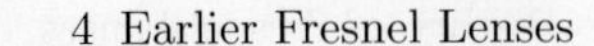

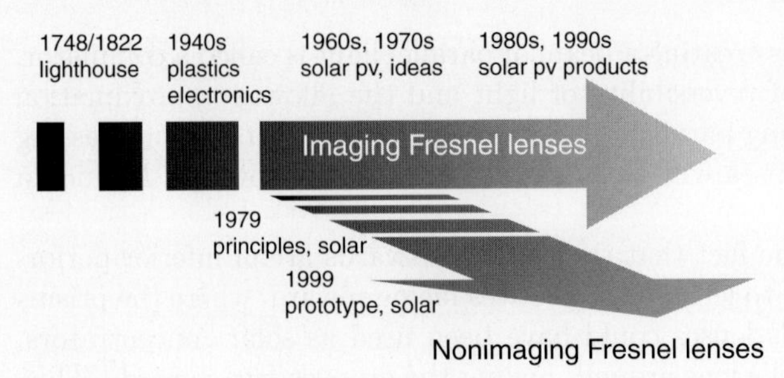

Fig. 4.3. History of Fresnel lenses

or PMMA became available in the 1950s. PMMA is resistant to sunlight, remains thermally stable up to at least 80°C, its spectral transmissivity matches the solar spectrum, and its index of refraction is 1.49, which is very close to the one of glass. Acrylic is the material of choice for most Fresnel lenses produced today.

Some Fresnel lenses used in solar energy applications were made of glass. Glass is an attractive option when lenses are to be used at high temperatures or when they double for glazing. Silicone has a history in Fresnel lens making, and can be used as a substrate under glass. Recent Fresnel lenses for solar concentration in space are manufactured from silicone. If a glass cover is used, it serves as a protective shield and structural element (for references see Sect. 4.2). Most lens designers of terrestrial applications choose PMMA for their lenses, as this material allows for high optical quality combined with less costly manufacturing technologies.

The history of Fresnel lenses has been visually presented in Fig. 4.3 with an emphasis on lens concentrators for the collection of solar energy and on the emerging separation of imaging and nonimaging optics.

4.2 Recent Developments

Initially, most Fresnel lenses selected for solar energy use had not been originally designed for the collection of solar rays. These lenses were imaging devices. Surprisingly, most of the major improvements in the design of Fresnel lenses listed in Table 4.1 were made with solar applications in mind. Nonimaging optics have recently begun to move from their solar origins into related fields like optoelectronics, e.g. the lenses of Miñano et al., and Terao et al.

The designs listed in Table 4.1 may be described by certain characteristics. Fresnel lenses can be designed according to the principles and assumptions itemized below. The number of designs in all tables in this section is by no means exhaustive.

Table 4.1. Major developments in Fresnel lenses suitable for solar applications

Year	Design reference	Application	Nonimaging optics	Numerical solution only	Cross-sectional angle θ	Perpendicular angle ψ	Minimum deviation prisms	Advanced/fashionable design	Thin lens, $f \gg d$	Prism size, $\Delta x \to 0$	Flat lens
1822	Fresnel [5]	lighthouse	–	–	–	–	–	–	–	–	○
1951	Miller et al. [107]	optoelectr.	–	–	–	–	–	–	–	○	○
1951	Boettner, Barnett [19]	optoelectr.	–	–	–	–	–	–	–	○	○
1977	Collares-Pereira et al. [35]	solar	–	–	–	–	–	–	○	○	–
1978	James, Williams [59], [60]	solar	–	–	–	–	–	–	–	○	○
1978	O'Neill [126]	solar	–	–	○	–	○	–	–	○	–
1979	Collares-Pereira [33]	solar	○	○	○	–	–	–	○	○	–
1979	Kritchman et al. [75], [76], [77]	solar	○	○	○	–	–	–	○	○	–
1981	Lorenzo, Luque [90], [91], [93]	solar	○	○	○	–	–	–	○	○	–
1984	Tver'yanovich [175]	solar	–	○	–	–	–	–	–	○	○
1997	Erismann [39]	optoelectr.	–	○	–	–	–	○	–	○	–
1999	Leutz et al. [83], [84]	solar	○	○	○	○	○	○	–	○	–
2000	O'Neill [129], [130], [133]	solar	–	–	○	–	○	○	–	○	–
2000	Miñano et al. [103]	optoelectr.	○	○	○	–	–	○	–	○	–
2000	Terao et al. [172]	optoelectr.	○	○	○	–	–	○	–	–	–

Design Principles

- Nonimaging optics: the lens in question is designed according to the principles of nonimaging optics, which were discovered in the 1970s, and found well suited for solar energy applications; see Sect. 2.1 for information on the edge ray principle and nonimaging concentration.
- Solution numerical only: the nature of the mathematics used in Fresnel lens design sometimes does not allow an analytical solution. If the number and extent of assumptions is to be reduced, numerical solutions may be called

for. As this generally requires a large number of iterations, computer-based simulations are needed.

- Cross-sectional angle θ: any nonimaging lens, whether of rotational symmetry (3D) or line focusing (2D), is designed with the help of a cross-sectional acceptance half-angle θ, which determines the part of the sky the lens may 'see'. Imaging lenses are generally designed to focus paraxial rays only, i.e. $\theta = 0$. However, the absorber, first, is larger than the focal *point*, and second, can be moved out of focus in such a way that some rays incident at an angle θ are collected. Another option is to create a 'red-edge' design of the imaging lens, i.e. use the refractive index of the longest desired wavelength to calculate the lens. Related is the absorber placement according to the circle of least confusion; see p. 71 for a description.
- Perpendicular angle ψ: the focal areas of line focusing, 2D lenses (imaging or nonimaging) are sensitive to changes of the angle of incidence not only in the cross-sectional plane, but also in the perpendicular plane. The significance of this has often been overlooked when designing linear Fresnel lenses for solar applications, where 2D characteristics are desired.
- Minimum deviation prism: optical losses and dispersion during refraction at a prism are least for the minimum deviation prism, where the angles of incident and exiting rays to the respective prism surface are equal.
- Fashionable design: solar energy applications, as well as manufacturing technologies usually require the outer side of the lens to be flat. Given this and the principle of minimum deviation prisms, an optimum convex shape of the Fresnel lens emerges from the simulation. By relaxing either principle, the design of lenses of other shapes is possible.
  'Fashionable' can mean the design of an aspherical lens in spherical shape (Erisman), for example. Any shape, provided that it is sensible enough not to obstruct the paths of refracted rays, is possible (Sect. 5.4). Again referring to Table 4.1, O'Neill's second lens is foldable and lightweight. Intended for powering satellites, it can be stowed during take-off. The nonimaging lens of Terao et al. uses total internal reflection (TIR) in the prisms for achieving a very low aspect ratio.

Assumptions

- Thin lens, $f \gg d$: if the thickness of the lens d becomes small in comparison to its focal length f, the sine theorem can be applied during the simplified simulation of the lens. With this assumption, it becomes possible to set up an equation describing the refracted rays emerging from a prism in relation to the prism center and absorber. The relocation of the rays during refraction in a prism is not considered.
- Prism size, $\Delta x \to 0$: the size of prisms in comparison to the absorber width is assumed to be small in all designs. Each prism is designed only once, but with some $\Delta x > 0$ extending from the design point, ideally in both directions. If that Δx becomes too large, the direction of the ray

emerging from one or both of the extremes of the prism may not point to the absorber. In combination with the edge ray principle of nonimaging designs, extreme rays from extreme ends of the prism will just miss the absorber, if the receiver size is matched to one ray from the center of the prism.
For imaging designs, any prism width greater than zero influences imaging quality, since the width of the ray emerging from the prism is always larger than the focal point, with zero width. In some imaging designs with very high optical quality requirements, one of the active prism surfaces is curved to account for this potential error, as was Fresnel's original center prism. Prisms cannot be made smaller than a multitude of wavelengths, if the rules of geometrical optics are to hold accurately (Sect. 8.2).

- Flat lens: Fresnel lenses, in particular imaging Fresnel lenses, are often assumed to be flat. This does not have to be the case, for imaging lenses can be shaped, just as nonimaging Fresnel lenses can be designed as flat lenses. Optimized nonimaging lenses with smooth outer surfaces must be shaped to allow for concentration of light incident from both the left and the right extremum given by the acceptance half-angle pair(s), if minimum deviation prisms are to be employed.
  Other advantages of the arched or domed shapes include the possibility of mounting the lens closer to the absorber, thus reducing chromatic aberrations, and a more rigid structure. Disadvantages are the more difficult design and production processes, and the chances of increased reflection losses at the lens surface and cosine losses at the absorber.

The usefulness of nonimaging optics for solar energy was demonstrated by the application of the (reflective, not refractive) compound parabolic concentrator (CPC) by Winston [186], refined by Rabl [142]. It did not take long, before the concept of nonimaging optics was applied to (refracting) Fresnel lenses. In the late 1970s, both Kritchman et al., and Collares-Pereira et al. designed bifocal Fresnel lenses, which are essentially nonimaging devices (for references, see Table 4.1). Both groups of designers intended to construct nonimaging lenses to enhance the ability of the lens to collect rays from a apparently moving sun within a cross-sectional pair of acceptance half-angles. O'Neil (1978) designed his successful bifocal lens, and realizes that this lens can compensate for tracking errors due to its prismatic structure. Lorenzo and Luque followed in 1981, and are followed by the work presented here, almost twenty years later, which is probably the first 'real' nonimaging design of a Fresnel lens, offering finite size prisms and a lens thick enough for manufacturing. These milestones of development have been listed in Table 4.1 to compare design principles and assumptions made.

Most probably, the late W. T. Welford deserves the honor to be credited with the initial idea for the design of curved lenses, based on his experiences with nonimaging optics that led to his and R. Winston's book [181], which has become a standard in the field. Collares-Pereira et al. [35] acknowledge a private communication with Welford in their first paper on the subject, which in turn is referred to in the first paper of Kritchman's group.

The developments in the design of Fresnel lenses for solar energy collection, and related applications has been summarized in Tables 4.2 and 4.3. Developments do not represent a continuous stream of improvements, as may be suggested by the keywords used in Tables 4.2 and 4.3 (refer also to Table 4.1). Not all authors were aware of previous or simultaneous advances. Some papers focus on the testing of conventional imaging Fresnel lenses; these have been included if the authors were able to draw conclusions of some significance for the application of Fresnel lenses in solar energy.

The distinction between imaging and nonimaging designs is strictly based on the definition given by Welford, previously related on p. 15, i.e. a classification that relates to the design approach taken. Bifocal, or 'red-edge' designs, are imaging designs, despite the fact that no image of the source is produced. These designs are presumingly achieved with a variation of the lens maker's formula, which is imaging design. This distinction is not always easy to make; also we are far from implying that imaging designs are inferior designs. Bifocal lenses are the most successful and advanced type of lens for the conversion of solar energy.

Typical flat Fresnel lenses are designed as focusing devices, i.e. the main parameter is the focal length of the lens. As long as certain assumptions are made, focusing lenses can be solved analytically. Such assumptions include the (reasonable) definition of the lens as thin, meaning that its focal length f must be much greater than its thickness d, $f \gg d$. Furthermore the lens is often assumed to have no thickness at all at the thinnest point, $d = 0$. The two refracting surfaces of the prisms constituting the body of the lens are usually not refined, although (with finite prism size) at least one of the surfaces must have an aspherical shape to account for focal aberrations [116].

Referring to Tables 4.1–4.3, two papers were published in 1951 (Miller et al.; Boettner and Barnett). The former group had designed and constructed Fresnel lenses for technical applications beginning in 1948, taking advantage of the new plastic materials. Their lens designs are of the imaging type; the papers show that the losses at a Fresnel lens are already well understood.

Research has focused on the development of evaluation techniques for Fresnel lenses under solar radiation. Ray tracing has been applied to find whether an incident ray hits an absorber or misses it. The foreshortening of the lens focus has destroyed many hopes of successfully using imaging Fresnel lenses as solar collectors. Tracking is an expensive option, and has only recently been developed to a stage of maturity. Harmon [51], who evaluates mass-produced focusing Fresnel lenses for their applicability in photovoltaics, probably captures what most results of simple lens analyses indicate, when he describes the lens as being

> ... an inefficient concentrator with losses that begin at 20 per cent and rise to about 80 per cent as the focal distance decreases. However,

Table 4.2. History of *imaging* Fresnel lenses for solar energy applications. Improved designs and related research by keyword. Representative publications

Year	Design reference	Design characteristics and remarks
1951	Miller et al. [107]	first plastic lens
1951	Boettner, Barnett [19]	plastic lens, flux calculation, production process
1961	Oshida [134, 135]	photovoltaic application, spectral considerations; also rotational prism sheets for daylighting
1973	Szulmayer [171]	extruded PVC strips
1975	Nelson et al. [121]	ray tracing, orientation of lens
1977	Collares-Pereira et al. [35]	curvature for very thin lens with high refractive index ($d \rightarrow 0$, $n \rightarrow \infty$), lens suggested by W.T. Welford, applied to two-stage concentrator
1977	Harmon [51]	evaluation for photovoltaic application
1978	Donovan et al. [36]	square lenses, large-scale application
1978	James, Williams [60]	flux uniformity for pv cells
1978	O'Neill [126]	bifocal, minimum deviation prisms, intentionally 'sloppy' design to cope with tracking errors and solar disk size
1978	Yatabe [190]	aberrations
1979	Nakata et al. [117]	bifocal design and test
1981	Shepard, Chan [158]	silicone substrate lens
1982	Moffat, Scharlack [110]	dome shape, injection molding
1984	Tver'yanovich [175]	five designs, including two-sided profile; coatings
1985	Mori [112]	multiple lenses and fiber optics for daylighting
1985	O'Neill [127]	three-dimensional lens in two-dimensional shape
1986	Franc et al. [41]	glass lens for use in buildings
1987	Mijatović et al. [105]	absorber iso-intensity problem
1989	Jebens [64]	uniform illumination, square lens
1989	Krasina et al. [73]	efficiencies
1991	Akhmedov et al. [1]	aspherical analytical solution
1992	O'Neill [128]	mass production of lens
1993	Soluyanov et al. [164]	imaging, accounts for solar disk size
1994	Kurtz et al. [78]	bifocal lens test for multijunction pv cells
1996	Grilikhes et al. [50]	imaging lens for space applications
1997	Erisman [39]	aspherical corrections in spherical shape
2000	O'Neill [130]	silicone stretch lens for space

Table 4.3. History of *nonimaging* Fresnel lenses for solar energy applications. Improved designs and related research by keyword. Representative publications

Year	Design reference	Design characteristics and remarks
1979	Colares-Pereira [33]	bifocal double-convex 'roof' lens, tracing of perpendicular rays in two-dimensional design
1979	Kritchman et al. [76]	bifocal design
1981	Lorenzo, Luque [89, 90, 91, 94]	ideal lens, evaluation and comparison
1984	Kritchman [75]	lens with secondary flow-line concentrator
1992	Miñano et al. [102]	lens–mirror integration
1994	Yoshioka et al. [191]	double elliptical rod lens, total internal reflection, not a Fresnel lens
1999	Leutz et al. [83]	cross-sectional and perpendicular acceptance half-angles, minimum deviation, optimum and other shapes
2000	Terao et al. [172]	total internal reflection

the lens is capable and adequate for low concentration purposes with photovoltaic systems. The most attractive aspects of using this lens as a solar concentrator are its availability and its potential low cost.

Designs can become more complicated than the typical focusing lens, when a cover or a coating is involved or when the lens sports grooves on both sides. In these cases analytical solutions can no longer be offered, and numerical calculations are introduced [175]. One reason why research has in the past often treated only the feasibility of existing focusing devices for solar energy applications might have been the lack of computing power for solving numerical problems for lenses with hundreds of prisms, which, done by hand, should be a mighty task indeed.

Shaped lenses (i.e. lenses with convex curvature) are usually not easily designed. Analytical solutions are not available for any lens other than the focusing dome lens, where the focal length f remains constant and the prisms are positioned along a curvature with radius f, measured from the focal point in the center of the lens system.

Contrary to imaging devices, nonimaging Fresnel lenses are designed starting not with the focal length of the lens, but with a pair of design angles, termed 'acceptance half-angles'. These acceptance half-angles open a 'window' in the hemisphere, facing the band of the relative movements of the sun, through which the collector 'sees' the sun, and through which solar rays can be received. In more technical terms, the collector's aperture is explicitly designed to accept rays incident from directions within certain angles.

As with to reflective mirrors, the concentrating behavior of lenses depends both on the incidence in the cross-sectional plane and on the incidence in the perpendicular plane. When designing collectors with line focus (2D), this makes a difference, as is seen from the discussion of the prism optics. Although Collares-Pereira [33] realizes the importance of the perpendicular fraction of the incoming ray when tracing rays through his lens, he did not include the perpendicular ray in the design of his lens. Neither did any of the other authors listed, even though lenses for solar energy applications of low and medium concentration should be used as 2D concentrators.

Collectors of rotational symmetry (3D) offer the possibility of higher concentration and flux ratios, but are usable only with accurate two-axis tracking. Line-focusing lenses are suitable for one-axis tracking, as the perpendicular design angle causes less severe focal foreshortening than the cross-sectional acceptance half-angle. Nonimaging Fresnel lenses can be used as stationary concentrators with low concentration ratios, thus overcoming the focal shortcomings of imaging devices.

Of the more sophisticated lenses mentioned, some have never left the computer of the designer, some prototypes have been built, and only the lens of O'Neill [126, 128, 131] has found its way into mass production. This is poised to change with the increasing availability of high-quality solar photovoltaic cells suitable for concentration.

A number of innovative prism/lens related designs for solar energy applications, e.g. in daylighting, should be mentioned to show the wide variety of possibilities for the utilization of the sun's power.

The 'Elephant Nose' of Oshida [134] consists of two prism sheets that can be rotated in such a way that the solar rays incident on the upper sheet, and refracted by both sheets a total of four times, become 'independent' of their original angle of incidence. This effect may simulate a sun at high altitude, even in morning or late afternoon hours.

One of the most interesting designs is the 'Sunflower' of a Japanese company [112]. A multitude of small imaging Fresnel lenses is assembled on a two-axis tracking mechanism. Solar rays are focused on the couplers of fiber-optic cables which transport the light with high efficiency ($\approx 90\%$) to a place where natural light is in demand, but not easily supplied. Refracted light is split into its colors, and the vertical positions of the foci of different colors are wavelength-dependent. The high accuracy of the 'sunflower's' design allows for splitting rays by adjusting the position of the coupler. This feature may be used in medicine, or agriculture, where some wavelengths are considered harmful.

The bulk, or rod, lens mentioned in Table 4.3 is of nonimaging design, but has little in common with Fresnel lenses, which are an assembly of prismatic structures. This bulk lens has the shape of two compounded inclined ellipses. Each one concentrates rays incident within some design angles. Seen in a cross-sectional plane, the edge ray is refracted onto the absorber edge that lies on the same side of the optical axis as the ellipse. In other nonimaging concentrators, the edge ray of one side is refracted towards the absorber edge of the opposite side of the optical axis. Furthermore, refraction in the rod happens only once before the ray reaches the absorber. Refraction in Fresnel lenses happens twice, and the number of reflections in a mirror-based design depends on the angle of incidence and the mirror shape. The concentration ratio of this type of lens is ≈ 2, depending on its design as a 2D or 3D concentrator ([191], [151], respectively).

A daylighting control mechanism utilizes the prism's property of total internal reflection for rays incident at an angle greater than the critical angle Φ_c. Lorenz [88] has proposed glazings with seasonally dependent solar transmittance based on the principle of total reflection. While active solar systems for the generation of heat and power, or light, are the focus of interest, passive systems should not be forgotten, as they serve to reduce energy loads where active systems are supplying energy.

4.3 Simple Fresnel Lenses

Imaging Fresnel lens design follows the same principles of geometrical optics that are also used for the design of other lenses, in particular the concepts of focal length and aperture. The design of some simple Fresnel lenses will be presented to enable the evaluation of focusing lenses for use as solar collectors.

Imaging Fresnel lenses are centered on the axis of the optical system. Imaging designs are generally realized for paraxial rays. There are no design angles or acceptance half-angles; in particular, the design processes for two-dimensional and three-dimensional lenses are identical, with the exception of the special treatment the corners of square 3D lenses receive if combined with circular absorbers.

The focal length of a plano-convex Fresnel lens condenser can be written with the lens maker's formula:

$$\frac{1}{f} = \frac{1}{i} + \frac{1}{o}, \tag{4.1}$$

where the conjugates, i.e. the distances of image and object points from the refracting surface of the thin lens, are denoted by i and o, respectively. In the case of the plano-convex collector lens, $o = \infty$, and $1/f = 1/i$.

The f/number is a measure of the aperture of the lens. It describes the ratio of the effective focal length to the diameter of the lens. The f/number is a measure of the flux concentration of the imaging lens. As the geometrical concentration ratio $C \to \infty$, the heat flux in the focal point is related to the amount of radiation concentrated:

$$\text{f/number} = \frac{f}{2R}, \tag{4.2}$$

where R denotes the distance of the extreme paraxial ray from the optical axis of the system. This is the ray that just passes through the aperture of the system. Smaller f/numbers mean larger apertures, and vice versa. Fresnel lenses are mostly free from spherical aberration since every step is designed separately for focusing. Due to their thinness, both absorption losses within the material and the change of those losses over the lens profile are small. Fresnel lenses can be designed as very 'fast' lenses, having a small f/number, and low aspect ratio. Only Fresnel lenses 'faster' than $f/0.5$, i.e. with a diameter twice their focal length are called 'impractical' [43].

Grooves In

A typical imaging Fresnel lens with grooves facing inwards will be presented here. The design has an analytical solution and follows Tver'yanovich [175]. In accordance with Fig. 4.4, three equations can be set up to describe the lens. The prism angle α is the goal of a simulation written as

$$n\,\sin\alpha = \sin\beta\ , \tag{4.3a}$$

$$\tan\omega = \frac{R}{f}\ , \tag{4.3b}$$

$$\beta = \alpha + \omega\ . \tag{4.3c}$$

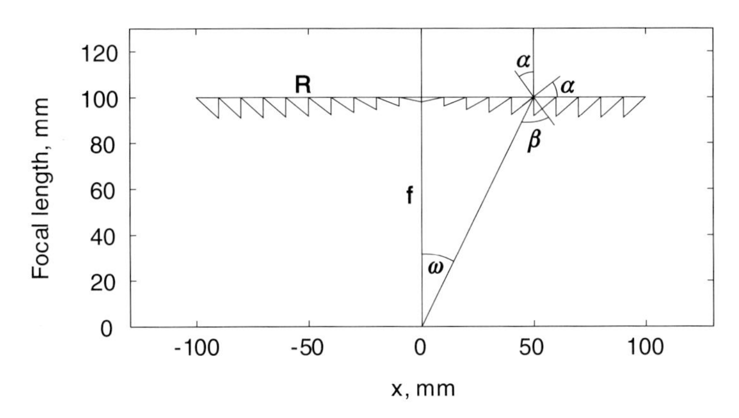

Fig. 4.4. Simple Fresnel lens with grooves facing inward. Analytical solution

Substituting β in (4.3a) with (4.3c) yields an expression for $\tan\alpha$:

$$n\,\sin\alpha = \sin(\alpha + \omega)\ ,$$

$$n\,\sin\alpha = \sin\alpha\,\cos\omega + \cos\alpha\,\sin\omega\ ,$$

$$\tan\alpha = \frac{\sin\omega}{n - \cos\omega}\ . \tag{4.4}$$

Since from (4.3b)

$$\sin\omega = \frac{R}{f}\cos\omega\ , \tag{4.5}$$

$\tan\alpha$ can be written as

$$\tan\alpha = \frac{R}{f}\frac{\cos\omega}{n-\cos\omega}\ . \tag{4.6}$$

Substituting $\cos\omega$ with the expression $f/(R^2+f^2)^{1/2}$ gives a final expression for the prism angle α in terms of focal length f and aperture R:

$$\tan\alpha = \frac{R}{n\sqrt{R^2+f^2}-f}\ . \tag{4.7}$$

The result of a simulation based on (4.7) is shown in Fig. 4.4. From the figure, it becomes obvious that the lens has at its thinnest points no thickness at all. The focal length is taken as $f = 100$ mm, and the aperture extends over $0 \le R \le R_{\max}$ by steps of $R/10$, where $R_{\max} = f$, to attain an f/number of 0.5 for this lens.

Grooves Out

A Fresnel lens fulfilling the same characteristics in terms of f and R can be designed with the grooves facing outward. Again, a simulation follows [175]. A set of four equations can be derived from Fig. 4.5:

$$\sin\alpha = n\ \sin\beta\ , \tag{4.8a}$$

$$-\alpha = \gamma - \beta\ , \tag{4.8b}$$

$$n\ \sin\gamma = \sin\omega\ , \tag{4.8c}$$

$$\tan\omega = \frac{R}{f}\ . \tag{4.8d}$$

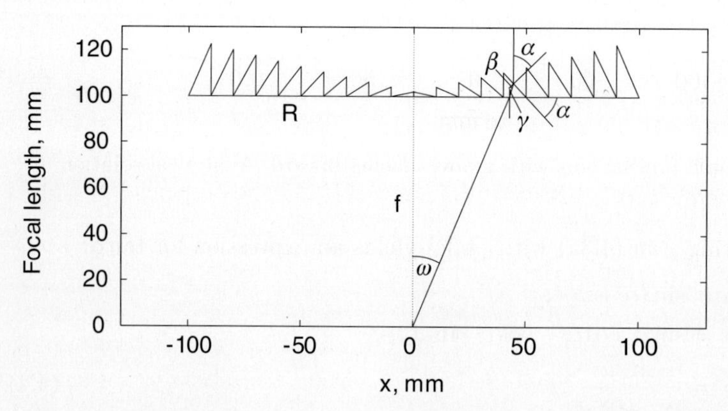

Fig. 4.5. Simple Fresnel lens with grooves facing outward. Analytical solution

In contrast to the lens with grooves facing inward, where the incoming paraxial ray is changed in its direction only once, in the case of the lens with grooves facing outward the paraxial ray is refracted twice. The angles β and γ account for this.

Beginning with the substitution of (4.8b) into (4.8a), an expression for $\tan\alpha$ can be found after some iterations:

$$\sin\alpha = n\,\sin(\gamma+\alpha)\ , \tag{4.9a}$$

$$\sin\alpha = n\,(\sin\gamma\,\cos\alpha + \cos\gamma\,\sin\alpha)\ , \tag{4.9b}$$

$$\tan\alpha = \frac{n\,\sin\gamma}{1-n\,\cos\gamma}\ . \tag{4.9c}$$

The numerator of (4.9c) is found from (4.8c) to be

$$n\,\sin\gamma = \sin\omega\ . \tag{4.10}$$

One may also want to express $1-n\,\cos\gamma$ in terms of ω. This is achieved starting from (4.10), which when squared can be written as

$$n^2\left(1-\cos^2\gamma\right) = \sin^2\omega\ . \tag{4.11}$$

Multiplying by n^2 and taking the square root results in an expression for the missing term of (4.9c):

$$n\,\cos\gamma = \sqrt{n^2-\sin^2\omega}\ . \tag{4.12}$$

The final equation reads as ($\sin\omega$ could be replaced by $R/\sqrt{R^2+f^2}$)

$$\tan\alpha = \frac{\sin\omega}{1-\sqrt{n^2-\sin^2\omega}}\ . \tag{4.13}$$

The prism angles resulting from the design of the two Fresnel lenses with grooves facing 'out' and 'in' can be compared. Figure 4.6 is a visualization of the calculus for lenses of different diameter. The f/number of the lens denotes the ratio of focal length to extreme paraxial ray for R, as previously shown in Figs. 4.4 and 4.5.

Figure 4.6 is evidence of the fact that Fresnel lenses designed in the presented way are aspherical. Were they merely reproductions of the spherical shape of, say, a plano-convex singlet, the prism angle would be equal to the tangent at the corresponding point on the aperture, similar to Buffon's echelon (see Fig. 4.1a).

It can be said that the lens with grooves facing outward is, due to the relatively larger prism angles, less prone to focal errors than the lens with grooves facing inward, where a small error in prism tilt close to the center of the lenses would lead to large errors due to off-axis focusing. The collection efficiency suffers. From this point of view, should point-focusing Fresnel lenses be used in the collection of solar energy, they should be lenses with the grooves facing outward.

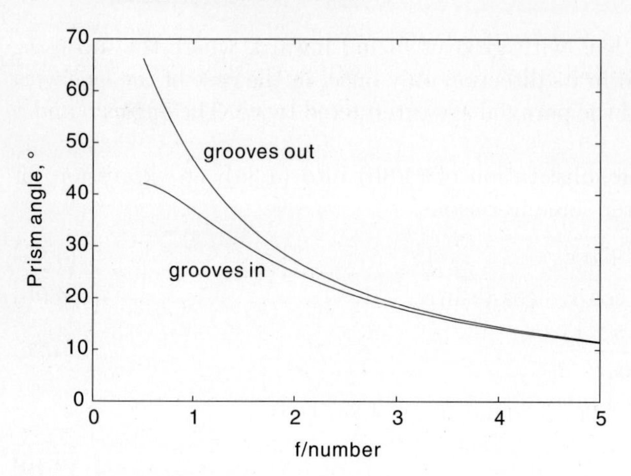

Fig. 4.6. Prism angles of Fresnel lenses with grooves facing 'out' and 'in'. The f/number denotes the ratio of focal length to diameter aperture of the lens. Paraxial rays, imaging collectors

Unfortunately, lenses sporting grooves 'out' are difficult to clean and are prone to shadowing losses, if the rays should not reach the lens in an exactly paraxial way. One might think of turning the lens around, so that the former grooves 'out' are now facing 'in'. This, however, may lead to total internal reflection at prisms distant from the optical axis due to the very large prism angles. Total internal reflection occurs if the refracted ray reaches the second surface at an angle greater than a critical angle. This critical angle was defined earlier for PMMA of $n' = 1.49$ to be $\Phi_c = 42.16°$ (see Fig. 3.2b).

Focal length shortening and off-axis shift becomes severe with larger incidence angles. There are cross-sectional θ-aberrations and perpendicular ψ-aberrations which depend on the respective angle of incidence and the refractive index of the lens material n. Off-axis lateral θ-aberrations happen in the cross-sectional plane, resulting in the focus shifting on a curve left or right and up. Longitudinal ψ-aberrations in the perpendicular plane foreshorten the back focal distance $B(\psi)$. The factors describing foreshortening in an imaging refractive concentrator are [99]

$$\ell_\theta \cong 1 - n\,(1 - \cos(\theta/n)) \ , \tag{4.14}$$

and for off-axis longitudinal ψ-aberrations in the perpendicular plane

$$\ell_\psi = \sqrt{1 - (1/n^2)\sin^2\psi} \ . \tag{4.15}$$

These equations for focal shortening as given are not accurate for all Fresnel lenses. There are strong differences according to which sides the grooves are facing; Meinel and Meinel [99] used flat cylindrical Fresnel lenses made of glass and fused silica for their analysis. The lenses are designed with grooves 'out', and are simply turned over for the measurements with grooves 'in'. In both cases, both off-axis and longitudinal errors are severe, and high precision tracking is imperative. Every imaging optical system has its own Petzval surface, where point images are formed for various incidence angles.

Circle of Least Confusion

Imaging Fresnel lenses produce inhomogeneous color flux in and near the focus. To minimize the effects, the absorber is usually placed at the circle of least confusion (CLC; [20, 52]), which is located closer to the lens than the actual focus. Dispersion becomes progressively greater in optical materials like glass or polymethylmethacrylate (PMMA) towards wavelengths shorter than that of yellow light used for the design of the system. The CLC is located where the refracted rays of the longest wavelength required to reach the absorber from the right-hand side of the lens and the refracted ray of the shortest wavelength needed on the absorber from the left-hand side of the lens (and vice versa) intersect, as shown in Fig. 4.7. 'Right' and 'left' are the sides on the right- and left-hand sides of the optical axis of the system defined in a cross-sectional view.

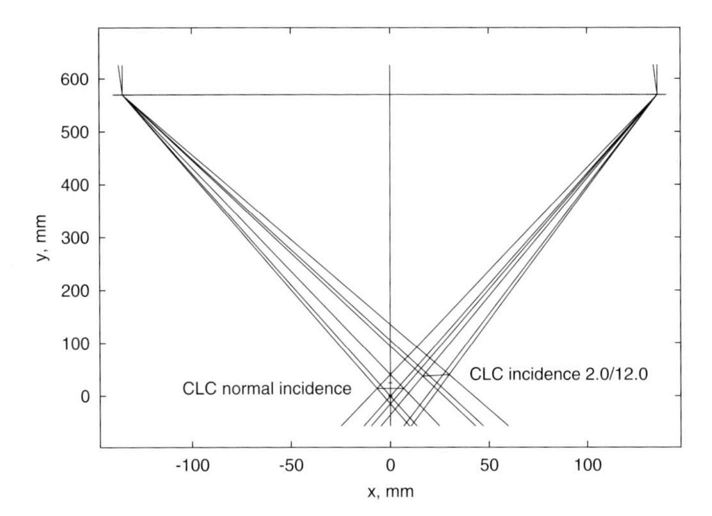

Fig. 4.7. Circle of least confusion (CLC) for two combinations of angles of incidence (cross-sectional θ/perpendicular ψ, in degrees) in a linear imaging Fresnel lens (f/number = 2.03). Not to scale

In the example the extreme wavelengths are determined by the response range of Si–photovoltaic cells, i.e. 300 and 1200 nm. Although the absorber size itself does not change in Fig. 4.7, focal errors are severe. For comparison, the absorber of a nonimaging Fresnel lens with similar aperture would be approximately 20 mm wide, and it would still catch most of the refracted

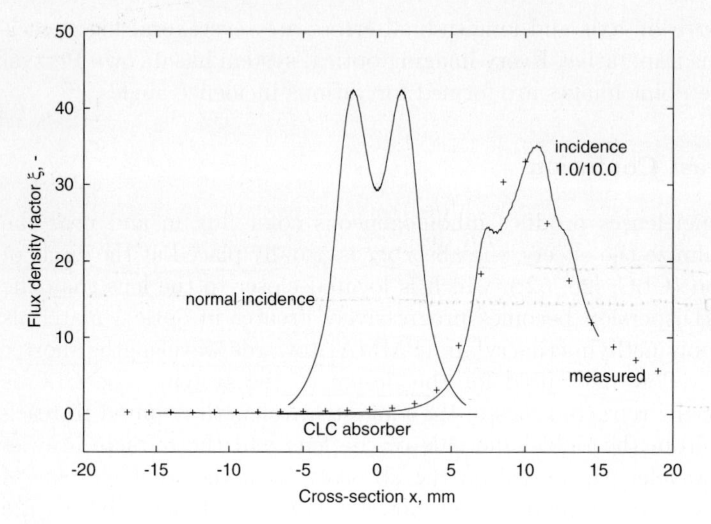

Fig. 4.8. Flux density factor of imaging lens for two combinations of angles of incidence (cross-sectional θ/perpendicular ψ, in degrees) in a linear imaging Fresnel lens (f/number = 2.03), compared with outdoor flux density measurement

rays at $\theta = \pm 2°/\psi = \pm 12°$. The imaging lens absorber must be three to four times wider to achieve similar tracking insensitivity.

The flux density for the same imaging lens under the sun (including solar disk size and terrestrial solar spectrum AM1.5D) is simulated and plotted in Fig. 4.8. The absorber designed by the CLC method, using normal incidence, is useful only for tracking accuracies smaller than 0.5° around the primary axis. At the nonimaging lens, the absorber edge would be touched by rays at an incidence of $\theta = \pm 2°/\psi = \pm 12°$. The flux density under the nonimaging lens of $\theta = \pm 2°/\psi = \pm 12°$ would be lower by a factor of about two. The focal distance or the aspect ratio of the nonimaging lens is half the value of that of the imaging lens.

A last remark (these topics are discussed in Sect. 9.7) concerns the color mixing in the flux delivered by the lens. Uniform illumination and homogeneous color distribution are important for optimum performance of multi-junction photovoltaic cells. The two flux peaks in Fig. 4.8 hint at insufficient color mixing. The lens described here is a commercially available Fresnel lens, originally designed for a copy machine, and not optimized for the collection of solar energy. Clearly, this imaging lens should not be used as a solar concentrator.

4.4 Domed or Arched Fresnel Lenses

Dome-shaped (3D) or arched (2D) lenses can be designed when prisms are chained along a semicircle centered at the focal point. The focal length is kept constant, and doubles as the radius of the semicircle. A shaped lens offers some advantages over flat lenses, namely:

- the shaped lens offers increased mechanical stability, and is more suited for use as a collector cover, as long as the grooves are located 'in' for easy cleaning;
- the shaped lens should reduce focal aberrations to some extent, as the distance between prisms and the designed focal point is kept constant, and smaller than in flat lenses;
- the shaped lens could gain some advantages under aesthetic considerations.

The disadvantages of the shaped lens include:

- its somewhat more complicated manufacturing process. The use of rollers after extruding the lens will be more difficult. For injection molding, the mold will have to be more sophisticated; when prisms are designed with undercut, the molding will use a collapsible core [110], and rolling will be impossible entirely (manufacturing technologies and problem circumnavigation techniques are discussed in Chap. 8.);
- the possibility of increased reflection losses due to the larger angles of incidence at the outer surface of the lens;
- increased cosine losses at the absorber for light refracted from the lower reaches of the lens.

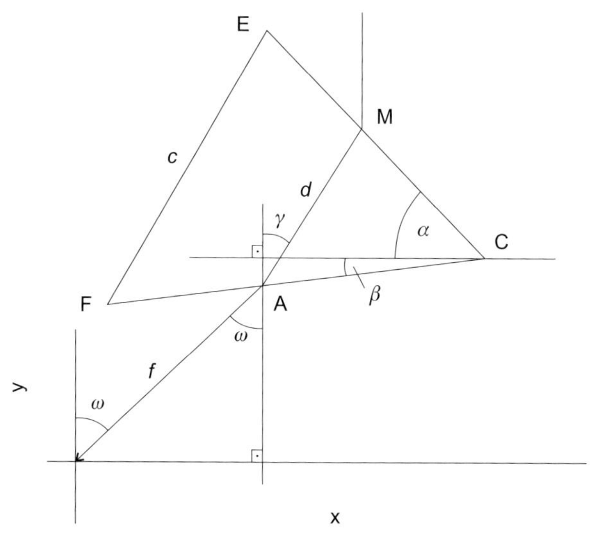

Fig. 4.9. Nomenclature for dome-shaped Fresnel lens, for enlarged prism from the right-hand side of the lens. For $f \gg d$: $f \approx f + d$; $\alpha \approx \omega$; $c||d$

Starting out from the parameters depicted in Fig. 4.9, the following two equations can be set up:

$$\sin\alpha = n\ \sin(\alpha - \gamma)\ , \tag{4.16a}$$
$$n\ \sin(\beta + \gamma) = \sin(\beta + \omega)\ . \tag{4.16b}$$

With $\alpha = \omega$, (4.16a) can be solved for γ via

$$\sin\omega = n\ \sin(\omega - \gamma)\ ,$$
$$\omega - \gamma = \arcsin\left(\frac{\sin\omega}{n}\right)\ . \tag{4.17}$$

Implicitly solved for this case, γ is

$$\gamma = \omega - \arcsin\left(\frac{\sin\omega}{n}\right)\ . \tag{4.18}$$

The second equation, (4.16b) is transformed to

$$n = \frac{\sin(\beta + \omega)}{\sin(\beta + \gamma)}\ . \tag{4.19}$$

With the theorem that

$$\frac{a}{b} = \frac{c}{d} \Leftrightarrow \frac{a+b}{a-b} = \frac{c+d}{c-d}\ , \tag{4.20}$$

it follows that

$$\frac{n+1}{n-1} = \frac{\sin(\beta+\omega) + \sin(\beta+\gamma)}{\sin(\beta+\omega) - \sin(\beta+\gamma)}\ . \tag{4.21}$$

From two theorems

$$\sin x + \sin y = 2\sin\frac{x+y}{2}\cos\frac{x-y}{2}\ , \tag{4.22a}$$
$$\sin x - \sin y = 2\cos\frac{x+y}{2}\sin\frac{x-y}{2}\ , \tag{4.22b}$$

it is found that (4.21) can be written as

$$\frac{n+1}{n-1} = \frac{2\sin\left(\frac{2\beta+\omega+\gamma}{2}\right)\ \cos\left(\frac{\omega-\gamma}{2}\right)}{2\cos\left(\frac{2\beta+\omega+\gamma}{2}\right)\ \sin\left(\frac{\omega-\gamma}{2}\right)}\ , \tag{4.23}$$

which can be simplified to

$$\frac{n+1}{n-1} = \tan\left(\beta + \frac{\omega+\gamma}{2}\right)\ \cot\left(\frac{\omega-\gamma}{2}\right)\ . \tag{4.24}$$

Implicitly, again, one finds the final expression for β as

$$\beta = \arctan\left(\frac{n+1}{n-1}\ \tan\left(\frac{\omega-\gamma}{2}\right)\right) - \frac{\omega+\gamma}{2}\ . \tag{4.25}$$

With (4.18) and (4.25), for γ and the prism inclination angle β, respectively, Fig. 4.10 has been derived. The design of the lens is determined by the fixed focal length, which in turn determines the prism angle α (as defined in Fig. 4.9).

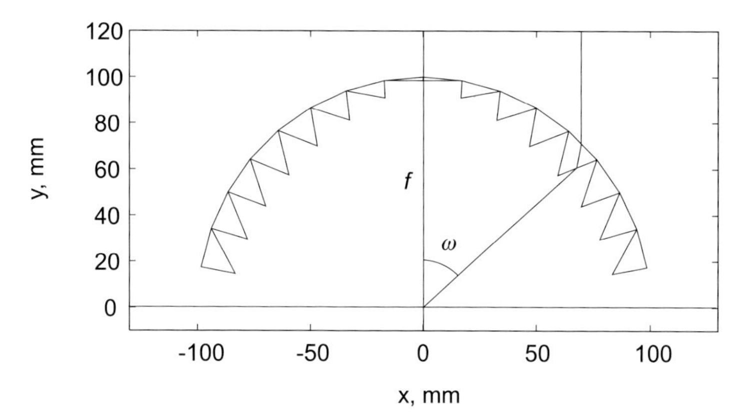

Fig. 4.10. Dome-shaped Fresnel lens with grooves facing inward. Analytical solution for constant focal length

This curved Fresnel lens is only slightly more difficult to calculate than the flat lenses presented above. The assumptions basically stay the same: the prism is thought to be thin, i.e. the distance $\overline{AM}$ is small, actually zero at point C. The prism is designed with the focal length f pointing towards point E for reasons of simplicity of simulation. Thus, the prism is not centered on the incident ray, but rather moved outwards from it. If the number of prisms is large enough, this method of calculation will have a negligible influence on the performance of this imaging lens.

Another result of the calculation of the dome lens has been presented by an unknown author [176]. Her or his final equation yields the same result as the calculus presented above, with a congruent outcome of Fig. 4.10. The formula, dating from 1975, reads

$$\delta = \arctan\left(\frac{\sin\omega}{n\,\cos\left(\arcsin\left(\frac{\sin\omega}{n}\right)\right) - 1}\right). \tag{4.26}$$

Note that here $\delta = \alpha + \beta$, as defined in Fig. 4.9. This equation is more elegant than (4.18) and (4.25). Since the results are the same, it can be understood that the same assumption of a thin lens underlies the two variations presented here.

5 Nonimaging Fresnel Lens Design

5.1 Applied Nonimaging Lens Design

A Fresnel lens is essentially a chain of prisms. Each prism represents the slope of the lens surface, but without the material of the full body of the conventional singlet. The imaging Fresnel lens refracts light from an object and forms an image in the focal plane. Aberrations have an impact on the quality of the image; the imaging lens should be free from such distortions. Due to the inaccurate manufacturing of the prism tips and grooves the image quality of Fresnel lenses is inferior to the images produced by aspherical full lenses.

The nonimaging Fresnel lens is designed with the objective of concentrating light rather than forming an image. Photographic accuracy is not the goal of a lens intended for solar energy applications. Means that help to create the perfect image are of secondary importance. The image forming requirements are relaxed to the extent that the incident light has to only reach the absorber, no matter where. The main goal of the design of a nonimaging Fresnel lens is to maximize the amount (energy) and quality (flux uniformity) of solar radiation concentrated by the lens.

Fresnel lens design often assumes thin lenses, i.e. thin prisms. 'Thin' stands for a thickness of zero. A thin imaging Fresnel lens can be designed analytically, as was shown previously with the design of flat as well as dome-shaped examples. This leads to considering the shape of the nonimaging Fresnel lens.

Kritchman et al. [76] dismiss the use of nonimaging flat Fresnel lenses on technical grounds. They require their lens to focus all rays of incidence θ_{in} onto one edge of the absorber; the focal length is required to be a constant. This is possible only for a spherical lens, but the flat lens can be a nonimaging concentrator, since the exactly defined focal point is a characteristic of imaging optics only (Kritchman et al. themselves suggest that this focal criterion be relaxed, and require the refracted rays to only reach a section of the absorber). The nonimaging flat Fresnel lens is possible ([90, 93, 94]), and represents the logical extreme of a curved lens, although the advantages of the convex shaped lens are significant ([33, 35]). A discussion of the shape of the ideal concentrator can be found in Sect. 6.5.

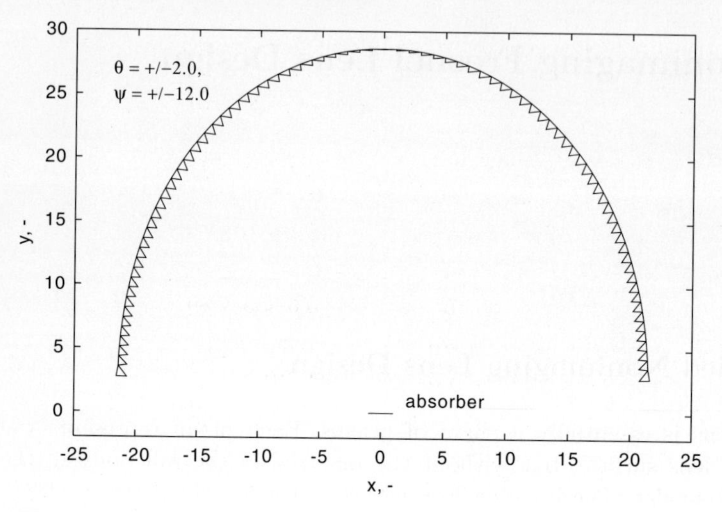

Fig. 5.1. The optimum nonimaging Fresnel lens. Cross-sectional acceptance half-angle $\pm\theta$; perpendicular acceptance half-angle $\pm\psi$. Oversized prisms

The nonimaging Fresnel lens is defined as 'optimum' if the following conditions are fulfilled, resulting in an arched linear lens (Fig. 5.1):

- the simulation has to yield minimum deviation prisms for minimum dispersion, and the smallest possible absorber size;
- the outer surface of the lens must be smooth to eliminate blocking losses on the outside. Also, this allows for cleaning of the lens.

The parameters to control during the design of the nonimaging lens are the following:

- the cross-sectional acceptance half-angle pair $\pm\theta$;
- the perpendicular acceptance half-angle pair $\pm\psi$;
- the angle ω that divides the aperture of the lens into equal segments (like the spokes of a wheel) and subsequently defines the pitch Δx by which the projected width of a prism is defined. This also limits the number of prisms constituting the lens. The designer may later set a fraction of this angle ω as a backstepping angle to start a new prism on the face of the one just finished, thus achieving a finite thickness of the lens;
- the (horizontal) prism inclination α for each prism;
- the prism angle β opening each prism;
- the average refractive index of the material n;
- an error margin ΔE for the numerical simulation;

These parameters can be set to fulfill the conditions given above, and define a two-dimensional, convex shaped, optimum nonimaging Fresnel lens of finite thickness. The lens is designed as the first aperture arched over

Table 5.1. Typical design parameters of the nonimaging Fresnel lens

Parameter	Designation	Typical values
Cross-sectional acceptance half-angle	θ	$\pm(0.5 \leq \theta \leq 30°)$
Perpendicular acceptance half-angle	ψ	$\pm(0.0 \leq \psi \leq 60°)$
Aperture segment	ω	$0.2 \leq \omega \leq 5.0°$
Prism inclination	α	$0.0 \geq \alpha \geq -20°$
Prism angle	β	$0.0 \leq \beta \leq 80°$
Refractive index	n_D	1.49
Error margin	ΔE	$1 \cdot 10^{-6}$

the second, or exit aperture, which is the absorber of unit width. Typical parameters for these design parameters are given in Table 5.1. Optimum values for the optimum lens, designed for specific solar radiation conditions, are given in Chap. 7.

The shape of the lens is defined by the condition of obtaining a smooth outer surface. Once this requirement is relaxed, the lens can assume any shape and still maintain minimum deviation prisms, as will be shown in Sect. 5.4.

5.2 The Optimum Linear Lens

The nonimaging lens of finite thickness cannot be completely described in analytical terms, and a numerical solution has to be found. With the help of a computerized design process the numerical design procedure can be run in an acceptable time for any level of accuracy desired.

Each prism in the chain forming the nonimaging lens is the result of many iterations. Figure 5.2 exemplifies this process for a prism in the upper right-hand half of a lens with the acceptance half-angle pairs $\theta = \pm 30°$ and $\psi = \pm 45°$. The units are multiples of the absorber half-length, which serves as a reference length of unity. The large number of iterations (at moderate accuracy requirements) can clearly be seen. The numerical algorithm, Newton's method, is used twice within some three infinity loops during the design process of the pictured prism, as it is the case for all other prisms until the design stops upon a new prism reaching the absorber level.

Design Process Overview

The design process will be explained in detail later. The following is a brief overview of the steps the program follows. The flow chart of the program for the design of the nonimaging lens is given in Fig. 5.3.

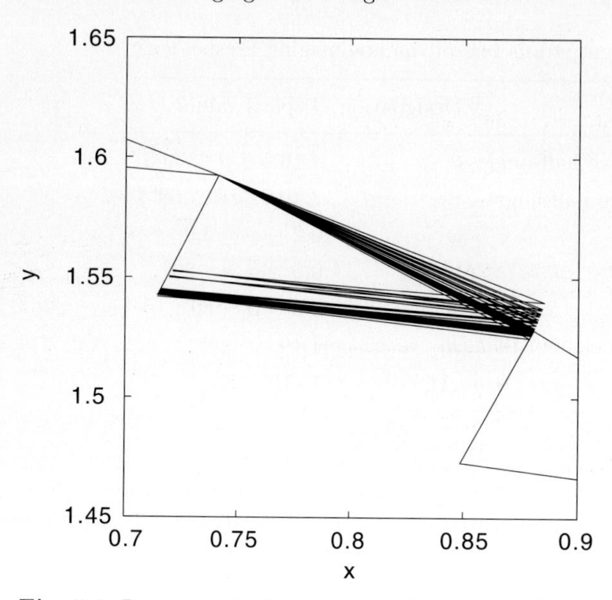

Fig. 5.2. Iterations during the numerical design of a prism in the upper right-hand of a nonimaging Fresnel lens with a cross-sectional acceptance half-angle $\theta = \pm 30°$ and a perpendicular acceptance half-angle $\psi = \pm 45°$. The axis lengths are multiples of the absorber half-width $d = 1.0$

The designer adjusts the variables the design depends on, i.e. the cross-sectional acceptance half-angle pair $\pm\theta$, the perpendicular acceptance half-angle pair $\pm\psi$, the angle ω that divides the aperture of the lens into equal segments, the prism inclination α, and the prism angle β, as well as the refractive index of the material n. An error margin ΔE is defined to determine the accuracy of the numerical solutions (Table 5.1).

After the design parameters are handed over to the program, the height of the lens over the absorber level is defined. The optical axis of the system is the starting point. From here on, the prisms on the right-hand side of the lens are calculated one by one until a next prism would reach the absorber level. Incidence within the acceptance half-angle pairs is simulated. The prisms are to be inclined and sized in order to be able to concentrate solar incidence within the edge rays $(+\theta\psi, -\theta\psi)$ onto the absorber.

The program sets an initial value for the prism inclination α. The edge rays are incident from both the left at $+\theta$, and from the right at $-\theta$, seen in the cross-sectional plane. The perpendicular acceptance half-angles $\pm\psi$ are taken into account later during the evaluation whether the refracted ray hits the absorber. The perpendicular incidence is symmetrical in the case of the 2D lens, and only one half-angle ψ has to be considered. In the case of the 3D lens, $\psi = 0$.

With the setting of the prism angle β the procedure enters two infinity-loops: only if a solution is found that satisfies the error margin in a comparison

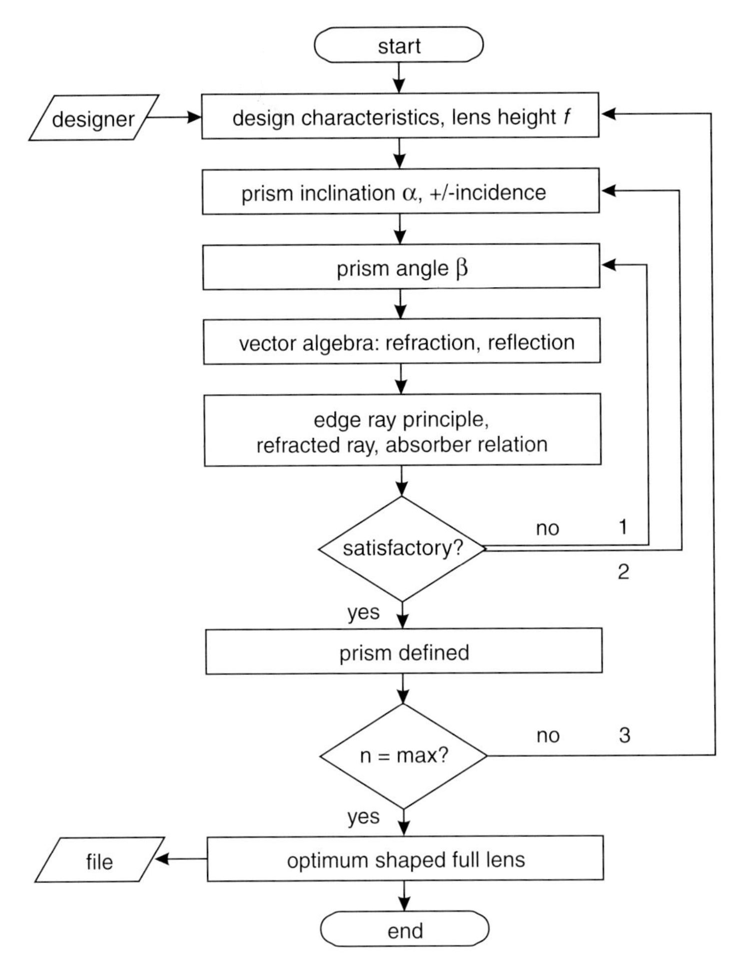

Fig. 5.3. Flow chart of the program that designs the shaped 2D nonimaging Fresnel lens. n denotes the number of prisms

of the path of the extreme refracted rays and the vectors pointing to the absorber's edges for *both* directions of incidence, do loops one and two (see Fig. 5.3) yield the final design of the prism. The numerical solutions in these loops are found with the help of Newton's method.

Within the infinity loops, the refraction is calculated for the extreme rays, and total internal reflection is checked. Every ray is represented by a vector. The edge ray principle is applied for evaluating the relation between refracted rays and the positioning of the prism over the absorber. If the edge rays for both directions of incidence (including the increased aberrations due to perpendicular incidence) reach the absorber within its edges, the prism is thought to satisfy the nonimaging design conditions, and its corner coordinates are finalized.

The angle ω turns the design towards a new prism, and the procedure is repeated until a next prism would reach the absorber level. The final product is the optimum-shaped nonimaging Fresnel lens, ready for evaluation.

Prism Inclination

A thorough explanation of the design procedure is as follows. To begin with, the height of the lens center over the absorber level f (the focal plane, or exit aperture) is determined. Initially, the first prism is assumed to be a horizontal plate of negligible thickness. Neither the shift of the twice-refracted ray normally observed during refraction at a thick plane-parallel plate nor the influence of the perpendicular angle ψ have to be considered. With the absorber half-length being the reference length $d = 1.0$:

$$f = \frac{d}{\tan\theta} \; . \tag{5.1}$$

From this reference point, the position and angles of the first prism to the right of the optical axis of the lens are determined. All succeeding prisms start where the previous one was finalized. Initial angles for the adjacent prisms are the final ones of the previous prism. Only the right-hand side of the lens is determined. Since the lens is symmetrical, the left-hand side of the lens is mirrored from the right-hand side.

The width of a prism is decided by the lens designer by setting the angle ω. Like the spokes of a wheel, the next prism will be set further to the right and further down towards the absorber level. Figure 5.4 explains how this is done.

The downward slope of the prism's inclination is defined as negative. The prism pitch Δx is found to be

$$\Delta x = \overline{AB}\cos\left(\beta - \alpha\right) \; . \tag{5.2}$$

From $\overline{BD}$ which is known from the position of the previous prism, and ω, we have

$$\overline{AB} = \overline{BD}\omega \; . \tag{5.3}$$

And so the prism pitch in x direction is given by

$$\Delta x = \cos\left(\beta - \alpha\right)\overline{BD}\omega \; . \tag{5.4}$$

The pitch of the prism in the y direction is then

$$\Delta y = \Delta x \tan\left(\beta - \alpha\right) \; . \tag{5.5}$$

At the first prism, the prism inclination angle α was set to zero. With two directions of incidence ($+\theta\psi$, $-\theta\psi$), two values for the prism angle β are found. Both angles are determined by Newton's method in infinity loops.

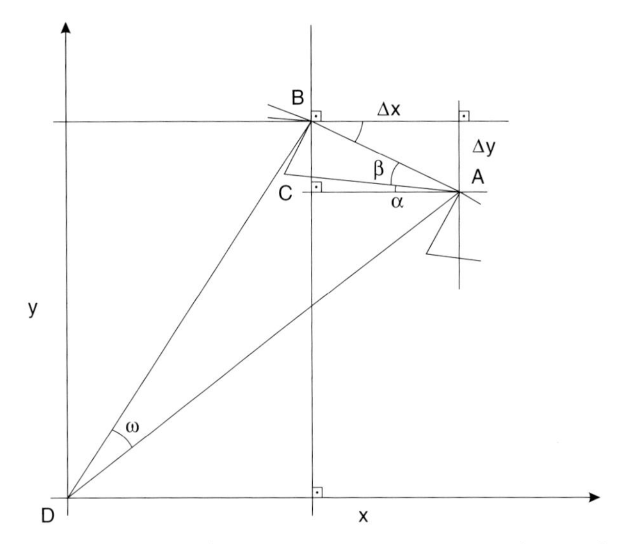

Fig. 5.4. Pitch of the lens. Each prism is moved towards the right and towards the absorber level by the angle ω

Imagine the result for the first iteration with $\alpha = 0$ (refer to Fig. 5.3, loop 1) does not yield two prism angles β_1 and β_2 for which

$$|\beta_1 - \beta_2| < \Delta E \ , \tag{5.6}$$

where ΔE is the error margin or a confidence level. If the condition in (5.6) is not fulfilled, the program enters into the infinity loop, and the prism inclination angle α is decreased by $\Delta\alpha = 1°$. Two new values (one for each direction of incidence) for the prism angle α are found. If the condition (5.6) remains unfulfilled, Newton's method is employed to find a new value for the prism inclination angle α.

Newton's Method

Newton's method [106] makes use of the derivative $f'(x)$ of a function $f(x)$ to determine a new value for x, thus approximating the root of a problem. Figure 5.5 is a sketch of the procedure generally used.

Formally, Newton's method requires a first solution to a function. This initial value x_1 is often a guess:

$$y_1 = f(x_1) \ . \tag{5.7}$$

The derivative of the function is (graphically) the tangent to this function at the point (x_1, y_1). From the slope of the tangent, a new value x_2 can be determined:

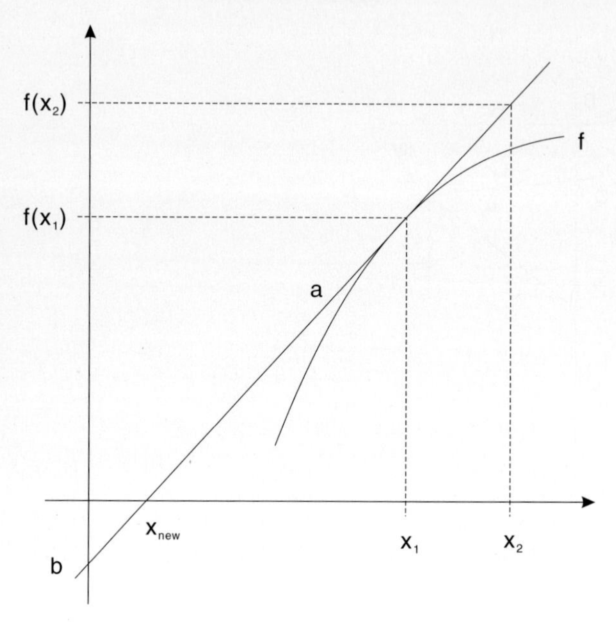

Fig. 5.5. Newton's method. The tangent (derivative) of a function $f(x)$ leads to a new root of the problem

$$f'(x_1) = \frac{y_1}{x_1 - x_2},$$
$$x_2 = x_1 - \frac{y_1}{f'(x_1)},$$
$$x_2 = x_1 - \frac{f(x_1)}{f'(x_1)}. \quad (5.8)$$

Or generally,

$$x_{i+1} = x_i - \frac{f(x_i)}{f'(x_i)}. \quad (5.9)$$

The new value for x is inserted into the algorithm, and yields a better solution as long as the function is well-behaved or changing slowly and smoothly. Although Newton's method works best for functions that have only one solution, it can be employed for functions with two roots, too. Guessing the initial values becomes more difficult then.

The procedure outlined in (5.7–5.9) is the general case. For the determination of the prism's inclination angle α, Newton's method has to be slightly modified. No formula for the function of α, and thus no derivative, can be given. Instead, two sets (x_1, y_1) and (x_2, y_2) have been calculated as results of the initial $\alpha_1 = 0°$, and the one decrease to $\alpha_2 = -1°$ representing the x_i. The y_i are the differences $\left|\beta_i^{+\theta\psi} - \beta_i^{-\theta\psi}\right|$ obtained for the prism angles β.

The indices here represent the direction of incidence $\pm\theta$. The slope a of the tangent can be determined to be

$$a = \frac{\Delta\beta_1 - \Delta\beta_2}{\alpha_1 - \alpha_2} \; . \tag{5.10}$$

As in the general version of Newton's method, the new value for x (here: α) is found as the point where the tangent intersects the abscissa. The form for a linear equation is

$$y = ax + b \; . \tag{5.11}$$

The y axis is intersected accordingly at a point b defined by

$$b = y - ax \; ,$$
$$b = \Delta\beta_1 - \frac{a}{\alpha_1} \; . \tag{5.12}$$

The intersection with the abscissa (the new value for x) is found for $y = 0$

$$x = -\frac{b}{a} \; . \tag{5.13}$$

This new value x_{new} for α should be closer to the optimum solution of the task to find a prism inclination angle α that suits incidence from both directions. Within the outer infinity loop of the program, Newton's method is applied for new pairs of (α_1, α_2) where the first is a result of Newton's method, and the second is the first one decreased by $\Delta\alpha$. After a few iterations, the prism inclination angle α is set; the criterion of (5.6) has been fulfilled.

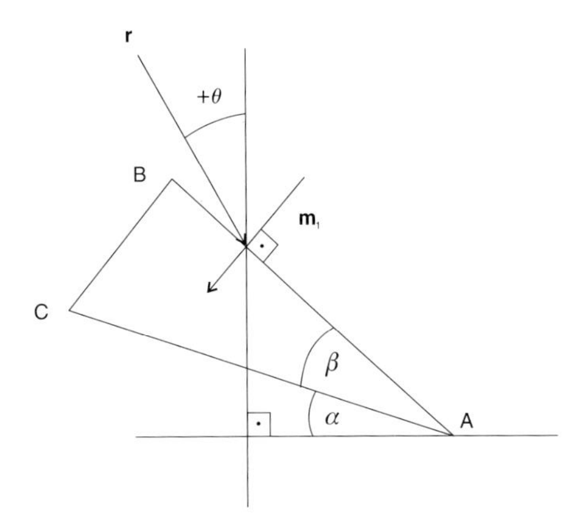

Fig. 5.6. Shading by a previous prism is the break condition in the infinity loop determining the prism inclination angle α

Shading

When the prism under consideration is rather low at the right-hand side of the lens, close to the absorber level, the prism angle β becomes large; the rays incident at a maximum angle $+\theta\psi$ from the left-hand side, could be shaded by a previous prism. Shading happens when the scalar product of the incident ray and the prism face normal becomes negative (see Fig. 5.6). If this condition

$$\boldsymbol{r} \bullet \boldsymbol{m}_1 < 0 \tag{5.14}$$

is true, then the infinity loop determining the optimum prism inclination angle α is broken, and the last value for α is kept constant and used for all subsequent prisms. The prism angle β, however, is increased further in order to optimize the prism for incidence $-\theta$ from the right-hand side of the lens, and for the perpendicular incidence ψ.

Prism Angle

A procedure very similar to the one outlined above for the optimization of the prism inclination angle α is employed for finding the optimum prism angle β. This time the condition to fulfill describes the position of the prism relative to the absorber in its relation to the direction of the refracted rays $\boldsymbol{q}^{+\theta\psi}$ and $\boldsymbol{q}^{-\theta\psi}$. These relations are shown in Fig. 5.7.

The prism position over the absorber is determined via two positioning vectors which describe the center point of the prism bottom in its position with respect to either end of the absorber. Incidence on the prism from the left and right should hit the absorber after refraction within the limits of the right or the left end of the absorber, respectively. The prism has to be designed in such a way, that this edge ray principle can be maintained. Thus, if the vector pair $\boldsymbol{q}^{-\theta\psi}$ and $\boldsymbol{d}^{-\theta\psi}$ as well as the vector pair $\boldsymbol{q}^{+\theta\psi}$ and $\boldsymbol{d}^{+\theta\psi}$ can be kept parallel, all rays formerly incident within the limits of the acceptance half-angles, and now leaving the prism after two refractions, will hit the absorber within its outer limits. It has to be noted that the vectors $\boldsymbol{q}$ in Fig. 5.7 are plotted to scale, and the vectors $\boldsymbol{d}$ are not.

The optimization criterion for Newton's method and the setting of the prism angle β (the setting of the prism inclination angle α in the outer loop was given in (5.6)) is

$$\left| \frac{\boldsymbol{q}.x}{\boldsymbol{q}.y} - \frac{\boldsymbol{d}.x}{\boldsymbol{d}.y} \right| < \Delta E \, . \tag{5.15}$$

Only when condition (5.15) is satisfied is the inner infinity loop broken, and the prism angle is found. This supplies a value for the outer infinity loop where the condition of (5.6) has to be met for the completion of the prism under design. It should be noted that the condition (5.15) includes the three-dimensional calculation of $\boldsymbol{d}$ and $\boldsymbol{q}$.

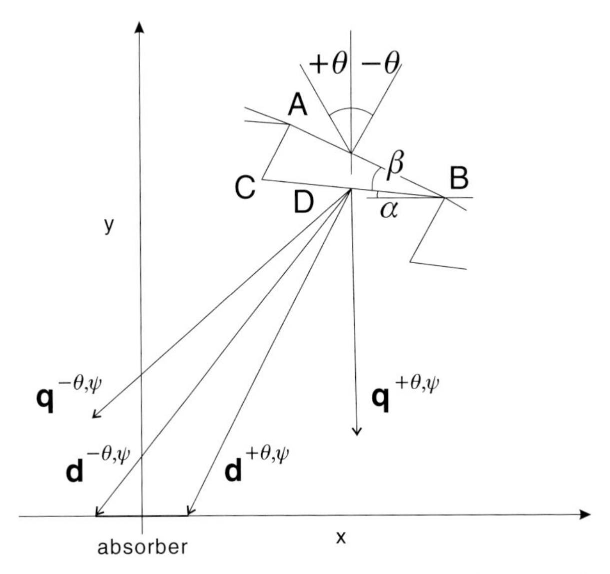

Fig. 5.7. Evaluating the prism position relative to the absorber. Refracted edge rays for incidence from both sides, depending on $\pm\psi$ (vectors $\boldsymbol{q}^{+\theta\psi}$ and $\boldsymbol{q}^{-\theta\psi}$), are compared with the prism position vectors relative to either end of the absorber $\boldsymbol{d}^{+\theta\psi}$ and $\boldsymbol{d}^{-\theta\psi}$. The vectors $\boldsymbol{q}$ are plotted to scale, but the vectors $\boldsymbol{d}$ are not. Projection onto cross-sectional plane

Minimum Deviation Prisms

The symmetry in the incidence angles $\pm\theta$ and ψ and in the condition determining the vectors $\boldsymbol{d}^{\pm\theta\psi}$ reaching the edges of the symmetrical absorber, as well as the quasi-symmetry in splitting the design of the prism into the calculation of two dependent angles α and β, results in the creation of prisms that are close to minimum deviation prisms.

Minimum deviation means minimum dispersion (Sect. 3.5). Minimum dispersion means the smallest possible absorber, if color aberrations are considered, or the least amount of losses due to color aberration, if the absorber is designed for yellow light.

The refractive power of the prism at incidence fulfilling the conditions of minimum deviation is as small as it can be. The symmetry expressed in the condition (5.15) leads the vectors $\boldsymbol{d}$ and $\boldsymbol{q}$ to enclose the imaginative vector of minimum deviation. Parallelity is required, and can be approximated by 'reversible' prisms, or refracted rays (vectors) surrounding the ray that would result from refraction at the minimum deviation prism.

Of course, one prism can have only one angle of minimum deviation, but the design described here yields paths for both edge rays that are reversible. For the maximum angle of incidence on the first surface from the left-hand side $\phi_1^{+\theta}$, an angle of refraction on the second surface $\phi_2^{+\theta}$ is recorded, where the latter approximately coincides with the angle of maximum incidence from the right-hand side $\phi_1^{-\theta}$, and the former roughly equals the angle of refraction for incidence from the right-hand side on the second surface $\phi_2^{-\theta}$. This behavior has previously been shown in Fig. 3.5.

Although minimum deviation happens only for one angle of incidence on each prism, symmetrical paths and the principle of the reversibility of light are the basic concepts of minimum deviation. Given the somewhat less stringent requirements of nonimaging optics, the 'reversible' prisms described here are to be called minimum deviation prisms.

Three-Dimensional Prism Design

One might think that the analysis presented is based on a two dimensional understanding of the prism, taking into account only the vector components that lie in the cross-sectional plane. Equation (5.15) could be presented to found this claim. However, the prisms are truly designed in a three dimensional manner. The vectors used to determine reflections and refractions do have three components, and are calculated in three dimensions. The following small experiment (Fig. 5.8) will clarify the importance of the third dimension in prism design.

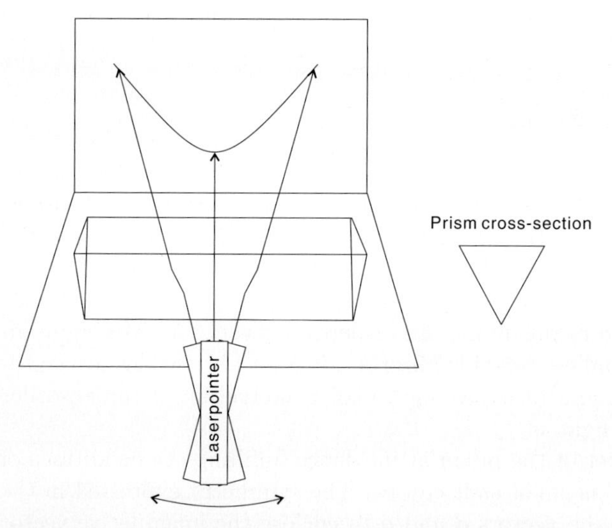

Fig. 5.8. Small experiment to clarify the three-dimensional nature of refraction at a prism

Monochromatic light from a laser enters a prism. The collimated beam is refracted twice and visualized at a screen. Initially the laser is oriented at an angle of 90° normal to the prism. As the laser is rotated sideways (not upwards or downwards), the light is refracted upwards more strongly than under initial conditions. The image of the refracted beam on the far screen traces out a 'u-shape' when the laser is rotated sideways in one plane. The effect is obvious and has strong implications on the design and performance of the linear Fresnel lens.

The three-dimensional design is best catered for by using vectors representing the incident and refracted rays. The two cases of reflection and refraction have been explained earlier with the help of vector algebra (3.1). It was found that a set of equations and constants can be found to calculate both reflection and refraction. The set of equations is (3.22a–3.22c)

$$\boldsymbol{w}.u = a\ \boldsymbol{v}.u + b\ \boldsymbol{u}.u\ , \tag{5.16a}$$

$$\boldsymbol{w}.v = a\ \boldsymbol{v}.v + b\ \boldsymbol{u}.v\ , \tag{5.16b}$$

$$\boldsymbol{w}.w = a\ \boldsymbol{v}.w + b\ \boldsymbol{u}.w\ . \tag{5.16c}$$

The factors a and b are given in Table 3.2. Three cases are significant for this evaluation: (i) refraction at the first surface of the prism, (ii) refraction at the second surface of the prism, and (iii) total reflection at the second surface of the prism. The refractive index changes from case (i) to cases (ii) and (iii) from $n' = n_{\text{acrylic}}$ to $n = n_{\text{air}}$. For the nomenclature used refer to Fig. 5.9.

Total reflection is prevented, it occurs only when the critical angle $\Phi_{\text{c}} = \phi_2' < 42.16°$. This value is found for the refractive index of the material (polymethylmethacrylate (PMMA)) of $n' = 1.49$, when substituting the angle of refraction in Snell's law with the maximum value of 90°.

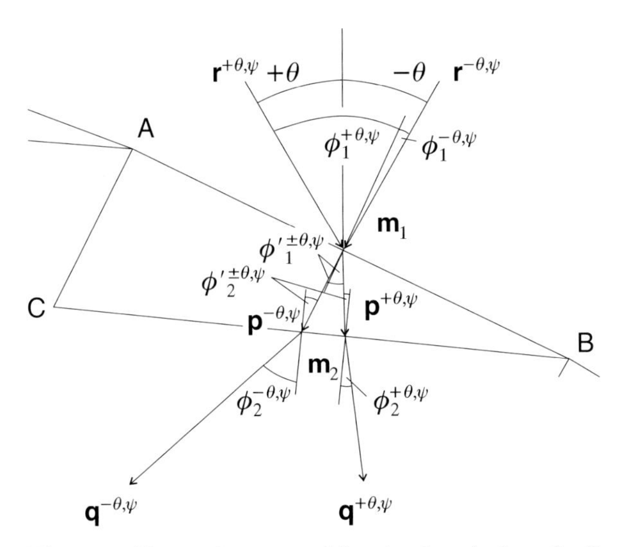

Fig. 5.9. Nomenclature used for the description of reflection and refraction at the prism. The incident rays $\boldsymbol{r}_{\text{i}}$, the rays refracted at the first surface $\boldsymbol{p}_{\text{i}}$, and the rays refracted at the second surface $\boldsymbol{q}_{\text{i}}$ are drawn to scale depending on the extreme acceptance half-angles $\pm\theta$ and ψ, the latter being symmetrical. Projection onto the cross-sectional plane

The angles ϕ_2' that are not subject to total reflection lie within a cone of rotational symmetry formed around the second surface normal $\boldsymbol{m}_2$.

The vector $\boldsymbol{p}$ is used for determination of the prism's tip (point C in Fig. 5.9). The tangent of the vector's projection onto the cross-sectional plane represents the slope of the vector. Point C can be calculated when the slopes and equations for the lines $l = \overline{AC}$ and $m = \overline{BC}$ are known. For the slope s_l of the line l we have

$$s_\mathrm{l} = \frac{\boldsymbol{p}.v}{\boldsymbol{p}.u} \, , \tag{5.17}$$

where u and v denote the horizontal and the vertical component of the vector $\boldsymbol{p}$, respectively. For the slope of the line m with the prism inclination α we have

$$s_\mathrm{m} = \tan\alpha \, . \tag{5.18}$$

The equations for the lines are derived from the normal form of the equation of a line:

$$y = ax + b \Leftrightarrow b = y - ax \, . \tag{5.19}$$

The components x and y of the points A and B are known. The intersection of the line l with the y axis at point l_y is given as

$$l_\mathrm{y} = A_\mathrm{y} - s_\mathrm{l} A_\mathrm{x} \, , \tag{5.20}$$

where the subscripts x and y denote the components of a point of the cross-sectional plane in the horizontal and vertical direction, respectively. The cross-sectional plane is the plane of the paper. It should be mentioned again that the vectors in Fig. 5.9 are three-dimensional vectors, although only their cross-sectional projections are plotted.

The intersection of the line m with the y axis at the point m_y is given as

$$m_\mathrm{y} = B_\mathrm{y} - s_\mathrm{m} B_\mathrm{x} \, . \tag{5.21}$$

The coordinates C_x and C_y are found when solving (5.19) for y and x, respectively. Two equations, one each for line $\overline{AC}$ and line $\overline{BC}$, are set equal and solved. For C_x (with (5.19) solved for y) we obtain

$$\begin{aligned} s_\mathrm{l} C_\mathrm{x} + l_\mathrm{y} &= s_\mathrm{m} C_\mathrm{x} + m_\mathrm{y} \, , \\ C_\mathrm{x} &= \frac{l_\mathrm{y} - m_\mathrm{y}}{s_\mathrm{m} - s_\mathrm{l}} \, . \end{aligned} \tag{5.22}$$

For C_y (having solved (5.19) for x) we obtain

$$\begin{aligned} \frac{C_\mathrm{y} - l_\mathrm{y}}{s_\mathrm{l}} &= \frac{C_\mathrm{y} - m_\mathrm{y}}{s_\mathrm{m}} \, , \\ C_\mathrm{y} &= \frac{l_\mathrm{y} s_\mathrm{m} - m_\mathrm{y} s_\mathrm{l}}{s_\mathrm{m} - s_\mathrm{l}} \, . \end{aligned} \tag{5.23}$$

The point C, marking the tip of the prism, has thus been found. The design of the prism is completed, and the prism's shape is set. The simulation calculates the coordinates for C for every iteration in the inner loop (the loop determining the prism angle β) in order to calculate point D in Fig. 5.7. From point D, the convergence criterion of (5.15) for the prism angle β is evaluated, before the criterion in (5.6) for the prism inclination angle α can be fulfilled in the outer infinity loop.

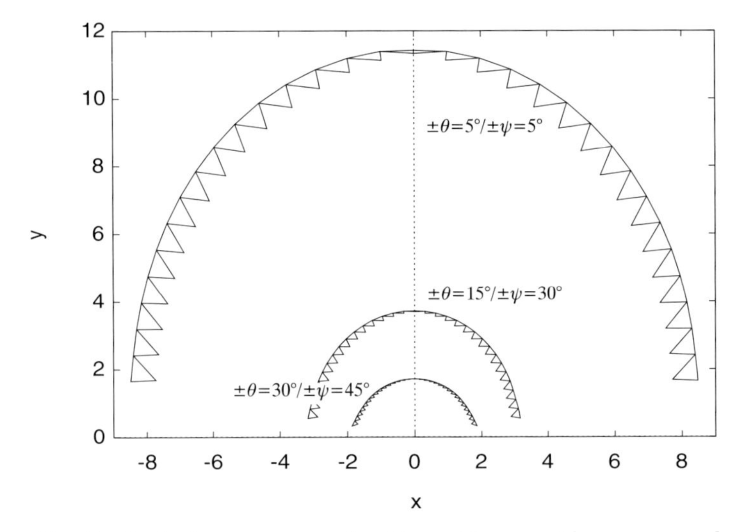

Fig. 5.10. Optimum nonimaging Fresnel lenses with acceptance half-angles $\pm\theta = 5°/\pm\psi = 5°$ (largest, highest concentration ratio); $\pm\theta = 15°/\pm\psi = 30°$; and $\pm\theta = 30°/\pm\psi = 45°$. Absorber half-width $d = 1.0$

The prism adjacent to the right of the one just finished is constructed in the same way. The pitch of the next prism and its size is defined by the angle ω. The design process continues, until the point where the next prism would reach down to the absorber level. Lenses can be designed with various acceptance half-angle pairs; a selection is plotted in Fig. 5.10. The size of the prisms has been chosen to be large to enable clear visualization.

It is possible to truncate the lens with the same reasoning as has been done for the CPC [142]. The truncated concentrator suffers little performance reductions while at the same time saving an economically significant amount of material. The lens becomes smaller, lighter, and easier to manufacture. A common truncation criterion is $f_t = 1/2f$. The lens is truncated at half-height over the absorber. Accordingly, the lens design is stopped when the next prism would cross the line given by $y = 1/2f$.

Although a rule of thumb, truncation of the nonimaging Fresnel lens at half-height is justified, as prisms on the lower half of the lens may be subject to shading for incidence from extreme angles within the design angles, as described on p. 86.

5.3 Rotational Symmetry

The nonimaging lens is usually designed as a linear lens. It produces a line focal area. Though only of medium concentration ratio, the tracking requirements are eased; when adopting orientation in a north–south direction, daily azimuthal tracking is sufficient. The line focal area furthermore is more economical in both solar thermal and photovoltaic applications, where the absorber is most easily produced with a linear shape (e.g. a heat pipe, or a photovoltaic panel, respectively).

The geometrical concentration ratio C of a three-dimensional nonimaging concentrator theoretically is much higher than that of a linear concentrator (see also p. 19, and for nonideal concentration p. 119):

$$C_{3D} = \frac{n}{\sin^2 \theta} \tag{5.24}$$

for the three-dimensional case, as opposed to

$$C_{2D} = \frac{n}{\sin \theta} \tag{5.25}$$

for the two-dimensional case. The cross-sectional acceptance half-angle θ is most decisive for obtaining the geometrical concentration ratio. The influence of ψ on C and the ambivalence of the concept of a geometrical concentration ratio will be discussed in Sect. 6.4. Note the influence of the medium that fills the concentrator in (5.24) and (5.25); the refractive index n is unity for vacuum and only very slightly different for air.

The three-dimensional nonimaging Fresnel lens can be obtained by using the same simulation used for the 2D concentrator. Only the perpendicular acceptance half-angle has to be set to $\psi = 0$. The result of the simulation is then the cross-section of a concentrator of rotational symmetry, instead of the linear lens obtained when the influence of ψ is taken into account. In order to calculate the performance values for the three-dimensional case, the cross-sectional values from the 2D case are squared.

The influence of ψ has been pictured in Fig. 5.11, where the cross-sections of two lenses of rather high concentration ratio are shown. The lenses have acceptance half-angle pairs of $\theta = \pm 0.5°$ and $\psi = 0.0°$ for the lens of rotational symmetry, and $\theta = \pm 0.5°$ and $\psi = 12.0°$ for the linear lens, respectively. The influence of the perpendicular acceptance half-angle ψ is clearly seen. The linear lens becomes slightly wider, and the prism angle adjusts to the effects of ψ.

Nonimaging Fresnel lenses of rotational symmetry can easily be designed, and may find interesting fields of application, such as lighting. For the collection of solar energy their employment could be advisable once concentration

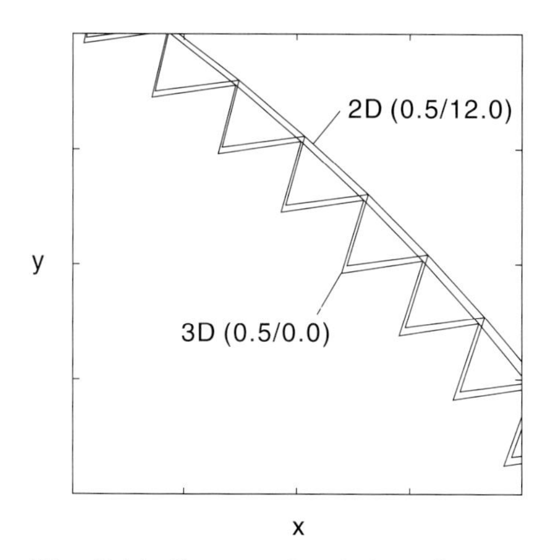

Fig. 5.11. Cross-sectional view of two nonimaging Fresnel lenses of $\theta = 0.5°$. One lens ($\psi = 0.0°$) is of rotational symmetry (3D), while the other ($\psi = 12.0°$) is a linear (2D) lens

ratios $C > 20$ are envisaged. Linear concentrators are practically restricted to concentration ratios below this value, due to the limit of one-axis tracking. Of course, linear Fresnel lens concentrators may track the sun over two axes; then the practical limit of concentration is set by the finite size of the sun ($\theta_s = 0.275°$) and color dispersion for the solar spectral irradiance. Concentration will not exceed 70 when measured as flux density.

5.4 Arbitrary Shapes

The nonimaging Fresnel lenses presented in this chapter are designed as optimum lenses for linear concentration of solar energy, and for any systems where the lens should roughly equal a collimator, such as lighting applications. The design incorporates any combination of acceptance half-angles θ and ψ, depending on what concentration ratio is to be achieved, accounting for tracking requirements and errors due to the size of the solar disk. Minimum deviation prisms of finite size constitute the Fresnel lens. The absorber has a finite size; a fraction of the sky 'seen' by the lens is refracted onto the solar receiver, or the size of the light source is refracted into the area to be illuminated. The lens can be used in a reversible way.

As opposed to imaging Fresnel lenses, where the distance between the lens and the focused image is determined by the *f/number* and the lensmaker's formula, the absorber half-width d and the height of the nonimaging lens above the absorber are related by $y_0 = d/\tan\theta$. This assumes the prism on the optical axis to be thin. In practice the innermost prisms are approximately flat plates.

From this point, a prism is determined with the help of the acceptance half-angle pair $\pm\theta$, describing (asymmetrical) incidence from the left and the right on the prism, and the symmetrical acceptance half-angle ψ. Only the right-hand side of the lens is determined, the left-hand side is later constructed as a mirrored version. A set of vectors representing the incidence from left and right on their paths of double refraction through the prism, $\boldsymbol{q}^{+\theta\psi}$ and $\boldsymbol{q}^{-\theta\psi}$, is derived.

Obviously, the vectors depend on the prism inclination α and the prism angle β. These are found in two nested infinity loops with the help of Newton's method. In the inner loop, the prism angle β is decided upon once both directions of incidence yield prism angles within an error margin. The outer loop compares the two vectors $\boldsymbol{q}^{+\theta\psi}$ and $\boldsymbol{q}^{-\theta\psi}$ pictured in Fig. 5.7 with the location of the prism defined by the vectors $\boldsymbol{d}^{+\theta\psi}$ and $\boldsymbol{d}^{-\theta\psi}$. Once the vectors $\boldsymbol{d}$ and $\boldsymbol{q}$ are found to be parallel within a confidence interval, the prism inclination α is fixed.

Subsequently, the next prism is found by feeding its optimization with the values of the previous prism. Once the next prism would reach the level of the absorber, or any other break criterion, the simulation is stopped. In order to achieve finite thickness of the lens, the prisms are designed to partly overlay the previous ones.

The starting point A for each subsequent prism can be chosen so that the prisms follow a predefined lens shape. Point A may be found as $y = f(x)$, where the mathematical function must fulfill two conditions. First, the slope of the function must allow the refracted light to reach the absorber; thus the

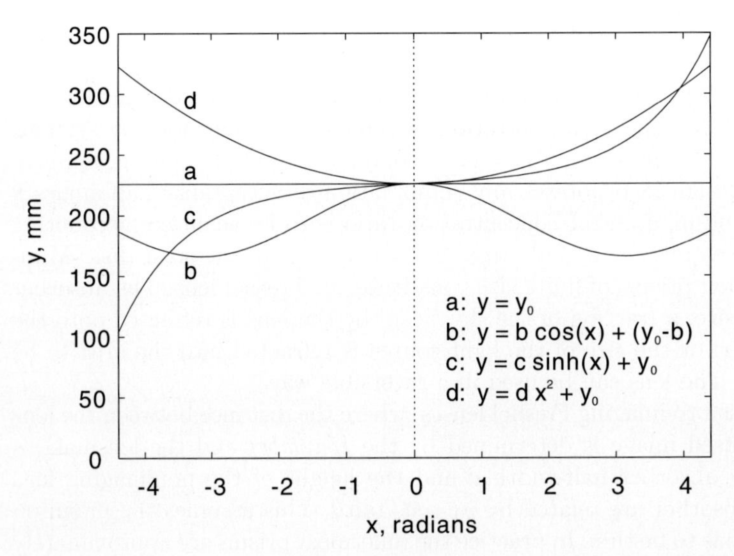

Fig. 5.12. Functions used to design shaped nonimaging Fresnel lenses

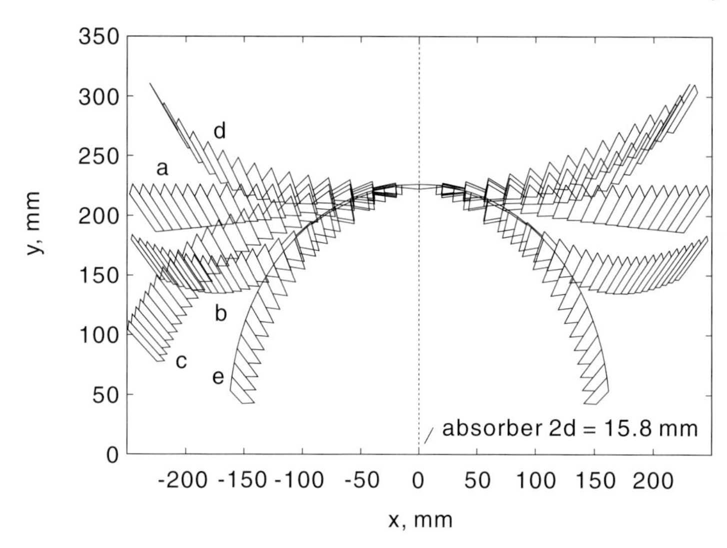

Fig. 5.13. Shaped nonimaging Fresnel lenses following mathematical functions. Acceptance half-angle pairs $\theta = \pm 2°$ in the cross-sectional plane (the plane of the paper), and $\psi = \pm 12°$ in the plane perpendicular to it. Oversized prisms. Lens e is the optimum lens shape, where a smooth outer surface is required. The previously manufactured prototype of e is truncated at $y = 1/2y_0$

tangent at any point must be smaller than one, with the absorber assumed to be small. The outermost prisms that follow function c in Fig. 5.12 and those in lens c in Fig. 5.13 are problematic in this respect. Second, the function must pass through the previously found y_0, according to the edge ray principle.

The previous requirement of a smooth outer surface must be given up, if minimum deviation prisms are to be employed. Consequently, nonimaging Fresnel lenses of arbitrary shapes carry grooves on both the inner and the outer surface.

Though not impossible, three-dimensional functions are to be avoided for practical reasons. Prisms would change their shape according to $y = f(x, z)$, and manufacturing would be a difficult task. One may understand the linear lenses in Fig. 5.13 as representing various cross-sections at different depths z of one lens wavering over a flat absorber like a blanket in the wind. Three-dimensional lenses of rotational symmetry can easily be obtained by setting the perpendicular acceptance half-angle $\psi = 0$.

The lenses in Fig. 5.13 are all designed taking into account the same acceptance half-angle pairs $\pm\theta = 2°$ in the cross-sectional plane (the plane of the paper), and $\pm\psi = 12°$ in the plane perpendicular to it. Acceptance half-angles and absorber size have been chosen to equal those of a prototype previously manufactured. This prototype has been plotted as lens e in Fig. 5.13, with the difference that the prototype is truncated at $y = 1/2y_0$.

For greater acceptance half-angles the lens moves closer to the absorber, i.e. if the distance between lens and absorber should be kept constant, the latter will become wider. Likewise, the curvature of an optimum lens with greater acceptance half-angles will become softer.

Shapes other than that of the optimum lens curved over the absorber, like flat or concave forms, are suboptimal in concentrating white light incident from both the left and right, since the spread of the dispersed beam becomes wider with increasing distance. The shape of a lens may be desirable on account of its smooth integration into existing shapes. Solar collectors may fit architectural designs such as daylighting, or lenses for lamps should suit lighting designs, for example. Erismann [39] designed a Fresnel lens, which is inherently aspherical, following a spherical shape for an infrared sensor, for reasons of fashion.

The principles of nonimaging optics have been successfully applied to the design of a novel class of nonimaging Fresnel lenses. The edge ray principle allows for the creation of nonimaging lenses with any combination of acceptance half-angles θ in the cross-sectional plane and ψ in the plane perpendicular to it.

Minimum deviation prisms constitute the lenses. Optimum nonimaging Fresnel lenses may have any shape. If the lens should have a smooth outer surface, its shape is convex. Non-convex shaped concentrator lenses with smooth outer surfaces are impossible, once minimum deviation prisms are used. The previous definition of 'reversible' minimum deviation cannot be fulfilled for entry and exit rays within the boundaries of the acceptance half-angles for prisms whose front faces are not inclined, so that their normal vector points towards the absorber.

The width of the nonimaging lens is restricted by the limit of the ideal concentration (and additionally by the refractive index n of the transparent material in between lens and receiver; see Sect. 6.5):

$$C = \frac{1}{\sin\theta} = \frac{R}{d}\,, \tag{5.26a}$$

$$R = \frac{d}{\sin\theta}\,, \tag{5.26b}$$

where R is the half-width of the loss-free lens and d is the half-width of the absorber. Real lenses may be wider, but their optical concentration ratio will not exceed the ideal limit, even if their geometrical concentration ratio may seem to suggest otherwise.

Nonimaging Fresnel lenses can be designed as 'fast' lenses. Besides the initial success of their use in solar energy applications as collectors of medium concentration for photovoltaics, lenses of various shapes can be integrated with architectural and technological requirements. A large field of anticipated applications is the use of the novel nonimaging lens as a collimator, or reversed concentrator, for lighting.

5.5 Diverger Lens for Lighting

The principle of the reversibility of light demands that the light source and the sink must be exchangeable. The nonimaging Fresnel lens concentrator in solar applications becomes a diverger in lighting applications. The absorber becomes the light source, and the acceptance half-angle pairs become the half-angle pairs defining the area to be illuminated. The profile of the reflecting CPC is a typical nonimaging device. The CPC is often used in lighting applications.

Light sources of finite size lead to strong aberrations in imaging optical systems. Imaging Fresnel lenses are point-focusing; a homogeneous flux distribution is difficult to achieve in lens designs due to the extended light source (the filament or cathode of the lamp) and dispersion. Nonimaging optics can overcome these limitations of imaging devices; once the flux distribution at the source is homogeneous, the lens diverger will provide homogeneous illumination. Color aberrations are minimized by the use of minimum deviation prisms with minimum dispersive power and the fact that light of different wavelengths mixes at some distance after refraction at the lens. The nonimaging lens does not provide an image of the source.

The area to be illuminated and the size of the light source are related by the half-angle pairs and, consequently, the width (for linear 2D lamps) or the diameter (for 3D lamps of rotational symmetry) of the nonimaging lens.

Figure 5.14 depicts a truncated, linear (2D) nonimaging Fresnel lens under a filament lamp. Light extends from the filament of the lamp a between A and B towards the novel nonimaging Fresnel lens e. The beams of light c and d are refracted into c' and d' at the lens e. The refracted beams extend over a pair of half-angles θ and ψ, where θ is the half-angle in the plane of the paper, ψ the half-angle in the plane perpendicular to the plane of the paper.

A parabolic reflector b is necessary to direct the light from the lamp towards the lens. This reflector extends to point C, in order to minimize losses due to rays that pass (after one or more reflections at the reflector) through the gap between C and D. While some rays are lost through this gap after one or more reflections at the reflector, none of the rays which have not been reflected misses the lens. Should the nonimaging Fresnel lens not be truncated, it may extend to point D', in which case the reflector can be truncated to C'.

The new lamp cover makes use of the optical properties of the nonimaging Fresnel lens, which consists of minimum deviation, minimum dispersion prisms for least geometrical losses and least color aberrations, to refract light extended by a light source of finite size into light within two pairs of half-angles.

In Fig. 5.14, the light source is a filament. Incandescent lights often call for diverger lenses of rotational symmetry (3D lenses), while the cathode of a fluorescent lamp calls for linear lenses (2D lenses) [85].

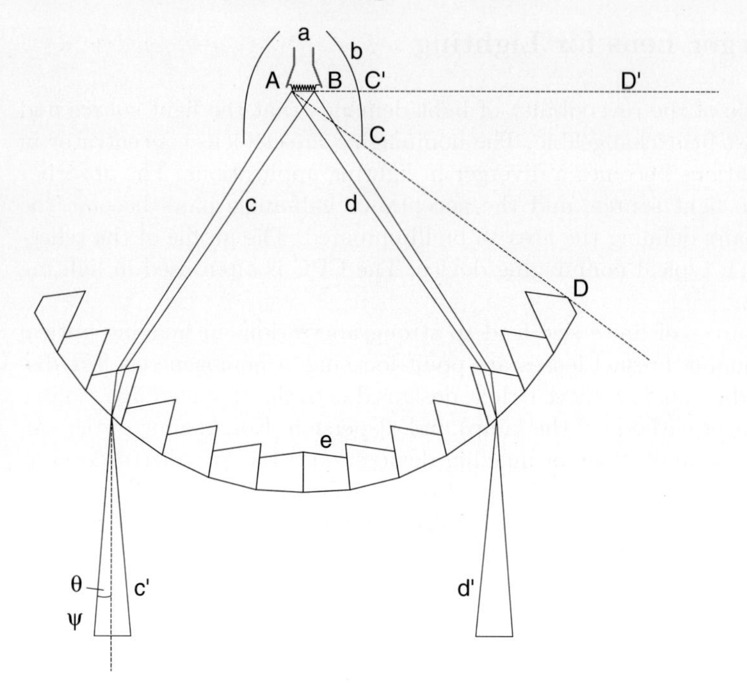

Fig. 5.14. Schematic of nonimaging diverger lens for lighting fixture

The prisms of the nonimaging lens are minimum deviation prisms. Their assembly, and thus the convex shape of the nonimaging Fresnel lens, is constrained by the requirement of a smooth outer surface of the lens. Should a smooth outer surface not be required, minimum deviation prisms can be assembled for any shape, and the nonimaging lens forming the lamp cover can be given any shape, as long as the body of the lens does not cross the rays between source and sink more than once. Fashionable lamp designs are possible without seriously affecting the optical performance of the nonimaging lens.

The edge ray principle states that for an ideal nonimaging optical system, the extreme rays at the first aperture (here the light source) will be the extreme rays also at the second aperture of the system (here the area to be illuminated). Graphically, this is demonstrated by the exemplary ray bundles c, c' and d, d' in Fig. 5.14. The ray bundles are diverged after the lens, filling a rectangle defined by the half-angle pairs θ and ψ. For the case of a lens of rotational symmetry, $\psi = 0$.

By modifying the half-angle pairs in relation to the constant size of the light source, the area to be illuminated can be controlled. This does affect the width or diameter of the nonimaging lens and the distance between the light source a and the lens e, as shown in Fig. 5.14. The smaller the divergence, the larger are both the width or diameter of the lens and the distance between the light source and the lens.

The flux distribution beyond the nonimaging lens approaches uniformity, as long as the light emitted by the source shows uniform characteristics. This is one of the characteristic properties of nonimaging optics.

6 Lens Evaluation

6.1 Losses

Solar collectors suffer from many types of loss. The collector is an energy system consisting of a concentrator, a receiver, some kind of support structure, and the means to connect it to a larger system such as a working fluid or an inverter in the cases of solar thermal energy conversion and grid-connected photovoltaic generation of electricity, respectively.

The solar concentrator is the first part of the system to interact with the incident radiation. The estimation and evaluation of losses at this point aims at two goals: first, the verification of the design process, and second, the facilitation of the concentrator's comparison with competitive designs. Explanation of losses leads to a closer understanding of the working principle of the concentrator, and may open ways for the optimization of the design. Comparing the concentrator with competing designs requires an honest characterization of the concentrator's losses based on an evaluation concept that finds widespread application.

A variety of numbers has been developed for the comparison of solar concentrators. The geometric concentration ratio and the optical concentration ratio are of prime importance, and will be discussed in detail in this chapter.

Losses at the concentrator can be categorized broadly according to the time when they become relevant during the evolving project as follows:

- Principal losses are related to the decision made regarding the type of concentrator. Most concentrators cannot collect diffuse radiation, for example. Many concentrators do not deliver homogeneous illumination of the receiver, and others may cut off a fraction of incident radiation, such as the previously mentioned skew rays in a 3D CPC. Principle losses can be reduced when the relation between concentrator type, its desired performance (application characteristics, e.g. desired process heat temperature or flux density on the absorber), and climatic conditions of the location where the system is to be installed are re-evaluated.
- Design inherent losses are the responsibility of the designer of the actual concentrator. A Fresnel lens, for example, can be designed in many ways, and most of them are suboptimal. Still, even when the particular design reaches an optimum, there are losses that cannot be avoided. If the prism

is of the minimum deviation type, tip and blocking losses will occur, when it is used in a nonimaging concentrator. Furthermore, the designer's choice of the concentrator material will influence its performance.
- Manufacturing and installation losses depend on the accuracy during manufacturing and the installation process of the concentrator. Losses of this kind may be regarded as caused partly by the decision on what type of concentrator to build, and partly on the design characteristics, which may make manufacturing difficult and result in losses due to not satisfying narrow tolerances. Some concentrators simply cannot be manufactured, an example being the Luneburg lens, an ideal imaging concentrator with continuously changing refractive index.
- Losses of wear and tear, which appear during the lifetime of the concentrator, but must be anticipated when the system is decided upon and designed. Solar radiation itself, humidity, extreme temperatures, sandstorms, etc. cause degradation of the optical materials involved and of the overall structure.

Other sources [49, 177] describe the total losses of reflective solar concentrators as a complex error function, which can be approximated by a Gaussian shape (standard deviation σ in rad):

$$\sigma_{\mathrm{D}}^2 = \sigma_{\mathrm{sun}}^2 + \sigma_{\mathrm{receiver}}^2 + \sigma_{\mathrm{tracking}}^2 + \sigma_{\mathrm{concentrator}}^2 , \tag{6.1}$$

where

$$\sigma_{\mathrm{concentrator}}^2 = \sigma_{\mathrm{specular}}^2 + \sigma_{\mathrm{contour}}^2 + \sigma_{\mathrm{mechanical}}^2 . \tag{6.2}$$

The concentrator mirror's losses (6.2) are a combination of the following factors:

- Microroughnesses in the range of the wavelength of visible light $\sigma_{\mathrm{specular}}$. Microroughnesses are due to material inhomogeneities, e.g. polishing marks or dust erosion. Calculations and material properties are given in [177], and [139].
- Local geometrical inaccuracies, or macroroughnesses, their dimension exceeding the wavelength of sunlight $\sigma_{\mathrm{contour}}$. Macroroughnesses are often caused by manufacturing deviations from the stipulated design.
- Global shape distortions $\sigma_{\mathrm{mechanical}}$, which are often caused by decisions about the concentrator operating principle and design. A typical example is a collector that is deformed by mechanical stress due to the windload or temperature fluctuation at a location it was not designed for.

The list above focuses on mirror-based concentrators. Refractive devices have different properties. Losses that occur because of optical interactions in the dimension of a few wavelengths will not be considered here. Ray tracing is a tool of geometrical optics, which is distinguished from physical or linear optics by its confinement to dimensions greater than those of the wavelengths of the radiation in question. The latter also stipulates that the Fresnel lens grooves and tips are set apart by at least a multiple of wavelengths. This 'multiple' in the actually manufactured lens is a factor of about one thousand, or the ratio of 0.5 mm to the wavelength of green light, 500 nm. Diffraction will briefly be discussed in Sect. 8.2.

Macroroughnesses and global shape distortions can be assessed by examining the error in the direction of the reflected or refracted ray. For mirror concentrators, the misorientation angle of the reflected ray is approximately twice the initial error of the incident ray (p. 9).

For the reader's convenience, a short course on the evolution of the angles tracing the ray during refraction shall be given here, with reference to Fig. 3.3. At the first surface, the ray incident at ϕ_1 is refracted to ϕ_1', according to Snell's law. Both angles are measured between the ray and the surface normal. At this point, it is useful to introduce an angle ϕ_t, which is used to translate the prism angle and the inclination into the calculation of the angles ϕ_2' and ϕ_2 at the second surface. The angle ϕ_t is measured as opened by the refracted ray within the prism and the (flat and horizontal) absorber normal:

$$\begin{aligned} \phi_1 &= \beta - \alpha - \theta_{\mathrm{in}} \ , \\ \phi_1' &= \arcsin\left(\frac{\sin\phi_1}{n'}\right) \ , \\ \phi_{\mathrm{t}} &= \beta - \alpha - \phi_1' \ , \\ \phi_2' &= \phi_{\mathrm{t}} + \alpha \ , \\ \phi_2 &= \arcsin\left(\sin\phi_2'\, n'\right) \ , \\ \theta_{\mathrm{out}} &= \phi_2 - \alpha \ . \end{aligned} \qquad (6.3)$$

The angle of incidence θ_{in} and the angle of refraction θ_{out} are measured with respect to the surface normal of a horizontal plate. The sign of θ_{in} is negative for incidence from the right, and positive for incidence from the left. The refractive index of the prism material is $n' = 1.49$ in this lens design. Note that these calculations are projections into the plane of the paper and exclude the behavior of the refracted ray due to incidence in the plane of the perpendicular angle ψ.

If we calculate (6.3) for $\theta_{\mathrm{in}} = -20.0°$ at a prism like the one pictured in Fig. 3.3, then $\theta_{\mathrm{out}} = 16.4°$. If we assume a surface error of 5° and $\theta_{\mathrm{in}} = -15.0°$, then $\theta_{\mathrm{out}} = 21.5°$. The lens, in contrast to the mirror, does compensate part of the slope errors. While the slope error must be multiplied by a factor of two when determining the beam error during reflection, the refracted beam direction changes only by an angle approximately equal to the slope error, for conditions near the minimum deviation angle. The reason is the nonlinear behavior of Snell's law.

The nonimaging Fresnel lens that is the theme of this book will be primarily evaluated according to inherent design losses. Manufacturers guarantee slope errors in the range of 1/100°, and the consequences for the performance of concentrators of medium concentration are marginal. The mechanical losses have to be carefully controlled choosing appropriate materials and secure fixtures.

Design losses can explain most of the differences between ideal and actual performance. The remainder are losses due to manufacturing and assembly, which can be estimated from the tolerances kept by the manufacturer and the performance of the lens in a test. Field tests will reveal losses of wear and tear.

6.2 Transmittance

The transmittance of a solar collector cover describes the relationship between the solar irradiance incident on the cover and the irradiance that passes the cover on its way to the absorber. The decision to use concentrators that work either on the basis of reflective or refractive optics, is a crucial one. According to this classification, losses resulting from transmittance of the respective concentrator type are considered principle losses.

Radiation entering a mirror-based concentrator, such as the compound parabolic concentrator (CPC), usually has to pass a transparent cover, and is then reflected by the mirrored walls towards the absorber. On the contrary, a refractive concentrator, such as the nonimaging Fresnel lens, combines in its lens body both cover and concentrating functions. Whereas the cover of a CPC may be a flat glass plate, the light entering the Fresnel lens has to pass through minimum deviation prisms.

The transmittance τ is the sum of reflectance losses ρ on each surface of the cover, reflectance losses ρ_{m} on a mirror surface, and absorption losses α in the cover material:

$$\tau = \tau_{\rho} \cdot \tau_{\rho_{\mathrm{m}}} \cdot \tau_{\alpha} \,. \tag{6.4}$$

Reflectance at each surface depends on the angle of incidence, as well as on the polarization of the radiation. Although sunlight is considered to be unpolarized, both the parallel and the perpendicular reflection fractions have to be calculated. Parallel and perpendicular refer to the plane spanned by the incidence and the surface normal. The reflection coefficients are calculated with the Fresnel equations:

$$r_{\parallel} = \frac{\tan^2(\Phi' - \Phi)}{\tan^2(\Phi' + \Phi)} \,, \tag{6.5}$$

for the parallel part, and

$$r_{\perp} = \frac{\sin^2(\Phi' - \Phi)}{\sin^2(\Phi' + \Phi)} \,, \tag{6.6}$$

for the perpendicular part of the reflection. The angles Φ and Φ' refer to the angle of incidence and the angle of refraction when radiation from a medium with refractive index n enters a medium with refractive index n'. The relation between these two angles is found by applying Snell's law:

$$\frac{n}{n'} = \frac{\sin \Phi'}{\sin \Phi} \, . \tag{6.7}$$

The reflection losses for unpolarized radiation at one surface are calculated as

$$\rho = \frac{1}{2} \left(r_{\parallel} + r_{\perp} \right) \, . \tag{6.8}$$

Note that the Fresnel equations (6.5), (6.6), and their combination in (6.8) yield the reflection losses for the crossing of one surface only. The second surface (a cover as well as a Fresnel lens has two surfaces) requires the use of (6.7), and recalculation of the Fresnel equations. For a number of surfaces n with ρ_{i}, the transmittance including reflectance losses will be

$$\tau_{\rho} = \tau_{\rho 1} \cdot \tau_{\rho 2} \cdots \tau_{\rho \mathrm{n}} \, , \tag{6.9}$$

where the individual transmittances satisfy

$$\tau_{\rho} = 1 - \rho \, . \tag{6.10}$$

For the prism, the angles of incidence on one surface are translated to the other surface by the prism angle β, and the above procedure will have to be respected. For multiple reflections of radiation passing through a slab of material with two parallel interfaces, the average transmittance of the two components $r_{\parallel}$ and $r_{\perp}$ can be calculated for unpolarized radiation as

$$\tau_{\rho} = \frac{1}{2} \left(\frac{1 - r_{\parallel}}{1 + r_{\parallel}} + \frac{1 - r_{\perp}}{1 + r_{\perp}} \right) \, , \tag{6.11}$$

which results in only slightly higher losses than the 'naive' (Rabl [143]) result of $\tau \approx 1 - 2r$. Reflection losses for smaller angles of incidence are usually below 10%. Figure 6.1 gives the results of (6.11), i.e. for a plate with multiple reflections at two parallel surfaces.

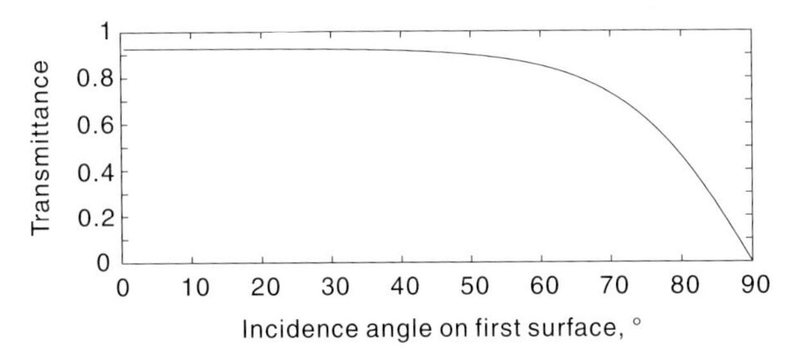

Fig. 6.1. Multiple reflections at two parallel surfaces: transmittance

From Fig. 6.1 it can be concluded that minimum deviation prisms cause least reflection losses for prisms. It is recalled that minimum deviation happens if $\phi'_1 = \phi'_2$, which are the angles between the surfaces normals and the refracted ray within the lens material. The sum of these two angles is a minimum for achieving the prism's optical power. In Fig. 6.1, it can be seen that their combined reflection losses are a minimum, since the slope of the transmittance curve becomes increasingly negative, ruling out optimum combinations of incidence angles other than twice the smallest one. The deviation of the slab is zero, of course.

Furthermore, the minimum deviation prism reduces the risk of total internal reflection at the critical angle ϕ'_2.

The reflectance losses of a mirror are calculated simply by taking into account the reflectivity of the material (ρ_{m}, around 0.85–0.95, for reasonably small angles) and the number of reflections n, which for the case of the CPC are available from graphs in [142, 143]:

$$\tau_{\mathrm{CPC}} = \rho_{\mathrm{m}}^{\mathrm{n}} \; . \tag{6.12}$$

The absorption of radiation in the cover is calculated by Bouguer's law. A fraction of radiation α is absorbed over a path of length s in a medium with extinction coefficient k:

$$\tau_{\alpha} = 1 - \alpha = \exp\left(-ks\right) \; . \tag{6.13}$$

Extinction coefficients for glass range from $4\,\mathrm{m}^{-1}$ for 'water white' glass to $32\,\mathrm{m}^{-1}$ for ordinary glass which appears greenish. Given the small absorption found in the previously mentioned spectral transmittance experiment (see Fig. 3.6, and thereafter), where reflectance for both samples of thicknesses $s =$ 1 mm, and $s =$ 10 mm, respectively, was absolutely dominant, it is probably safe to assume extinction coefficients for PMMA towards the lower end of the range.

It can be concluded that the transmittance of a thin Fresnel lens acting as a cover of the concentrator is higher than the transmittance of a CPC, where reflectance losses strongly increase for small acceptance angles. For all concentrators involving refraction at one point or another, the angular dependence of reflectance is a cause of concern for efficient design. The steep slope of the outer reaches of the Fresnel lens, for example, are problematic in this sense; truncation of the lens receives some justification from this point of view.

6.3 Geometrical Losses

Losses due to the optical design of the lens are evaluated with the help of ray tracing. Ray tracing is a procedure of geometrical optics and has become the standard method for evaluating solar collectors.

Nonimaging concentrators, in particular, are to be evaluated by ray tracing, as there is no other meaningful tool for this task; Welford [180] notes that geometrical vector flux formalisms can be used for simply shaped concentrators only. The étendue, or optical throughput, used for calculating the maximum theoretical geometrical concentration ratio along the lines of the edge ray principle does not provide any information on the flux density, i.e. the radiation flux crossing a unit of area.

Ray tracing denotes the process of calculation of a ray's path through an optical system. This process had been very time-consuming in the past, before the age of computers. Jenkins and White [66] explain different methods of ray tracing for various kinds of rays. A major problem used to be the calculation of logarithmic and trigonometric functions, a process that was time-consuming and prone to errors when done by hand.

Both the introduction of the very first computers to ray tracing and the use of formulas involving quadratic equations requiring Newton's method for solving instead of trigonometry is anticipated in an 1951 article [53]. More recently, ray tracing has been adopted for the automatic calculation of nonimaging concentrators [48]. Nowadays, software for optical design on personal computers usually includes ray tracing procedures, and often nonimaging design examples.

For the evaluation of the prototype Fresnel lens, rays are created forming a hemisphere over the linear lens. Since the lens is symmetrical in the direction of the angle ψ, only half the hemisphere has to be simulated. If rays are created in steps of one degree, 180 rays (90 rays simulating incidence from the left side, 90 rays for incidence from the right) have to be used in the cross-sectional plane of the lens, and 90 rays perpendicular to it. One half of the prototype lens consists of some 300 prisms, and two extreme rays are traced for each prism. The other half of the lens can be assumed to have the same properties for reasons of symmetry, as long as the rays do not carry any value, i.e. the simulation is not based on a radiation model. Then, during the ray tracing of the nonimaging lens, some 9.7 million rays have to be followed trough the lens and down to the absorber level, or at least through three intersections. Both the design and the evaluation of nonimaging concentrators are strongly dependent on computing power.

Three types of losses can be evaluated by tracing rays through each prism of the lens. Passing through one prism, the incident rays may interfere with the back of this prism, if, generally speaking, the angle of incidence of light entering the prism from the right (the prism being part of the right-hand side of the lens) is larger than the cross-sectional design angle. This type of loss is termed unused tip losses; confusingly the 'unused' tip of the prism appears empty at the top, but overfull of light towards the bottom of the prism.

The incident ray may be refracted by the prism in such a way that its path toward the absorber could be blocked by the front of the adjacent, lower prism. This type of loss is called blocking loss. It usually happens for incidence at angles much greater than the design angle θ for rays entering the lens from the left-hand side (the prism still being part of the right-hand side of the lens).

Both the unused tip losses and the blocking losses are geometrical losses that occur at the level of the prisms. A third loss, also geometrical in nature, occurs due to the refractive influence of the perpendicular incidence angle ψ. This variation of the geometrical losses is design inherent, and will be discussed in connection with the homogeneous illumination of the receiver and the 'hot spot' problem in Sect. 9.1.

The following considerations refer to prisms at the right-hand side of the lens (and, applied symmetrically, to the left-hand side of the lens, of course). The nomenclature of the prism and the refraction at its first surface (front) and its second surface (bottom) have been depicted in Fig. 5.9. Additionally, one may recall the prism angle β between the prism front and bottom, and the prism inclination α as the angle between the prism bottom and a horizontal line through the rightmost corner of the prism. This inclination angle is defined to have a positive sign for counterclockwise rotation around this point.

Figure 6.2 may serve to explain the procedure for ray tracing at the nonimaging lens. The prisms are placed in the context of a three-dimensional coordinate system. This system is the same as that used in the design of the lens (Fig. 2.4):

$$\begin{aligned} \boldsymbol{r}.u &= \sin\theta_{\text{in}}\ \cos\psi_{\text{in}}\ , \\ \boldsymbol{r}.v &= -\cos\theta_{\text{in}}\ \cos\psi_{\text{in}}\ , \\ \boldsymbol{r}.w &= -\sin\psi_{\text{in}}\ . \end{aligned} \tag{6.14}$$

The refracted ray is generated as a vector, again following the procedure which resulted in Table 3.2, using (3.22a–3.22c). A number of vectors are created for each prism to analyze its geometrical losses.

Tip Losses

In Fig. 6.2, the vector $\boldsymbol{p_1}$ is created, depicting the refracted ray within the prism. The vector is aligned through point C and intersects the prism front at point H. If point H is not located on the actual prism front, i.e. not between points A and E, an intersection of $\boldsymbol{p_2}$ running through point A with the prism bottom at point U is created. If that point U lies on the actual prism bottom, i.e. within the boundaries of points C and E, no unused tip losses are found.

If $\boldsymbol{p_1}$ intersects the prism front within its actual range, between points A and E, the rays incident between A and H at the prism front are refracted in such a way that they are subject to refraction or total internal reflection (as in Fig. 6.3b, although very small, and clearly in Fig. 6.4b). All rays incident at the angle in question and between points A and H are qualified as unused

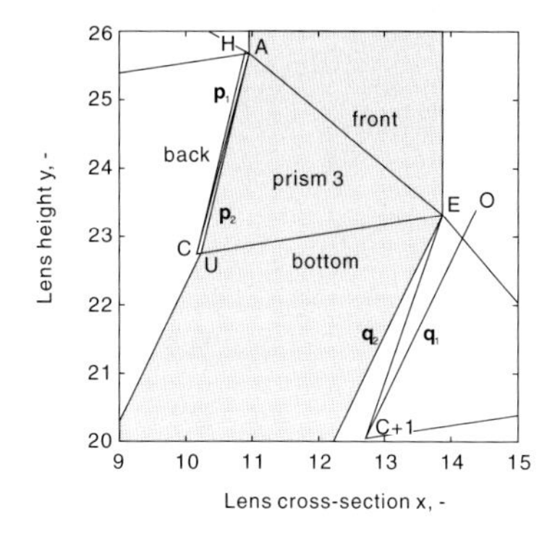

Fig. 6.2. Ray tracing at prism 3 of the nonimaging lens with $\theta = \pm 2°$ and $\psi = \pm 12°$. Normal incidence $\theta_{\text{in}} = \psi_{\text{in}} = 0°$. Projection into the plane of the paper

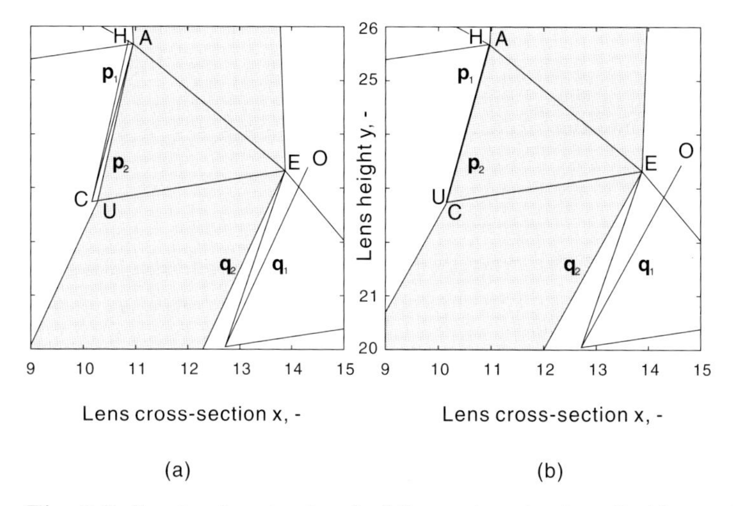

Fig. 6.3. Ray tracing at prism 3 of the nonimaging lens. Incidence at design angles (**a**) $\theta_{\text{in}} = +2°$, $\psi_{\text{in}} = 12°$; (**b**) $\theta_{\text{in}} = -2°$, $\psi_{\text{in}} = 12°$. Projection into the plane of the paper

tip losses by discounting the ratio of $\Delta\overline{AH}/\Delta\overline{AE}$ from the effective width of the prism. The lengths may be calculated as the thickness of the ray bundle incident between points A and E, referring to the beam's angle of incidence (i.e. the minimum thickness of the beam), or as the projection of the beam thickness into the horizontal line. The symbol Δ shall indicate the use of one of the modification methods described:

$$L_{\text{tip}} = \frac{\Delta\overline{AH}}{\Delta\overline{AE}} \,. \tag{6.15}$$

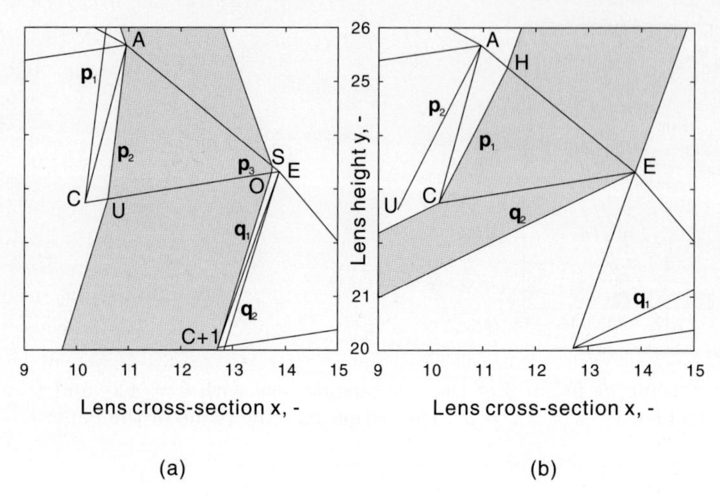

Fig. 6.4. Ray tracing at prism 3 of the nonimaging lens with $\theta = \pm 2°$ and $\psi = \pm 12°$. Incidence exceeding design angles (**a**) $\theta_{\text{in}} = +20°$, $\psi_{\text{in}} = 30°$; (**b**) $\theta_{\text{in}} = -20°$, $\psi_{\text{in}} = 30°$. Projection into the plane of the paper

Blocking Losses

Blocking losses occur when the twice-refracted ray hits the back of the adjacent, lower prism, here prism 4 to the right of prism 3. A vector $\boldsymbol{q_1}$ is generated, which states the direction of the twice-refracted ray passing through the prism tip of prism 4, point $C + 1$. This vector $\boldsymbol{q_1}$ is intersected with the bottom of prism 3. No losses occur if that intersection, at point O, happens outside (to the right) of point E in Fig. 6.2.

Once vector $\boldsymbol{q_1}$ intersects the actual prism bottom at point O as in Fig. 6.4a, the simulation assumes blocking losses, because the rays exiting the prism between O and E are blocked by the back of prism 4. These losses are translated via a vector $\boldsymbol{p_3}$ from O to an intersection S with the prism front. The point S lies then within points A and E. Blocking losses are discounted as the ratio $\Delta\overline{SE}/\Delta\overline{AE}$ from the effective width of the prism, in one of the two ways described above:

$$L_{\text{block}} = \frac{\Delta\overline{SE}}{\Delta\overline{AE}} \, . \tag{6.16}$$

Clearly, from Figs. 6.2–6.4, blocking losses occur for incidence from the left for $\theta_{\text{in}} \gg \theta$, and unused tip losses occur for rays incident from the right for $\theta_{\text{in}} \geq \theta$, for the right-hand side of the lens. Since the lens is symmetrical, losses occur on both sides of the lens, but as a 'mirrored' version. In general, the prisms behave quite well, though, and unused tip and blocking losses hardly ever occur for rays entering the lens at angles smaller or equal to those of the design acceptance half-angles.

Usually, all rays entering the prism front exit through the prism bottom. The thickness of the ray bundle on both sides may be different, though. The cause for this is prism magnification, or concentration, which is a phenomena independent of lens magnification. It can be explained by deviation [107].

Absorber Misses

Rays entering the lens at angles greater than the design angles are not supposed to hit the absorber. They have been presented in Fig. 6.4 for the purpose of illustrating the simulation's assessment of geometrical losses at the prisms. Should a refracted bundle of rays miss the absorber, the losses L_{miss} are set to one. If the ray bundle hits only part of the absorber losses become

$$L_{\mathrm{miss}} = \frac{\Delta s}{s} \, , \tag{6.17}$$

where s is the horizontal projection of the thickness of the ray bundle at the absorber level and Δs is the part of the beam's horizontal width that extends over any of the two absorber edges.

Optical Efficiency

For any one combination of incidence $\theta_{\mathrm{in_i}}/\psi_{\mathrm{in_j}}$ (the subscripts i and j count the rays in their respective plane), the actual width of incidence over the lens over all its prisms k can be calculated by summarizing the minimum thickness of the incident beam between points A and E, or by summarizing the beams' projective lengths. Thus, the width of the beam of incidence on the lens W_{b} becomes

$$W_{\mathrm{b}}\left(\theta_{\mathrm{in_i}}/\psi_{\mathrm{in_j}}\right) = \sum_{\mathrm{k}=0}^{\mathrm{n}} \Delta\overline{AE}_{\mathrm{k}} \, . \tag{6.18}$$

To obtain the effective ('optically working') width W_{e} of the lens for any one combination of incidence, the width of the incident beam W_{b} in (6.18) has to be modified by the geometrical losses described in (6.15), (6.16), and (6.17):

$$W_{\mathrm{e}}\left(\theta_{\mathrm{in_i}}/\psi_{\mathrm{in_j}}\right) = \sum_{\mathrm{k}=0}^{\mathrm{n}} \left(\Delta\overline{AE}_{\mathrm{k}} \, \left(1 - L_{\mathrm{tip_k}} - L_{\mathrm{block_k}} - L_{\mathrm{miss_k}}\right)\right) \, , \tag{6.19}$$

where the losses L have to be calculated for one combination of incidence. With this the geometrical properties of the optical efficiency of the lens have been found. The optical efficiency of the lens η is found by taking into account the transmittance of the lens τ from (6.4). Naturally, the optical efficiency of the lens is independent of how the width of the incident ray is defined:

$$\eta\left(\theta_{\mathrm{in_i}}/\psi_{\mathrm{in_j}}\right) = \sum_{\mathrm{k}=0}^{\mathrm{n}} \frac{W_{\mathrm{e_k}} \, \tau_{\mathrm{k}}}{W_{\mathrm{b_k}}} \, . \tag{6.20}$$

Although it is impractical to define one single optical efficiency for a nonimaging lens, a value $\overline{\eta}$ may be found by calculating the average efficiency of a lens for incidences $-\theta \le \theta_{\text{in}} \le +\theta / -\psi \le \psi_{\text{in}} \le +\psi$, or the 'plateau' defined by the design angles:

$$\overline{\eta}\,(\theta_{\text{in}}/\psi_{\text{in}}) = \frac{1}{m}\sum_{-\psi}^{+\psi}\sum_{-\theta}^{+\theta}\sum_{\text{k}=0}^{\text{n}} \frac{W_{\text{e}_\text{k}}\,\tau_{\text{k}}}{W_{\text{b}_\text{k}}}\,, \tag{6.21}$$

where m is the number of summation steps over the plateau, or the number of rays traced.

The optical efficiency of a truncated lens of acceptance half-angles $\theta = \pm 2° / \psi = \pm 12°$ has been plotted in Fig. 6.5 for a range of incidence combinations. It is important to describe the nonimaging lens in terms of the incoming light reaching the absorber. This value, the optical efficiency, incorporates geometrical losses due to unused prism tips and blocking at adjacent prisms' backs, absorber misses, and optical losses (transmittance) accounting for reflections on both surfaces. Figure 6.5 shows a peak of the optical efficiency in the range of the design angles. It is remarkable that there is a plateau of high optical efficiency in spite of occasional blocking losses. Its sharp drop after exceeding the design angles is a typical feature of nonimaging concentrators due to their design according to the edge ray principle: incidence at angles greater than the design angles generally misses the absorber.

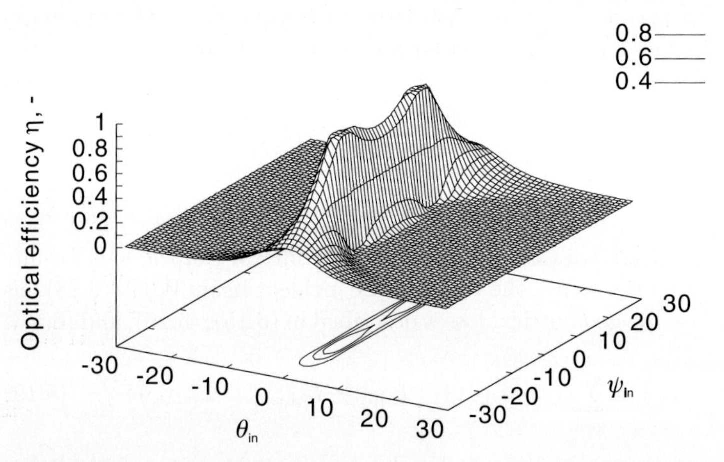

Fig. 6.5. Optical efficiency of the nonimaging lens with $\theta = \pm 2°$ and $\psi = \pm 12°$ for a range of incidence angles. The levels of the contourlines are given in the upper right-hand corner. Lens truncated

6.4 Concentration Ratios

The geometrical concentration ratio C for a lens with $\theta = \pm 2°$ and $\psi = \pm 12°$ at normal incidence is calculated to be 19.13, as shown in Fig. 6.6. This result is the ratio between the 'thickness' of the incident radiation W_{b} (from (6.18), measured between the outermost rays for a lens that would reach the absorber level) and the absorber width d:

$$C\left(\theta_{\mathrm{in_i}}/\psi_{\mathrm{in_j}}\right) = \frac{W_{\mathrm{b}}}{d} \,. \qquad (6.22)$$

For lenses of rotational symmetry, this has to be modified to

$$C_{\mathrm{3D}}\left(\theta_{\mathrm{in_i}}\right) = \frac{W_{\mathrm{b}}^2}{d^2} \,. \qquad (6.23)$$

The concentration ratio should be given for one combination of incidence only or for the combinations of incidence within the design angles. Fig. 6.6 gives the concentration ratios for a range of incidence combinations in 1° steps.

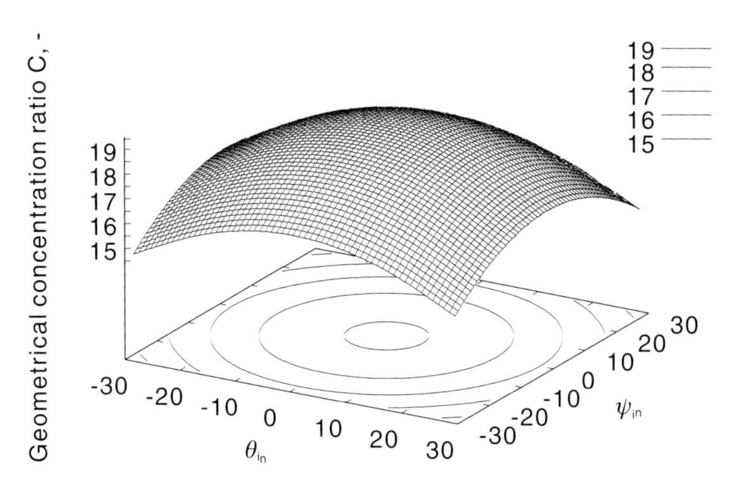

Fig. 6.6. Geometrical concentration ratio of the nonimaging lens with $\theta = \pm 2°$ and $\psi = \pm 12°$ for a range of incidence angles. The levels of the contourlines are given in the upper right-hand corner. Lens truncated

Stating a concentration ratio for a shaped Fresnel lens is an ambiguous concept, as one could also define a concentration ratio calculated as the ratio of the horizontal projection of the incident beam and the receiver width (Fig. 6.7). The latter concept resembles the one used for describing the edge ray principle, as this is based on flat horizontal apertures, as at a CPC. For incidence angles close to 90°, however, this concentration ratio becomes infinity. Nevertheless, for reasons of comparability, this concept is preferred

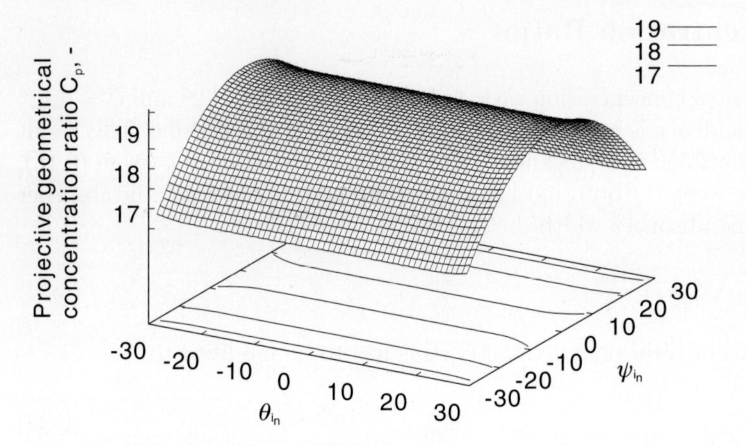

Fig. 6.7. Projective concentration ratio of the nonimaging lens with $\theta = \pm 2°$ and $\psi = \pm 12°$ for a range of incidence angles. The levels of the contourlines are given in the upper right-hand corner. Lens truncated

here. The discussion follows the concepts of concentration ratios outlined by Welford and Winston [182], with the following clarifications.

The theoretical geometrical concentration ratio for a linear concentrator is, of course,

$$C_{2\mathrm{D}} = \frac{1}{\sin\theta} \,. \tag{6.24}$$

For the exemplary lens with the acceptance half-angles $\theta = \pm 2°$ and $\psi = 12°$, we obtain $C = 28.65$, using (6.24). The elevation angle ψ leads to a higher concentration ratio for a 'quasi-3D' concentrator:

$$C_{\mathrm{q-3D}} = \frac{1}{(\sin\theta \;\sin\psi)} \,, \tag{6.25}$$

but limits the acceptance of radiation from perpendicular directions. We are considering (projected) lengths instead of areas, and recommend the use of the concentration ratio $C_{2\mathrm{D}} = 1/\sin\theta$.

The idea behind this way of proceeding becomes more clear when comparing the Fresnel lens with a CPC collector. Imagine the CPC to be short, i.e. having a perpendicular cut-off angle due to shading from both ends of the tube, which has to be considered during evaluation of the CPC's performance. Likewise, the perpendicular angle ψ in the lens is an angle cutting off solar rays, rather than an angle useful in concentration. The elevation angle should be understood as limiting the performance of the refractive system; see also Table 6.2.

The optical efficiency η multiplied with the geometric or the projective concentration ratios C or C_p, respectively, results in an optical concentration ratio of the lens given by

$$\eta_C = \eta C \; . \tag{6.26}$$

The optical concentration ratio η_C can be understood as the ratio between the radiation intensity with the concentrator and the radiation intensity without the concentrator. It is again calculated for combinations of angles of incidence. The two optical concentration ratios are pictured in Figs. 6.8 and 6.9. The difference between the optical concentration ratio based on the optical efficiency η and geometrical concentration ratio C and its projectively based counterpart resulting from η and C_p is insignificant, because the projective concentration ratio becomes exceedingly large only when the optical efficiency of the lens approaches zero.

Nonimaging concentrators like the CPC or this Fresnel lens can be truncated at half-height ($f_t = 1/2f$). The parabolic mirrors or the prisms are cut with minor impact on concentrator performance but significant cost reduction due to material savings. A comparison of truncated and nontruncated lenses has been pictured in Fig. 6.10 for a stationary lens of acceptance half-angles $\theta = \pm 25°$ and $\psi = \pm 35°$, where a wider plateau offers better visualization of the effect.

Even for the truncated lens, the range of incidence angle combinations giving almost uniform optical concentration is surprisingly wide before sharply dropping. Calculated with the projective concentration ratio, this range is slightly higher, and its maximum is shifted towards greater angles of incidence. Lenses designed with greater acceptance half-angles further widen this plateau, but its elevation is reduced.

The optical concentration ratio resembles an energy efficiency. It makes the performance of a lens comparable to the performance of other solar concentrators. Also, this ratio can be integrated into an overall efficiency of the

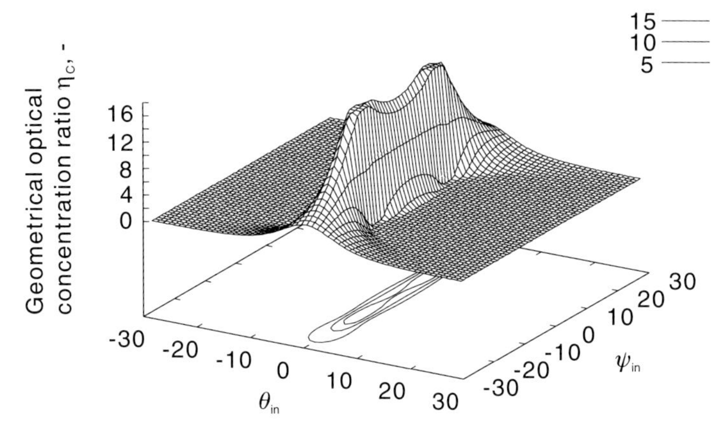

Fig. 6.8. Geometrical optical concentration ratio of the nonimaging lens with $\theta = \pm 2°$ and $\psi = \pm 12°$ for a range of incidence angles. The levels of the contourlines are given in the upper right-hand corner. Lens truncated

collector (multiplying it with the absorber's efficiency) and into system evaluations.

Table 6.1 illustrates that it should make sense to truncate the lens at half-height above the absorber. Performance values are not seriously affected by truncation. The table contains values for the optical efficiencies η, geometric concentration ratios C, and projective optical concentration ratios η_C for some nonimaging Fresnel lenses with various combinations of acceptance half-angles.

Nonimaging lenses of rotational symmetry reach higher efficiencies and concentration ratios than linear lenses. The truncated 0.5/0.0 lens ($\psi = 0.0$, rotational symmetry) sports the highest optical efficiency in the field: 0.91. Given the transmittance losses of ≈ 0.07, the geometrical losses of this lens are very small. Stretching the rotational symmetry into a line-focusing concentrator with $\psi > 0.0$ illustrates the influence of the perpendicular acceptance half-angle on the performance of the lens. A ratio C_{lin} describing the relation between perpendicular and cross-sectional acceptance half-angles for the linear nonimaging lens is defined by

$$C_{\text{lin}} = \frac{\psi}{\theta} \,. \tag{6.27}$$

This value should only increase in a reasonable manner. Optical efficiencies and optical concentration ratios of lenses for which C_{lin} exceeds one order of magnitude decrease by around 10%, and 20%, respectively. Table 6.1 clearly shows this trend. Large values of C_{lin} lead to an increase to absorber misses due to the refractive contribution of the perpendicular angle ψ.

Nonimaging devices with an acceptance half-angle $\theta = \pm 0.5°$ are designed for applications aimed at accurately tracking the sun with the help of electro-

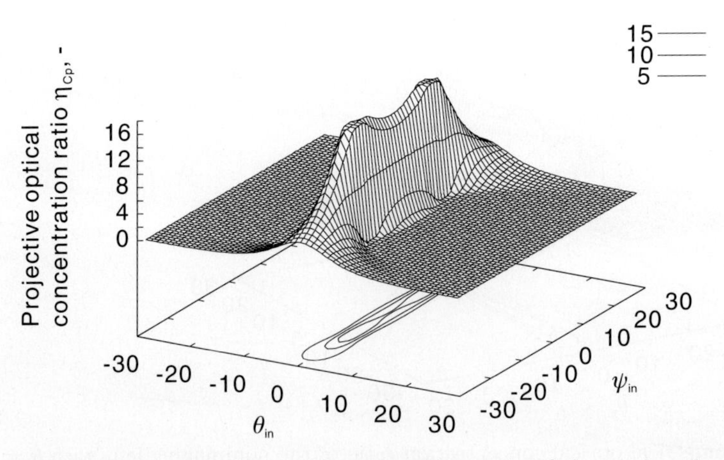

Fig. 6.9. Projective optical concentration ratio of the nonimaging lens with $\theta = \pm 2°$ and $\psi = \pm 12°$ for a range of incidence angles. The levels of the contourlines are given in the upper right-hand corner. Lens truncated

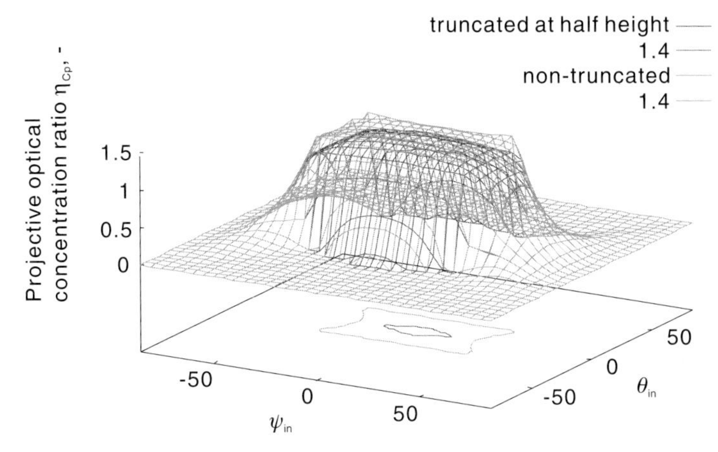

Fig. 6.10. Optical concentration ratio based on the projective concentration ratio of a nonimaging Fresnel lens with acceptance half-angles $\theta = 25°$, $\psi = 35°$, and angles of incidence ψ_{in} and θ_{in}. The same lens is truncated at half-height above the absorber for comparison, and is shown as an inner grid along with the center contour line (contourline levels in the upper right-hand corner)

nically controlled trackers. Passive, simple trackers achieve accuracies of $\pm 2°$, in one axis only. Stationary collectors have cross-sectional acceptance half-angles of 25° and greater. The optical efficiencies η of lenses with increasing acceptance half-angles do not fall to below 0.65 in Table 6.1, but following the decrease in geometrical concentration ratios C, the optical concentration ratios η_C do rapidly drop.

For some cases, e.g. for incidence at design angles at the 27/45 lens, the optical concentration ratio may be smaller than unity; the losses exceed the gains of the concentrator at these incidence angles. Still, the purpose of the stationary lens is fulfilled, and incidence at design angles at least *contributes* to the efficiency of the concentrator, described as averaged efficiency on the 'plateau' visible in Figs. 6.6–6.10, for incidence angle combinations $-\theta \leq \theta_{\text{in}} \leq +\theta / -\psi \leq \psi_{\text{in}} \leq +\psi$.

Normal incidence and incidence at design angles lead to highest efficiencies; the lens should, if possible, for practical reasons face the light source directly. The intercepted beam of light is 'thickest' in this condition, and cosine losses are reduced.

Truncation of the lenses (here at half-height) leads to an increase of efficiencies over the full lens, but a drop in the area of incidence W_{b} causes the geometrical concentration ratio to drop, and subsequently is responsible for the slightly lower optical concentration ratio of the truncated lens. The shape of the optimum nonimaging Fresnel lens is the reason why the decrease in concentration ratios of the truncated lens is accompanied by a rise in its optical efficiency. Geometrical losses at prisms towards the outside of the lens

Table 6.1. Average efficiency ratios of exemplary nonimaging Fresnel lenses over their respective acceptance half-angles. 'Truncated' stands for the truncation of the lens at half-height over the absorber. η: optical efficiency; C: geometrical concentration ratio; η_C: projective optical concentration ratio

Lens (θ, ψ), incidence[a]	η	C	η_C
Full lens			
2.0, 12.0 plateau	0.65	21.30	13.93
2.0, 12.0 $\perp$	0.71	21.47	15.27
2.0, 12.0 design	0.44	20.99	9.17
Truncated lenses			
2.0, 12.0 plateau	0.71	18.98	13.47
2.0, 12.0 $\perp$	0.80	19.13	15.27
2.0, 12.0 design	0.45	18.70	8.42
27.0, 45.0 plateau	0.76	1.48	1.21
27.0, 45.0 $\perp$	0.81	1.71	1.38
27.0, 45.0 design	0.41	1.08	0.51
0.5, 0.0 rotational[b]	0.91	5751.71	5234.05
0.5, 3.0 plateau	0.91	75.84	68.87
0.5, 6.0 plateau	0.79	75.84	59.97
0.5, 12.0 plateau	0.46	75.88	35.13
2.0, 0.0 rotational	0.76	329.86	273.49

[a] *plateau* means value averaged over incidence combinations within the design angles ($-\theta \leq \theta_{\text{in}} \leq +\theta/-\psi \leq \psi_{\text{in}} \leq +\psi$); $\perp$ stands for normal incidence; *design* specifies incidence at any design angle combination ($|\theta_{\text{in}}| = |\theta|/|\psi_{\text{in}}| = |\psi|$).
[b] Lenses of *rotational* symmetry are three-dimensional concentrators.

are greater than those towards its center. Furthermore, the former intercept less radiation. Truncation of the nonimaging lens makes practical sense.

Based on the discussion of different ways of calculating the geometrical concentration ratio of a 'quasi-3D' nonimaging concentrator like the nonimaging Fresnel lens, the concentration ratios are calculated following theoretical concepts. These are compared with the actual geometrical and optical concentration ratios of the lenses in Table 6.2.

The lenses in Table 6.2 are truncated, but the effect of truncation on the concentration ratios is weak for two-dimensional lenses (e.g. see Fig. 6.10). Clearly, the theoretically found concentration ratios do not accurately describe the performance of the lens; the equation $1/(\sin\theta\ \sin\psi)$, in particular, is misleading in its assumption that the perpendicular angle ψ would be an angle of concentration, and not the cut-off angle that it actually is.

Table 6.2. Average concentration ratios. Theoretical concepts[a] and actual values. Ideal concentration ratios $C_{2D} = n/\sin\theta$ and $C_{q-3D} = n/(\sin\theta\ \sin\psi)$, with the refractive index $n_{air} = 1.0$. Optical efficiency $\overline{\eta}$, actual geometrical concentration ratio $\overline{C}$, projective optical concentration ratio $\overline{\eta}_C$, all averaged over incidence within design angles. Actual lenses truncated

Lens (θ, ψ)	C_{2D}	C_{q-3D}	$\overline{\eta}$	$\overline{C}$	$\overline{\eta}_C$
2/12	28.65	137.82	0.71	18.98	13.47
27/45	2.20	3.12	0.76	1.48	1.21
0.5/0.0	–	[b]13131.56	0.91	5751.71	5234.05

[a] See also Sect. 6.5.
[b] $n/\sin^2\theta$, likewise the actual concentration values are squared results of the cross-sectional case.

If the lens is of rotational symmetry, or a three-dimensional concentrator, the gap between theoretical and simulated performance cannot be explained by the, actually very high, optical efficiency. The 'lost half', which cannot be explained by geometrical losses, must be interpreted as the influence of non-ideal concentration due to the refractive index of the lens. The next section will show that for an ideal lens the refractive index must approach infinity.

6.5 Nonideal Concentration

A concentrator is called 'ideal' when all rays entering the first aperture of the concentrator system within two pairs of acceptance half-angles θ and ψ exit through the second aperture of the concentrator system over a solid angle [182].

Four issues are to be discussed when determining whether the nonimaging Fresnel lens may be an ideal concentrator, namely

- does the lens act as an ideal light source illuminating the receiver, and under what conditions may the lens be considered a Lambertian radiator? The refractive index of the lens, and not only the refractive index of the transparent material between lens and receiver, plays a decisive role in this matter;
- is the idealness of the lens affected by its design as a two-dimensional or three-dimensional concentrator, and what influence does the perpendicular acceptance half-angle ψ have?
- to what extent is the performance of the lens practically restricted by total reflection, whether internal or on the outer surface?
- how does truncation of the lenses outer reaches curtail its performance?

If the light source is ideal ('Lambertian'), being characterized by constant flux over all directions, the absorber must equally be a Lambertian radiator if the concentrator should be called ideal. The sun is an almost ideal radiator, but, unfortunately, the nonimaging Fresnel lens is not.

Refractive Index of Infinity

In [93] the nonlinear behavior of Snell's law is made responsible for asymmetrical input and output angles (ϕ_1 and ϕ_2 in Fig. 3.3 and (6.3)) at the prisms of the lens. Thus, the absorber would not see the lens (and the sun) as a Lambertian light source, unless the lens was made out of material with refractive index $n \to \infty$. This explanation assumes that the symmetry of these angles is perfect, when the prism surfaces are almost parallel, as allowed by the asymptotic Snell's law with a refractive index approaching infinity. Obviously, ϕ_1 and ϕ_2 are not symmetrical for all prisms of the lens, even though ideal. Still, the ideal nonimaging lens must have $n \to \infty$.

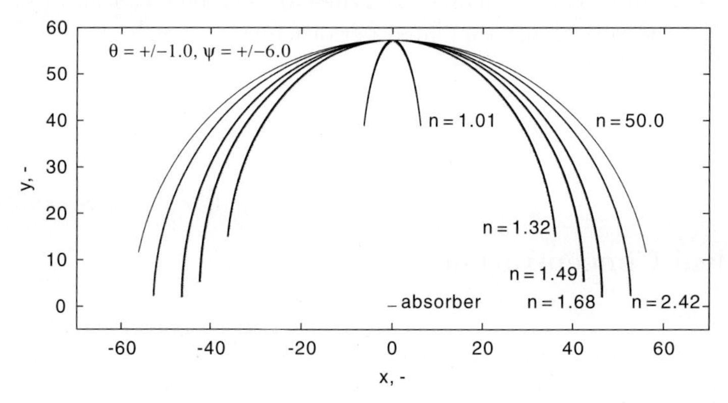

Fig. 6.11. Influence of the refractive index n on the width of the nonimaging Fresnel lens of $\theta = \pm 1^\circ$ and $\psi = \pm 6^\circ$, approaching ideal concentration. Units are multiples of the absorber half-width d

Figure 6.11 pictures linear nonimaging Fresnel lenses of various refractive indices. The design angles are kept constant at $\theta = \pm 1^\circ$ and $\psi = \pm 6^\circ$. The cases of $n = 1.01$ and $n = 50.0$ are the extreme examples; the latter is almost identical to $n \to \infty$. Of more practical value are the lenses with refractive indices close to that of water ($n = 1.32$), PMMA ($n = 1.49$), flint glass ($n = 1.68$), and diamond ($n = 2.42$).

The shape of the ideal nonimaging Fresnel lens ($n \to \infty$) can be determined based on the following two-dimensional concepts:

- The prism should be a prism of minimum deviation, to keep the spread angle of the outgoing beam as small as possible. The spread of the beam directed towards the absorber of small size determines how far from the absorber the prism can be installed, in order that the extreme rays of the beam meet the edges of the absorber. The front and bottom surfaces of the ideal prism with $n \to \infty$ become almost parallel. For the plane parallel plate, and normal incidence, the deviation can be $\delta = 0$, representing the ideal case.

- The refractive power of the prism depends on its refractive index and on its geometry, i.e. the prism angle β. If the prism angle can be kept small due to a high index of refraction, the prism inclination α can be strongly negative: the face of the prism can see the light source and the bottom of the prism can see the absorber. Again, this leads to the possibility of moving the prism further away from the absorber; the slope of the lens is less steep for larger n at a given height over the absorber.

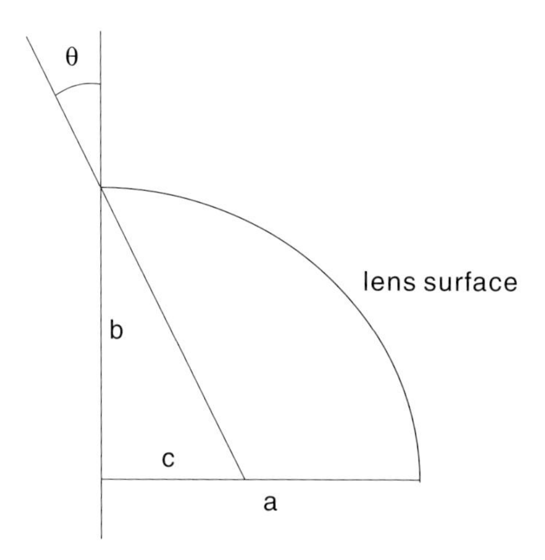

Fig. 6.12. Elliptical shape of the ideal nonimaging Fresnel lens

We assume that the ideal lens will have an elliptical shape [94]. We determined the starting point for the lens design as (see Fig. 6.12)

$$b = \frac{c}{\tan\theta} , \tag{6.28}$$

where c is the absorber half-width, which is set to unity. The foci of the ellipse are located at the absorber's edges. The general equation describing the ellipse is

$$c = \sqrt{a^2 - b^2} , \tag{6.29}$$

where a and b are the half-axes of the ellipse. From (6.29) and (6.28), a can be derived as

$$a = \frac{c}{\sin\theta} . \tag{6.30}$$

This, clearly, is the definition of the geometrical concentration ratio of the ideal two-dimensional nonimaging concentrator:

$$\frac{a}{c} = \frac{1}{\sin\theta} \,. \tag{6.31}$$

Table 6.3. Geometrical concentration ratios C_g for lenses of different refractive indices n_D, and their comparison with the ideal $C = 1/\sin\theta$. Lenses of $\theta = \pm 1°$ and $\psi = \pm 6°$

Material	n_D	C_g	C_g/C_{ideal}
Air-like	1.01	6.3	0.11
Water	1.32	36.3	0.63
PMMA	1.49	42.5	0.74
Glass	1.68	46.5	0.81
Diamond	2.42	52.8	0.92
Ideal[a]	50.0	56.2	0.98

[a] The refractive index $n = 50.0$ is taken in our simulation as representing $n \to \infty$. Due to its numerical approach, the prisms can be designed by the program only to a set confidence level, here $\Delta E = 10^{-6}$.

Looking at the lenses in Fig. 6.11 reveals that only the lens of $n \to \infty$ is an ideal concentrator. All other lenses are not wide enough. Table 6.3 lists the geometrical properties of the lenses in Fig. 6.11. The comparison shows that material choices can make some difference: the expected performance of water-filled lenses and lenses made from glass differs strongly enough to necessitate a detailed optical, and possibly an economic, analysis.

Note the difference between the ideal nonimaging concentration ratio $n/\sin\theta$ (derived in Sect. 2.1, where n is the refractive index of the material between the thin lens and the absorber, which here is $n_{air} = 1.0$) and the concentration ratio $1/\sin\theta$ referring to the refractive index of the thin lens itself. We find that two refractive indices have an influence on the concentration ratio of the nonimaging lens:

- the refractive index of the thin lens itself, fulfilling the conditions of maximum concentration $1/\sin\theta$ for $n \to \infty$. The concentration ratio of the real lens ($n \approx 1.5$) is discriminated by about 20–25% when compared to the ideal lens of $n \to \infty$;
- the refractive index of the transparent material filling the gap in between thin lens and absorber, leading to the familiar ideal concentration ratio $n/\sin\theta$.

Unfortunately, we are not able to examine further the performance of lenses of higher refractive index, because the wavelength-dependent dispersive

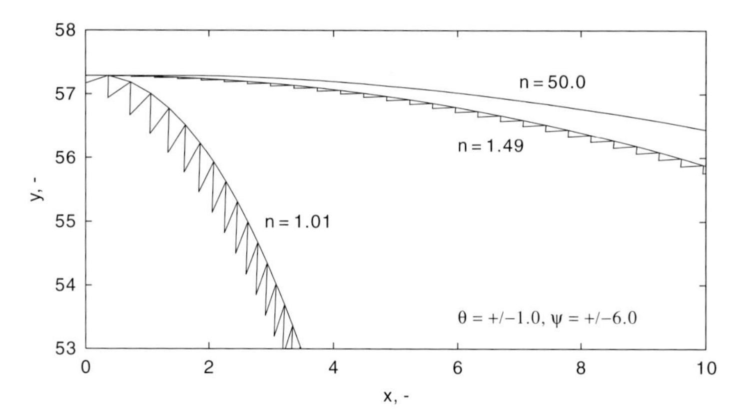

Fig. 6.13. Prism angles and inclinations at prisms of lenses of three refractive indices

properties of these materials are not available to us, even though the refractive power of the material is given. Diamonds have very strong dispersive power, and properly cut, let colors sparkle around those whom they are decorating. Strong dispersion may make them unsuitable for concentrating wavelengths across the solar spectrum. At the ideal lens concentrator, the refractive indices for all wavelengths must be $n(\lambda) = \infty$.

Prism angles and inclinations are shown on an enlarged scale in Fig. 6.13, where it is clearly shown that prisms with low a refractive index have to compensate for this lack of refractive power with a large prism angle β, which can only be achieved with a steep slope of the prism front.

Figure 6.14 shows some of the outermost prisms of two lenses with different secondary acceptance half-angle ψ. The prism angles are small, while the prism inclinations are extremely negative, which is facilitated by the high refractive index.

Design Angle ψ

The incorporation of the perpendicular acceptance half-angle ψ into the design of the nonimaging Fresnel lens does change the shape of the ideal linear lens, as seen in Fig. 6.14. In contrast to the 2D lens, the three-dimensional Fresnel lens concentrator ($\psi = 0$) is not influenced by any effects of the perpendicular design angle.

At the linear lens, incidence from $0 < \psi_{\text{in}} \leq \psi$ is considered in the three-dimensional design process. If $\psi \neq 0$ and $\theta = \textit{constant}$, the absorber is not covered by the designed flux. Focal shortening happens in both the cross-sectional and in the perpendicular direction of incidence. When the perpendicular incidence angle is smaller than the perpendicular acceptance half-angle, the focal plane moves from the position found for the perpendicular design angle further down below the lens. For a prism located in the right part of

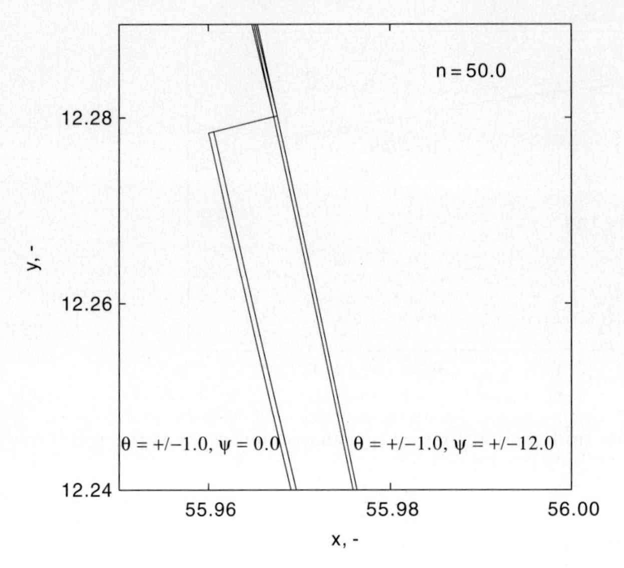

Fig. 6.14. Prisms of the ideal nonimaging Fresnel lens with refractive index $n \to \infty$, showing the small but real influence of the perpendicular acceptance half-angle ψ on lens width and prism angle

the lens, some rays incident at angles smaller than ψ miss the absorber on its right-hand side.

Total Reflection and Truncation

Total internal reflection may occur for lower prisms, effectively limiting the width of the lens due to the resulting minimum required prism inclination. This is an inherent design problem and limits the performance of the first aperture. While total internal reflection is related to the limit of concentration of the ideal lens, near-total reflection on the outer surface of the lens for prisms towards the absorber level is a problem limiting the performance of the practical lens.

For the ideal nonimaging concentrator, the exit aperture angle must be 90° (Fig. 2.3). This is impossible with a truncated lens, which is cut at some height over the absorber, unless secondary concentration is employed. The truncated lens cannot be an ideal concentrator.

In conclusion, the ideal nonimaging Fresnel lens must be

- a Lambertian radiator, with a refractive index of infinity for all incident wavelengths, from which it follows that the lens has an optimum elliptical shape with smooth outer surface;
- a 3D concentrator ($\psi = 0$), to eliminate the losses induced by the perpendicular acceptance half-angle.

We note that the relation between the concentrators' dimensions (2D or 3D) and their ideal behavior is reversed for the CPC and for the nonimaging Fresnel lens. The CPC is ideal as two-dimensional concentrator; but when designed with rotational symmetry, some skew rays within the acceptance half-angle are rejected [182]. The nonimaging Fresnel lens is ideal as a three-dimensional concentrator, when there is no influence of the perpendicular acceptance half-angle.

The fundamental difference between the CPC and the nonimaging Fresnel lens is that, while an ideal CPC can actually be constructed, the ideal nonimaging lens cannot, as a material with refractive index approaching infinity does not exist.

Total reflection, internal and at the outer surface, as well as the prism angle and inclination are determined by the refractive index of the prism material, and effectively limit the width of the lens. Truncation further degrades the ideal performance of the lens. Nonetheless, we would like to define the novel nonimaging Fresnel lens as a concentrator with ideal behavior.

7 Optimization of Stationary Concentrators

7.1 Choice of Stationary Collector

In this chapter, the main characteristics of stationary nonimaging solar concentrators are optimized in respect of the latitude of their installation and the direct fraction of solar radiation incident on the collector plane at the location.

Nonimaging concentrators can be characterized and optimized according to their

- acceptance half-angles: $\pm\theta$ for 3D concentrators, and θ, ψ for symmetrical 2D concentrators;
- collector tilt: inclination of the collector towards south. The tilt angle depends on the latitude of the location;
- orientation: west-east or north–south, or in between.

The objective function of a simulation is to maximize the intercepted yearly cumulated solar irradiance, while reaching a geometrical concentration ratio as high as possible. If only the total energy yield was to be considered, a flat plate collector would be the logical maximum, as it intercepts beam and diffuse radiation from all directions.

Values for insolation on a tilted plane are calculated by means of a simple, yet universal radiation model. The compound parabolic concentrator (CPC) and the Fresnel lens concentrator (FLC) are used as examples for the subsequent optimization of the acceptance half-angles of the stationary concentrator.

Concentrator technologies should be used worldwide in applications such as solar-driven air conditioning, low-pressure steam generation, or electrical power supply. For predicting the approximate performance of solar concentrators for a typical year it is necessary to obtain the yearly average direct fraction of total solar radiation from as little input data as possible, e.g. from the latitude of the proposed location and the monthly average clearness index $\overline{K_{\mathrm{T}}}$. The latter describes cloud cover, as well as the atmospheric transparency, with a weak influence on results. Data sets for $\overline{K_{\mathrm{T}}}$ are readily available in handbooks on solar energy, notably in [37] for many locations around the globe.

Simple radiation models deriving the direct and diffuse fraction of global irradiation have been proposed by various authors. A recent overview [58] proves their general viability. The disadvantage of more complicated models is the need for large amounts of data. Recent models focus on deriving irradiance values on the ground from satellite images. Satellite data [24, 144] is becoming increasingly available, but the computational efforts in storing, handling and calculating the data are substantial. The challenges lie with the prediction of short-term data of high resolution, and with the extraction of beam radiation [96].

Nonimaging solar concentrators are defined by acceptance half-angles. Any direct radiation incident within a pair of these angles is concentrated onto the collector's receiver. Line-focusing, two-dimensional (2D) concentrators require the definition of one acceptance half-angle when symmetrical, and the definition of two acceptance half-angles when asymmetrical. Point-focusing, three-dimensional (3D) concentrators are symmetrical, thus one acceptance half-angle is sufficient. The well-known compound parabolic concentrator is a 2D concentrator with a symmetrical pair of acceptance half-angles. Refractive concentrators can be described in similar terms as their counterparts based on the principle of reflection of incident rays. The Fresnel lens analyzed here has two pairs of perpendicular acceptance half-angles, due to the fact that refraction occurs both in the cross-sectional plane (acceptance half-angle θ) and perpendicular to it (acceptance half-angle ψ).

Stationary solar concentrators will be installed at any given latitude. They should be tilted towards the equator and oriented east–west, or north–south, or possibly another direction. The radiation model described in this chapter simulates a solar concentrator, resulting in proposed acceptance half-angles, tilt, and orientation. These values are optimized according to the cumulated solar irradiance, balanced by the concentration ratio of the collector. The radiation model described can be used for any location on earth where monthly average $\overline{K_{\mathrm{T}}}$ values are recorded. If measured solar data is available, this can be incorporated with slight modifications of the input section of the model.

It has to be emphasized that the optimization of a solar concentrator in the described way makes sense only for stationary collectors which do not follow the apparent movements of the sun. Concentrators that track the sun, on the other hand, ideally are always oriented towards the sun. Their optimization is subject to considerations of the type and the accuracy of tracking, the size of the solar disk, which limits the acceptance half-angle, and the theoretical limit of concentration.

The performance of the stationary concentrator depends on

- the concentrator's characteristics, i.e. optical properties developed with ideal radiation in mind;
- the insolation: where do rays come from, and at what intensity, when seen by the concentrator?

An insolation model can show the need to redesign a concentrator for a specific location and insolation. For large systems, a site specific concentrator may be developed.

For the case of the stationary nonimaging Fresnel lens concentrator, two pairs of acceptance half-angles, along with the optimum tilt and azimuth orientation of the collector, can be derived. It will be shown that, although differences do exist, the approach yields acceptance half-angle pairs which are quite insensitive to changes in the climate or the solar radiation characteristics.

The results of the radiation model based concentrator optimization are a decisive factor in the process that leads to the design, fabrication, and testing of a stationary prototype of the nonimaging Fresnel lens concentrator.

7.2 Solar Radiation Model

The computer simulation used to describe the radiation model has three parts, all set up in loops for all combinations of position of the sun in the hemisphere, azimuth and tilt towards south and east of the collector, and its acceptance half-angles. The three parts are:

1. In the first part of the program, possible positions of the sun in the sky are developed for locations at any latitude.
2. Parameters of the solar concentrator are introduced – can the collector with some azimuth, tilt, and acceptance half-angles 'see' the sun?
3. The third part of the program deals with the amount of direct and diffuse insolation on an inclined surface. The average solar fraction on a horizontal plane that is diffuse is calculated from a set of monthly average clearness indices $\overline{K_{\mathrm{T}}^{\mathrm{m}}}$. A reference value for the total insolation is found. The amount of radiation collected by the concentrator is accumulated and balanced by the collector's concentration ratio.

The combination of azimuth, tilts, and acceptance half-angles for the superior collector are stored to be analyzed in detail. The flow chart of the radiation model is shown in Fig. 7.1.

Positions of the Sun in the Sky

The sun stands directly over the Tropic of Cancer at 23.45° during the summer solstice in the northern hemisphere, and directly over the Tropic of Capricorn at −23.45° during the summer solstice in the southern hemisphere. At equinoxes, the sun moves just over the equator. Seen from locations on the equator, the positions of the sun cover a band between the corresponding ±23.45° in the celestial hemisphere.

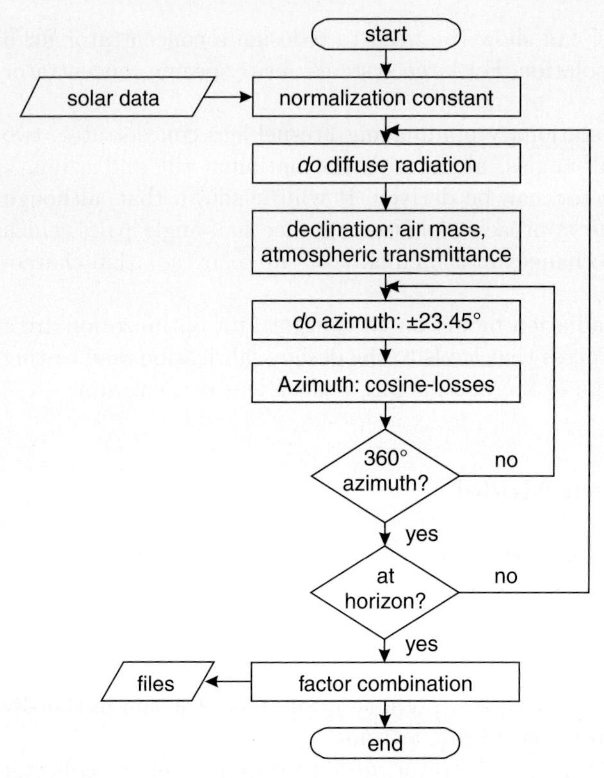

Fig. 7.1. Flow chart of the solar radiation model

The simulation of the position of the sun covers either the northern or the southern hemisphere. They are symmetrical. A coordinate system is set up, with u pointing towards south, v towards west, and w towards the zenith of the sky. As shown in Fig. 7.2, two angles α and β describe the complete northern hemisphere by a vector $\boldsymbol{h}$, is given by

$$\boldsymbol{h} = \begin{pmatrix} \sin\alpha\ \cos\beta \\ -\sin\alpha\ \sin\beta \\ \cos\alpha \end{pmatrix} . \tag{7.1}$$

Due to the declination δ, the sun can occupy only a band of positions in the sky. The sun will lie over the normal to the earth's surface only over locations within the borders of the Tropics of Cancer and Capricorn. The apparent position of the sun depends on the latitude of the location from where the sun is seen. Thus, the coordinate system for $\boldsymbol{h}$ has to be corrected for latitude ϕ by a rotation around the v axis, as shown in Fig. 7.5, where the resulting vector $\boldsymbol{s}$ is shown:

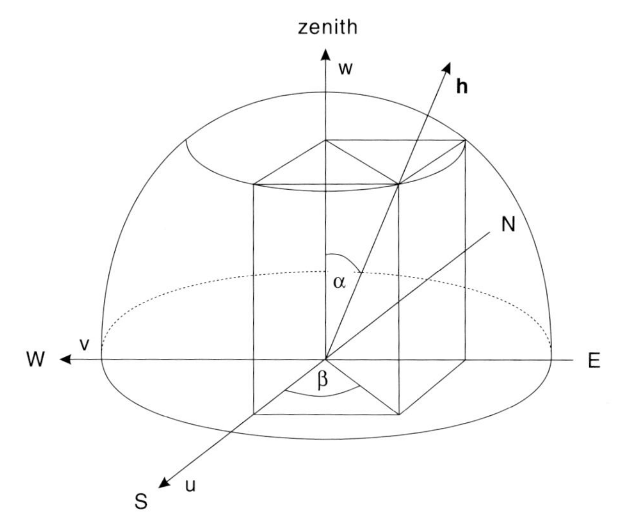

Fig. 7.2. The northern hemisphere described by a vector $\boldsymbol{h}$ set off the zenith by an angle α, with the angle β running from north $(-\pi)$ via west, south, and east to north $(+\pi)$

$$\boldsymbol{s} = \begin{pmatrix} \sin\alpha \ \cos\beta \ \cos\phi - \cos\alpha \ \sin\phi \\ -\sin\alpha \ \sin\beta \\ \sin\alpha \ \cos\beta \ \sin\phi + \cos\alpha \ \cos\phi \end{pmatrix} . \tag{7.2}$$

Contrary to our first thoughts, the coordinates are not turned upward in Fig. 7.5 in order to have the sun appear lower over the horizon u, but instead are turned downward. A collector at latitude ϕ is imaginatively moved to the equator – what elevation above the horizon must the sun now have in order that the collector can still see it? The sun appears higher above the horizon seen from the equator than seen from a certain latitude ϕ. A schematic is shown in Fig. 7.3, the imagined sun would be already in the southern hemisphere.

The rotation of the coordinate system around the v axis by an angle ϕ is achieved by the following procedure. Facing the plane spanned by u and w from the positive side of v, the new lengths u' and w' are to be expressed by the existing lengths u and w, as well as by the angle of rotation ϕ. Fig. 7.4 illustrates the process.

Since the coordinate system is rotated around the v axis, the rotation calculations can be done for the two-dimensional plane spanned by u and w. The result should be a transformation in matrix form. From Fig. 7.4 we have

$$u = a + b \, ,$$

$$a = \frac{u'}{\cos\phi} \, ,$$

$$b = w \ \tan\phi \, ,$$

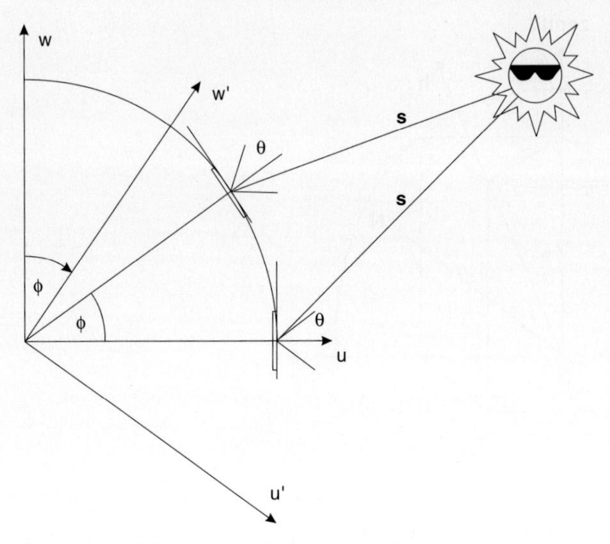

Fig. 7.3. Schematic of changing elevations of the sun when a collector with a pair of acceptance half-angles is moved to the equator by rotating the coordinate system. In order to still be seen by the collector within the acceptance half-angle θ, the elevation of the sun above the equator must be higher, similar to rotating the system (u, w) into the system (u', w')

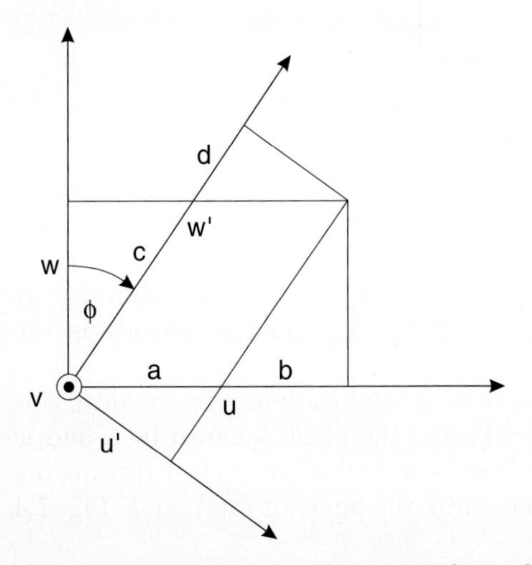

Fig. 7.4. Turning a two-dimensional coordinate system. Schematic

$$
\begin{aligned}
u' &= (u - w\ \tan\phi)\cos\alpha \\
&= u\ \cos\phi - w\ \sin\phi\,,
\end{aligned}
\tag{7.3}
$$

and

$$
w' = c + d\,,
$$

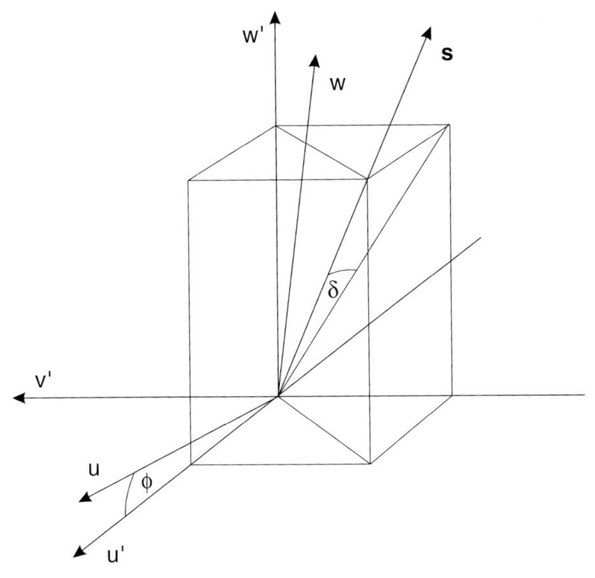

Fig. 7.5. Coordinate system used to describe the position of the sun (vector $\boldsymbol{s}$) in the hemisphere. Corrected for the latitude ϕ of a location by rotation around the v axis. The declination δ is used for excluding impossible positions of the sun

$$
\begin{aligned}
c &= \frac{w}{\cos\phi} , \\
d &= a \; \sin\phi , \\
w' &= \frac{w}{\cos\phi} + \sin\alpha \, (u - w \; \tan\phi) \\
&= u \; \sin\phi + \frac{w}{\cos\phi} - \frac{w \; \sin^2\phi}{\cos\phi} \\
&= u \; \sin\phi + \frac{w \left(1 - \sin^2\phi\right)}{\cos\phi} \\
&= u \; \sin\phi + \frac{w \; \cos^2\phi}{\cos\phi} \\
&= u \; \sin\phi + w \; \cos\phi .
\end{aligned} \tag{7.4}
$$

Therefore, the 2×2 matrix for the plane of u' and w' is

$$
\begin{pmatrix} u' \\ w' \end{pmatrix} = \begin{pmatrix} \cos\phi & -\sin\phi \\ \sin\phi & \cos\phi \end{pmatrix} \begin{pmatrix} u \\ w \end{pmatrix} . \tag{7.5}
$$

The three-dimensional 3×3 matrix is obtained by accounting for the v axis which has not changed, as

$$
S = \begin{pmatrix} u' \\ v' \\ w' \end{pmatrix} = \begin{pmatrix} \cos\phi & 0 & -\sin\phi \\ 0 & 1 & 0 \\ \sin\phi & 0 & \cos\phi \end{pmatrix} \begin{pmatrix} u \\ v \\ w \end{pmatrix} . \tag{7.6}
$$

Multiplying the 3 × 3 matrix S (7.6) with the vector $\boldsymbol{h}$ (7.1) results in the vector $\boldsymbol{s}$ (7.2) describing the position of the sun in a hemisphere over a location at latitude ϕ:

$$\boldsymbol{s} = S \times \boldsymbol{h}\,. \tag{7.7}$$

This vector $\boldsymbol{s}$ and the declination δ are used as exclusive parameters to determine possible positions of the sun in the sky. Whenever δ from Fig. 7.5 exceeds 23.45°, there is no possibility for the sun to be observed directly overhead. The direct radiation is zero.

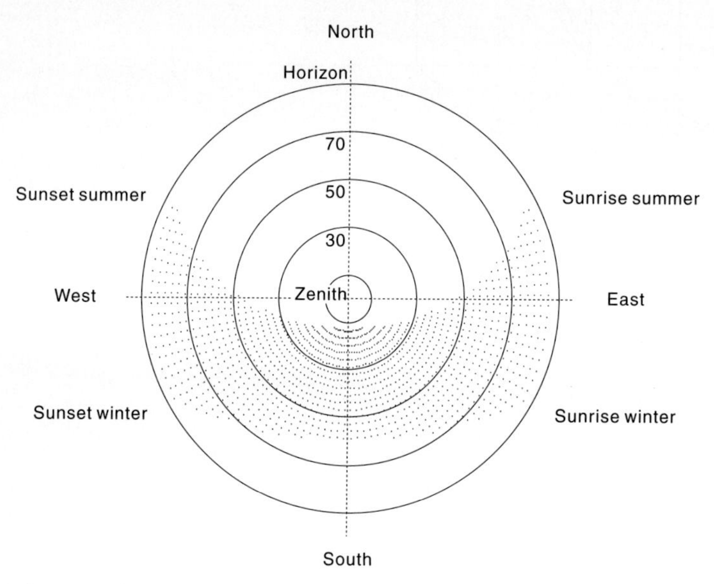

Fig. 7.6. Simulated positions of the sun as seen from Tokyo, Japan at 35.5°. Since the maximum summer solstice declination of the sun is 23.45°, the sun never stands directly over this latitude

The hemisphere over Tokyo at 35.5° northern latitude is shown in Fig. 7.6, centered around the true northern zenith. Standing highest during the summer solstice, the sun follows a path that reaches 12° below the zenith. At the winter solstice the sun covers the opposite low edge of the band.

Approximating the Direct Solar Fraction

Since a radiation model for initial approximations of the typical performance of solar concentrators should be applicable worldwide, this model is based on a single set of data input, i.e. the monthly average $\overline{K}_{\mathrm{T}}$ values for any location. The desired result gives the average yearly solar insolation on a horizontal plane that is diffuse, $\overline{H}_{\mathrm{d}}^{\mathrm{y}}/\overline{H}$, as a fraction of the total insolation.

Collares-Pereira and Rabl [34] have described a monthly correlation between $\overline{K_\mathrm{T}}$ and $\overline{H_\mathrm{d}^\mathrm{m}}/\overline{H}$. With all angles in degrees [37] the equation is

$$\frac{\overline{H_\mathrm{d}^\mathrm{m}}}{\overline{H^\mathrm{m}}} = 0.775 + 0.00606\,(\omega_\mathrm{s}^\mathrm{m} - 90) - (0.505 + 0.00455\,(\omega_\mathrm{s}^\mathrm{m} - 90))\cos\left(115\overline{K_\mathrm{T}^\mathrm{m}} - 103\right) . \quad (7.8)$$

The monthly sunset hour angle $\omega_\mathrm{s}^\mathrm{m}$ is calculated from a list of monthly declinations δ^m and the latitude of the location ϕ by (7.9) [37]. Monthly declinations δ^m can be found with the monthly average clearness indices for Shimizu, Japan, in Table 7.1

$$\omega_\mathrm{s}^\mathrm{m} = \arccos\left(-\tan\phi\ \tan\delta^\mathrm{m}\right) . \quad (7.9)$$

Table 7.1. Monthly average clearness indices K_T^m for Shimizu, Japan[a], and monthly declinations δ^m

	K_T^m	δ^m, deg
January	0.48	−20.9
February	0.45	−13.0
March	0.46	−2.4
April	0.41	9.4
May	0.41	18.8
June	0.38	23.1
July	0.47	21.2
August	0.49	13.5
September	0.44	2.2
October	0.46	−9.6
November	0.48	−18.9
December	0.49	−23.0

[a] Source: [37].

The arithmetic average of the twelve monthly values yields the yearly diffuse fraction on a horizontal plane:

$$\frac{\overline{H_\mathrm{d}^\mathrm{y}}}{\overline{H^\mathrm{y}}} = \frac{1}{12}\sum_{\mathrm{m}=1}^{12}\frac{\overline{H_\mathrm{d}^\mathrm{m}}}{\overline{H^\mathrm{m}}} . \quad (7.10)$$

For Tokyo, Japan, at 35.5° northern latitude, using the data for Shimizu, the procedure outlined in (7.8–7.10) yields a diffuse fraction of 0.43, close to the actual measured value of 0.46. The direct fraction $\overline{H_\mathrm{b}^\mathrm{y}}$ follows as

$$\overline{H_{\mathrm{b}}^{\mathrm{y}}} = 1 - \overline{H_{\mathrm{d}}^{\mathrm{y}}} = 0.57 \,. \tag{7.11}$$

Whenever global and either diffuse or direct radiation data is not available (e.g. from the World Radiation Data Center [189]), the approximation of direct radiation with $\overline{K_{\mathrm{T}}}$ and $\overline{H}$ data sets from [37] is a good substitute.

7.3 Radiation on a Tilted Plane

Direct radiation can only be incident on the concentrator's receiver if the collector's aperture 'sees' the sun. The earth moves in such a way that the sun is observed as being directly overhead in the celestial hemisphere at all latitudes between the Tropics of Cancer and Capricorn at $\pm 23.45°$. In order to be seen by the concentrator, the sun has to lie within the acceptance half-angles of the nonimaging concentrator.

The cumulated energy of the ray bundle coming from a particular direction of the sky incident at a location at a particular latitude depends on

- the air mass to be passed by the sunlight. The value for the air mass AM equals one when the sun is at the zenith, and increases with the zenith angle ϑ. To a close approximation at sea level [37]

$$\mathrm{AM} = 1/\cos\vartheta \,, \qquad 0° \leq \vartheta \leq 70° \,; \tag{7.12}$$

- the declination of the sun, or its altitude over the horizon. The declination is a measure of the sun's apparent velocity of altitude changes.

The energy of a bundle of rays is modified by the lengths of their paths through the earth's atmosphere, and is a function of their angles of incidence ϑ and the atmospheric transmissivity τ [170]:

$$w_1(\vartheta) = \tau^{1/\cos\vartheta} \,. \tag{7.13}$$

Atmospheric transmissivity depends on the amount of humidity, particles, and aerosol droplets in the air. A constant value of 0.7 may be adopted due to its limited influence in (7.13); but it should be noted that the transmissivity of air over potentially large areas affected by volcanic eruptions or heavy smog might reduce the cumulative energy of direct rays by 40% while affecting the global solar radiation by a moderate 15–25% (on the effects of the Philippine Mt. Pinatubo's eruption on radiation over Spain, see [125]). Extreme transmissivity values do influence the performance of concentrating solar collectors.

The second factor affecting the energy accumulated for the beam radiation from a segment of the sky is the velocity with which the sun moves along its path at different declinations. As can be seen from the declination δ [185], we can write

$$\sin\delta = \sin\delta_0 \, \sin 2\pi t \,, \tag{7.14}$$

where $\delta_0 = \pm 23.45°$ and t is the fraction of the year ($0 \leq t \leq 1$). The sun's vertical movement over the horizon is fastest around equinoxes and appears almost stationary near solstices, resulting in energy peaks when cumulating energy from portions of the sky for the latter. The energy of a direct ray bundle is [170]

$$w_2(\delta) = \frac{A}{\pi} \frac{\cos\delta}{\sqrt{\sin^2\delta_0 - \sin^2\delta}}, \qquad |\delta| < \delta_0 , \tag{7.15}$$

where A is a normalization constant. Thus, the beam radiation incident on a collector aperture is the sum of the accumulated energy of the ray bundles coming from a segment of the sky within $\pm 23.45°$ of the celestial hemisphere.

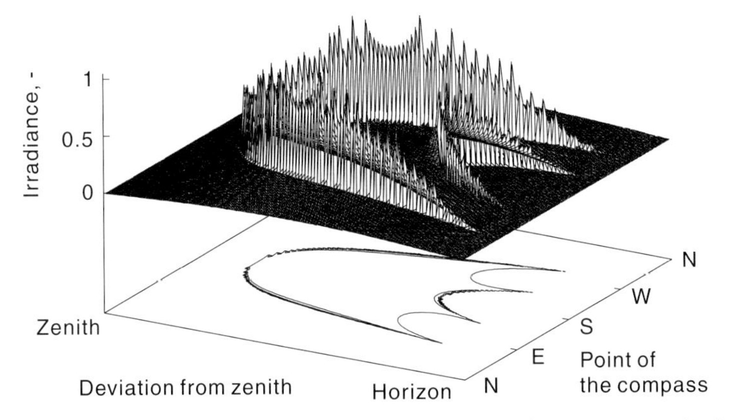

Fig. 7.7. Total normalized energy of ray bundles originating in hemisphere segments incident on a horizontal plate in Tokyo, Japan at 35.5°. The highest peaks depict beam radiation on and near the solstices; the winter solstice path is lower but as clearly marked as the summer solstice path. Diffuse radiation is largely responsible for the shape of the base

Figure 7.7 clearly shows the irradiance peaks that can be attributed to the energy of beam radiation on and near solstice days. The graph has been calculated for beam and diffuse radiation incident on a horizontal plate in Tokyo, Japan at 35.5° northern latitude, using the data from Table 7.1. Higher peaks constitute the summer solstice chain; the lower and shorter chain further to the right is the winter solstice chain.

A similar procedure has been used to simulate the yearly cumulated irradiance on a moving plate (Fig. 7.8). The plate is rotated along the azimuth angle from north to south, and to north again, completing a full circle. The collector is tilted between 0°, its horizontal position, and 180°. Only after the plate has been turned upside down, and away from the sun, does the incident radiation becomes zero. This is due to the diffuse radiation present. No albedo (ground reflection) is considered here.

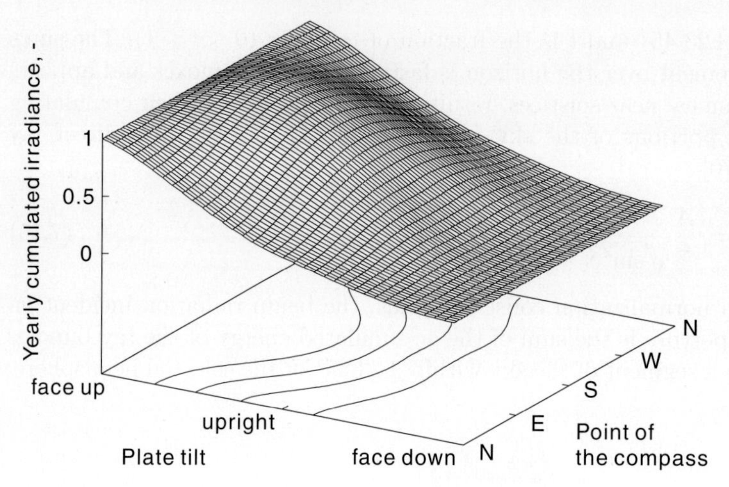

Fig. 7.8. Total solar irradiance incident on a plate in Tokyo, Japan at 35.5°. The plane is rotated by 0°–360° and tilted by 0°–180°

Diffuse radiation is assumed to be uniformly distributed over the whole hemisphere. One speaks of isentropic distribution of diffuse radiation. In Fig. 7.7 diffuse radiation adds the base to the total radiation pictured. The height of the base has been determined by the fraction of solar radiation that is diffuse as derived in (7.8–7.10).

The irradiances in Figs. 7.7 and 7.8 are normalized for the case of the horizontal plate. Accordingly, the irradiance may differ from unity. It may exceed unity in locations north of the Tropic of Cancer or south of the Tropic of Capricorn, if the plate is tilted towards the equator. The effect is caused by the possibility of reducing cosine losses (Fig. 7.9) of the direct fraction of the incident solar radiation.

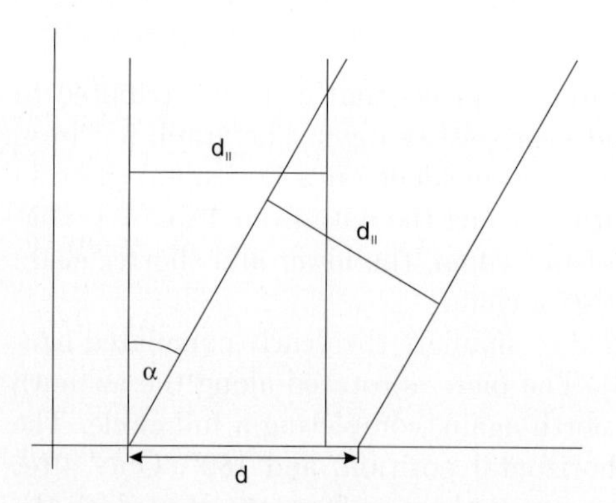

Fig. 7.9. Solar radiation incident on a plane at an angle α off the normal covers a larger area in the beam's projection, $d > d_{\|}$. The effect is termed cosine losses

This is also the reason why tracking plates, like solar concentrators (which always face the sun) can have a higher solar radiation utilizability η_r than tilted flat plate collectors, even though the concentrators are unable to collect the diffuse part of the radiation. The gains of the concentrator resulting from reduced cosine losses depend on the amount of the direct fraction of sunlight at any given location. Only in solar-rich areas is the direct fraction large enough to offset the lost diffuse fraction's energy, and obtain a radiation utilizability $\eta_r > 1$.

7.4 Acceptance by a Solar Concentrator

The amount of direct and diffuse radiation on a tilted plate is an approximation for the total radiation incident on a flat plate solar collector. Concentrating nonimaging solar collectors require an enhanced model due to their ability to concentrate direct solar radiation within symmetric angles while light incident from directions outside the acceptance half-angles is cut off. Uniformly distributed diffuse radiation is collected within the angles, but this fraction does not significantly contribute to the performance of the concentrator, unless the latter's acceptance half-angles are wide.

Acceptance half-angles are used to define the concentration ability of a nonimaging solar collector. An example of a line-focusing 2D collector is the compound parabolic concentrator, where the concentration ratio C based on the acceptance half-angle θ, is $C = 1/\sin\theta$.

Concentrators may contain two pairs of acceptance half-angles. Line-focusing Fresnel lenses are described with a primary cross-sectional pair of acceptance half-angles, and a secondary pair perpendicular to that. Fresnel lenses are prone to focal foreshortening in both planes spanned by the two pairs of acceptance half-angles. Both the CPC and the Fresnel lens can be designed as a nonimaging concentrator, where the mirror or the lens, respectively, does not picture an image of the sun on the absorber. The linear CPC is an ideal concentrator: the pair of acceptance half-angles accurately describes the amount of light collected as incidence at angles larger than the acceptance half-angles is discretely cut off.

For the optimization aimed at with this radiation model, the collector has to be freely moved in three dimensions under the atmospheric hemisphere. All possible positions must be obtained. Of prime importance is the part of the sky covered by the sun's apparent movements, in particular the solar positions near solstices, where the sun's elevation changes slowest. There, the sun remains longest, and energy from these directions is highest. Two pairs of acceptance half-angles are defined (only the cross-sectional one is important in the case of the CPC) to evaluate the amount of radiation collected by the concentrator.

A collector coordinate system in pictured in Fig. 7.10. The collector moves around the y axis by ξ, around the x axis by γ, and rotates around the z axis

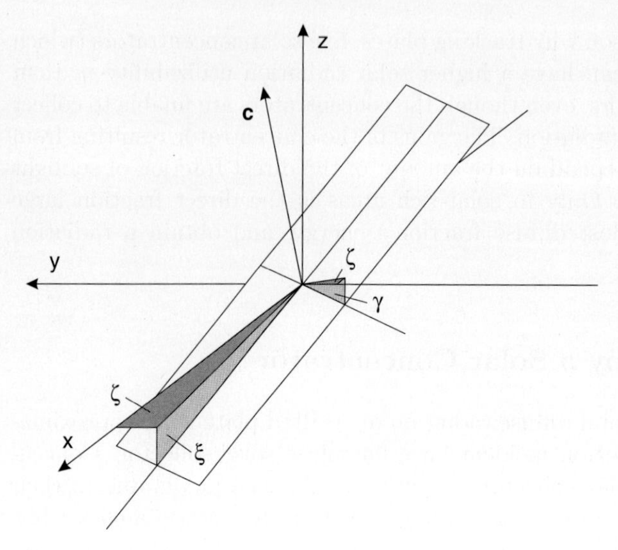

Fig. 7.10. The collector normal c in a coordinate system, where the collector is tilted around the x axis by γ and around the y axis by ξ, as well as being rotated around the z axis by the azimuth angle ζ

by the azimuth angle ζ. Table 7.2 summarizes the three degrees of freedom for a full collector transformation.

The respective coordinate transformations result in the following matrix for the rotation around the y axis by ξ:

$$A = \begin{pmatrix} x' \\ y' \\ z' \end{pmatrix} = \begin{pmatrix} \cos\xi & 0 & -\sin\xi \\ 0 & 1 & 0 \\ \sin\xi & 0 & \cos\xi \end{pmatrix} \begin{pmatrix} x \\ y \\ z \end{pmatrix} . \tag{7.16}$$

Rotation of the collector around the new x' axis by γ is described by

$$B = \begin{pmatrix} x'' \\ y'' \\ z'' \end{pmatrix} = \begin{pmatrix} 1 & 0 & 0 \\ 0 & \cos\gamma & \sin\gamma \\ 0 & -\sin\gamma & \cos\gamma \end{pmatrix} \begin{pmatrix} x' \\ y' \\ z' \end{pmatrix} . \tag{7.17}$$

Finally, the azimuth ζ around the z'' axis yields

Table 7.2. Three degrees of freedom for the collector transformation in order of application

	Angle	Transformation
1.	ξ	tilt around y axis
2.	γ	tilt around x' axis
3.	ζ	azimuth around z'' axis

$$C = \begin{pmatrix} x''' \\ y''' \\ z''' \end{pmatrix} = \begin{pmatrix} \cos\zeta & -\sin\zeta & 0 \\ \sin\zeta & \cos\zeta & 0 \\ 0 & 0 & 1 \end{pmatrix} \begin{pmatrix} x'' \\ y'' \\ z'' \end{pmatrix} . \tag{7.18}$$

Once the collector is described, all its positions can be simulated. The matrix D is the result of the coordinate transformation:

$$D = A \times B \times C . \tag{7.19}$$

How does the collector 'see' the sky? How does the sun move relative to the collector normal $\boldsymbol{c}$? The resulting matrix D is multiplied by the vector $\boldsymbol{h}$ describing the position of the sun in the original hemisphere:

$$\boldsymbol{c} = D \times \boldsymbol{h} . \tag{7.20}$$

The process yields for the vector $\boldsymbol{c}$ (in its components)

$$\begin{aligned} \boldsymbol{c}.u = {} & (\cos\xi\ \cos\zeta + \sin\xi\ \sin\gamma\ \sin\zeta)\,\boldsymbol{h}.u \\ & + (-\cos\xi\ \sin\zeta + \sin\xi\ \sin\gamma\ \cos\zeta)\,\boldsymbol{h}.v \\ & + (-\sin\xi\ \cos\gamma)\,\boldsymbol{h}.w , \\ \boldsymbol{c}.v = {} & (\cos\gamma\ \sin\zeta)\,\boldsymbol{h}.u \\ & + (\cos\gamma\ \cos\zeta)\,\boldsymbol{h}.v \\ & + (\sin\gamma)\,\boldsymbol{h}.w , \\ \boldsymbol{c}.w = {} & (\sin\xi\ \cos\zeta - \cos\xi\ \sin\gamma\ \sin\zeta)\,\boldsymbol{h}.u \\ & + (-\sin\xi\ \sin\zeta - \cos\xi\ \sin\gamma\ \cos\zeta)\,\boldsymbol{h}.v \\ & + (\cos\xi\ \cos\gamma)\,\boldsymbol{h}.w . \end{aligned} \tag{7.21}$$

The collector can now be freely moved under the sun. To be able to determine whether a ray is incident within the acceptance half-angles, these have to be expressed in terms of the components of $\boldsymbol{c}$. The projection of the cross-sectional acceptance half-angle θ into the vw plane yields

$$\theta_{\mathrm{p}} = \arctan\left(\frac{\boldsymbol{c}.v}{\boldsymbol{c}.w}\right) , \tag{7.22}$$

while the projection of the perpendicular acceptance half-angle ψ onto the uw plane results in

$$\psi_{\mathrm{p}} = \arctan\left(\frac{\boldsymbol{c}.u}{\boldsymbol{c}.w}\right) . \tag{7.23}$$

By using the normal $\boldsymbol{c}$ and the projection of the acceptance half-angle(s), radiation incident on the aperture of the collector is detected after it is assured that the collector normal points towards the sky. This is done by calculating the scalar product Δ of $\boldsymbol{c}$ and $\boldsymbol{h}$:

$$\Delta = \boldsymbol{c} \bullet \boldsymbol{h} . \tag{7.24}$$

The scalar product becomes negative if $\boldsymbol{c}$ and $\boldsymbol{h}$ point in opposite directions. The number Δ depends on the angle between the two vectors and accounts for the cosine losses (see Fig. 7.9). In fact, it is a correction for ray bundles incident at angles other than normal on a collector aperture. Projected onto a horizontal plate, the projected area (its diameter) of the ray bundles becomes larger with increasing angles of incidence.

With the collector freely moving in the simulated space, and the incident angles determined, the stage is set for tilt and orientation optimization of the solar collector.

7.5 Compound Parabolic Concentrators

Three parameters of stationary linear CPC are to be optimized:

- acceptance half-angles $\pm\theta$;
- tilt γ; towards the south, seen from the northern hemisphere; and
- orientation (N–S, E–W, or in between).

Two steps during optimization can be distinguished. The first step is the maximization of the global solar radiation incident on the collector within its acceptance half-angles. This value, S, is a function of the time the collector's aperture is exposed to direct solar incidence. This value is calculated as a fraction of the hemisphere that is 'seen' by the collector, i.e. the yearly integrated global irradiance

$$S = \sum \text{radiation}\,(w_1, w_2) \ . \tag{7.25}$$

The second step is the maximization of the collector's concentration ratio (2.8), or its radiation concentration ratio R, defined as the product of the concentration ratio and the sum of incident solar radiation:

$$R = \frac{1}{\sin\theta} \sum \text{radiation}\,(w_1, w_2) \ . \tag{7.26}$$

Both values, the integrated irradiance S and either the concentration ratio C or the radiation concentration ratio R, can be added to form a combined optimization criteria. Both values are ratios with strong dependence on the acceptance half-angle. Both values take opposing courses when drawn over the acceptance half-angle, as will be seen later. While the concentration ratios become smaller with increasing acceptance half-angle, the integrated irradiance becomes larger. Where the sum of both values is greatest, the collector acceptance half-angle is optimum.

Repeating the procedure for all possible combinations of tilt and azimuth, and comparing the results, yields the optimum collector orientation, tilt, and acceptance half-angle for any location.

Optimization Simulation

In contrast to previous sections, where the collector normal $\boldsymbol{c}$ was calculated to evaluate how the collector 'sees' the sun, the following optimization considers a collector that is tilted towards the south, and rotated between a north–south and an east–west orientation. Now the fixed hemisphere coordinates 'see' the collector's normal.

The new vector $\boldsymbol{f}$ is calculated as the product E of the inverse matrices A^{-1}, B^{-1}, and C^{-1}. These are derived from the matrices in (7.19) in the following way. A matrix T

$$T = \begin{pmatrix} t_{11} & t_{21} \\ t_{12} & t_{22} \end{pmatrix} , \tag{7.27}$$

becomes its inverse

$$T^{-1} = \begin{pmatrix} t_{11} & -t_{21} \\ -t_{12} & t_{22} \end{pmatrix} . \tag{7.28}$$

The matrix E is calculated as

$$E = C^{-1} \times B^{-1} \times A^{-1} . \tag{7.29}$$

The fixed coordinates of the hemisphere are introduced by means of the hemisphere vector $\boldsymbol{h}$, which is now fixed pointing towards the zenith (its w component becoming unity):

$$\boldsymbol{h} = \begin{pmatrix} 0 \\ 0 \\ 1 \end{pmatrix} . \tag{7.30}$$

The matrix E and the fixed hemisphere vector $\boldsymbol{h}$ are multiplied, resulting in the vector $\boldsymbol{f}$ describing how the hemisphere 'sees' the collector:

$$\boldsymbol{f} = E \times \boldsymbol{h} . \tag{7.31}$$

The resulting vector $\boldsymbol{f}$ is given here for reasons of completeness in its components:

$$\begin{aligned}
\boldsymbol{f}.u &= (\cos\zeta \ \cos\xi + \sin\zeta \ \sin\gamma \ \sin\xi)\, \boldsymbol{h}.u \\
&\quad + (\sin\zeta \ \cos\gamma)\, \boldsymbol{h}.v \\
&\quad + (\cos\zeta \ \sin\xi - \sin\zeta \ \sin\gamma \ \cos\xi)\, \boldsymbol{h}.w , \\
\boldsymbol{f}.v &= (-\sin\zeta \ \cos\xi + \cos\zeta \ \sin\gamma \ \sin\xi)\, \boldsymbol{h}.u \\
&\quad + (\cos\zeta \ \cos\gamma)\, \boldsymbol{h}.v \\
&\quad + (-\sin\zeta \ \sin\xi - \cos\zeta \ \sin\gamma \ \cos\xi)\, \boldsymbol{h}.w , \\
\boldsymbol{f}.w &= (-\cos\gamma \ \sin\xi)\, \boldsymbol{h}.u \\
&\quad + (\sin\gamma)\, \boldsymbol{h}.v \\
&\quad + (\cos\gamma \ \cos\xi)\, \boldsymbol{h}.w .
\end{aligned} \tag{7.32}$$

This procedure allows for the evaluation of the collector's tilt, described by the angle between $\boldsymbol{h}.w$ and $\boldsymbol{f}.w$. The collector's azimuth off south (still facing south, though) allows for evaluation of the collector's radiation concentration ratio in other orientations than N–S, or E–W.

The acceptance half-angles of the collector are projected into the vw plane as in (7.22). Every ray whose incidence angle exceeds the limits of

$$\theta_{\mathrm{p}} = \arctan\left(\frac{\boldsymbol{c}.v}{\boldsymbol{c}.w}\right) \tag{7.33}$$

is cut off in the simulation, and does not contribute to the integration of incident irradiation.

Fortunately for this optimization of the stationary CPC, there is a clear maximum stating the one best combination of collector acceptance half-angle, tilt, and orientation.

This optimization is carried out for the Tokyo latitude (35.5°) and the Tokyo climate with a diffuse fraction of solar radiation of 0.43. As expected, the optimum tilt of the stationary collector (with a reasonable acceptance half-angle) lies in the range of the latitude. Figure 7.11 shows radiation concentrations for tilted collectors of E–W orientation, based on (7.34).

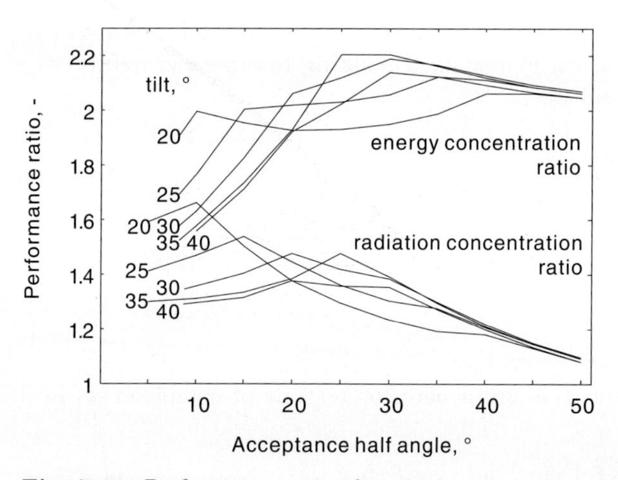

Fig. 7.11. Performance ratios for tilted compound parabolic concentrators ($\gamma = 20°, 25°, 30°, 35°, 40°$, facing south), E–W orientation. The energy concentration ratio is the sum of the radiation concentration ratio and the normalized yearly solar irradiation

The maxima at various tilt angles can clearly be seen. For their interpretation, taking a look at Figs. 7.6 and 7.7 proves useful. Tilting a collector with certain acceptance half-angle towards the south increases the amount of incident radiation, as long as the increase is not offset by having the collector facing parts of the sky with little or no direct radiation at all. Due to the energetical dominance of the solstice chains, they must be included for superior radiation concentration ratios.

The optimum CPC has been found by defining an energy concentration ratio E, based on the radiation concentration ratio R (see (7.26) and (2.8)) of the CPC with its acceptance half-angle θ, and a normalized yearly irradiation S, which is the integrated global irradiance accepted by the collector aperture. The irradiation is expressed as the ratio between the irradiance accepted by the collector and the irradiance falling upon a horizontal surface at the location in question. Thus, S can exceed unity if the collector is tilted towards south, and has a large acceptance half-angle in order to receive as much solar radiation as possible. The energy concentration ratio E is simply calculated as

$$E = S + R\,. \tag{7.34}$$

The performance of the optimum compound parabolic concentrator is pictured in Fig. 7.12. The normalized yearly irradiation and the radiation concentration ratio are added up resulting in the energy concentration ratio. The optimum peak is recorded for a tilt of 35° (in Tokyo at a latitude of 35.5°) and a CPC with an acceptance half-angle $\theta = 27°$.

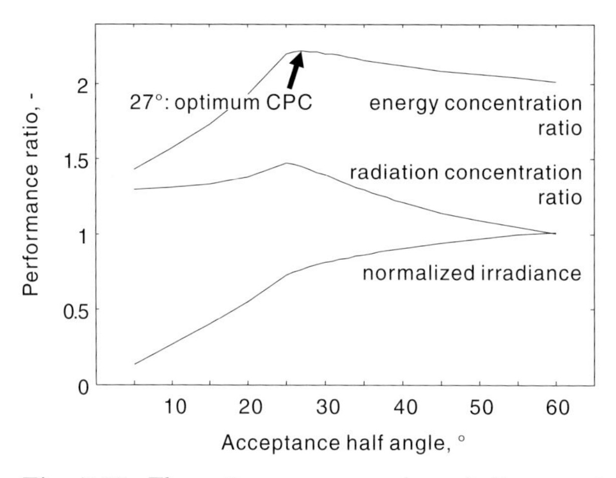

Fig. 7.12. The optimum compound parabolic concentrator for Tokyo, Japan at 35.5° northern latitude is found by maximizing the energy concentration ratio E, the sum of the radiation concentration ratio, and the normalized yearly irradiation. The optimum CPC's acceptance half-angle is 27°

The maxima wander towards lower radiation concentration ratios as well as larger acceptance half-angles for larger tilts off zenith (Fig. 7.11). One is not limited in choice to the obvious peaks; a smaller tilt sometimes offers a better radiation concentration ratio for the same acceptance half-angle than a larger tilt. Thus, it is very useful to include the aperture exposure fraction in the evaluation process. The aperture exposure fraction is defined as the ratio of solar positions 'seen' by the collector to the total number of positions taken by the sun during one year.

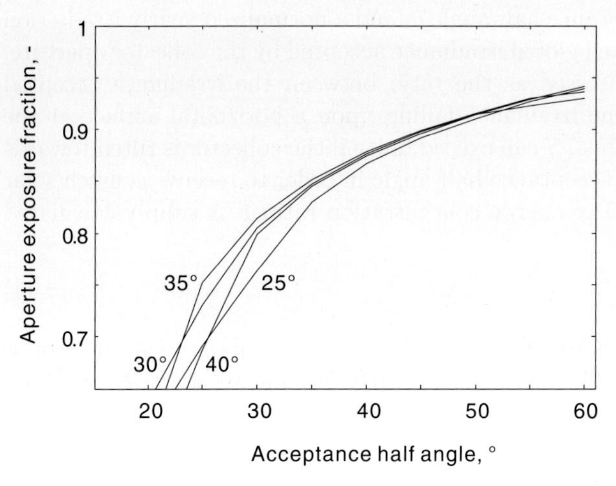

Fig. 7.13. Aperture exposure fraction for tilted compound parabolic concentrators ($\gamma = 25°$, $30°$, $35°$, $40°$, facing south), E–W orientation, for Tokyo at 35.5° northern latitude

The aperture exposure fraction of collectors with various tilts is pictured in Fig. 7.13. Collectors installed at a tilt slightly smaller or equal to the latitude of the location offer the highest aperture exposure fraction for a constant acceptance half-angle. Figure 7.13 is drawn for stationary ideal compound parabolic concentrators with E–W orientation.

Applications for solar concentrators pose different demands. A high radiation concentration ratio opens the possibility of relatively high driving temperatures, whereas a large aperture exposure fraction is desirable where operating time, rather than operating temperature, is the imperative. Radiation concentration ratio and aperture exposure fraction form a tradeoff expressed in Fig. 7.12 and (7.34).

As for its orientation, a collector installed in the N–S direction is not competitive for the acceptance half-angles envisaged in this optimization. Such installations prove superior over E–W orientations only when one-axis tracking is available. In this case, the collector would always face the sun. As the solar band is slimmer in the N–S direction than it is in the E–W direction, reflection losses at the cover and absorber for a N–S directed device will be negligible, as opposed to the one in the E–W direction, where morning and afternoon rays enter the collector at large angles.

A stationary collector (tilted at 35°) for various orientations between N–S and E–W is shown in Fig. 7.14. The collector in the E–W direction exhibits a large maximum for an acceptance half-angle of 27°, which coincides with

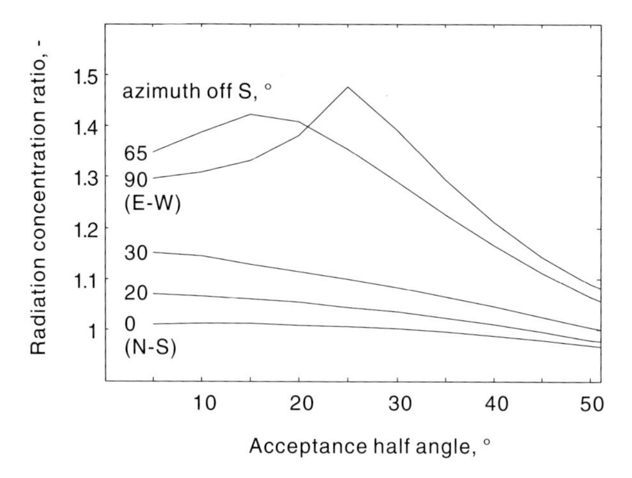

Fig. 7.14. Radiation concentration ratios for tilted compound parabolic concentrators (35°), rotated from N–S to E–W orientation. The collector normal points south (±5°)

previous results [170], where the optimum acceptance half-angle for a collector at latitude 35.5° was determined to be 26°.

The same cluster of collectors is shown in Fig. 7.15 to evaluate the effect of the azimuth on the aperture exposure fraction. The collector normal faces south (within ±5°) at a tilt of 35° for all cases. There is a clear superiority for the collector in the E–W orientation; other orientations of stationary collectors are not competitive.

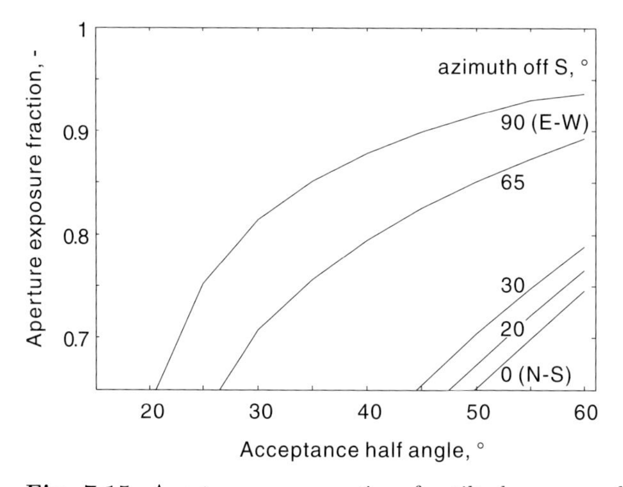

Fig. 7.15. Aperture exposure time for tilted compound parabolic concentrators (35°, facing south) rotated from N–S to E–W orientation

Taking into consideration both the radiation concentration ratio and the aperture exposure fraction by means of an energy concentration ratio, the optimum stationary collector in Tokyo, Japan, has the following characteristics:

- east–west orientation;
- tilt slightly smaller or equal to latitude;
- acceptance half-angle $\theta = 25$–$40°$ as desired, the theoretical maximum being $\theta = 27°$.

Other locations with other latitudes and climates, as well as the availability of one-axis tracking, will slightly change the above choice; the influence of climates (and solar radiation availability) on the optimum acceptance half-angle of the CPC is small. The performance of the concentrator, however, does vary with location.

The energy concentration ratio E from (7.34) is calculated for various locations around the globe from

$$E = S + R\,\rho_{\mathrm{m}}\,, \tag{7.35}$$

with a slight modification accounting for the reflectance losses on the side walls of the CPC, ρ_{m}. The sum of the intercepted radiation S and the radiation concentration ratio R of the CPC are defined in (7.25), and (7.26), respectively. The results are printed in Table 7.3.

Table 7.3. Specifications of optimum 2D compound parabolic concentrators in the east–west orientation and for different locations. Normalized performances in terms of intercepted radiation S, radiation concentration ratio R, and energy concentration ratio E are presented, along with the CPC tilt γ, and its acceptance half-angle θ

Location	latitude ϕ	tilt γ	θ	S	R	E
Tokyo, Japan	35.5	-34	27	0.767	1.453	2.219
Lanzhou, China	36.1	-33	27	0.736	1.394	2.130
Nancy, France	48.7	-46	27	0.775	1.468	2.243
Santiago de Cuba	20.5	-19	27	0.768	1.455	2.224
Singapore	1.0	-1	27	0.688	1.302	1.990
Pretoria, South Africa	-25.8	+25	28	0.785	1.439	2.224

In Table 7.3, it should be noted that the individual performances of the concentrators are normalized according to the amount of radiation each location receives. A direct comparison of the values S, R, and E can only indicate which fraction of radiation is concentrated. As the CPC concentration ratios at all locations are very similar, it can be concluded that the peaks of radiation around the summer and winter solstices fill the 'window' opened by the concentrator rather well, which is true for higher latitudes; otherwise the radiation and the selective concentration do not coincide, as is the case for Singapore, where the seasons are less defined due to its equatorial location.

Since the CPC does intercept part of the diffuse radiation, i.e. a fraction of isentropic diffuse radiation congruent to the CPC's acceptance half-angles, the lower performance values in Singapore can be taken only as a weak indication that the direct fraction of radiation may be lower. Relatively more irradiance reaches the ground at angles greater than θ than in other locations, but in absolute terms, the irradiance within the design angles may well be higher than in other locations.

7.6 Quasi-3D Concentrators

In the previous section possible criteria for optimizing solar concentrators have been explained for the example of the compound parabolic concentrator (CPC). Optimum acceptance half-angles, tilt, and orientation of the collector can be determined by adding the collector's radiation concentration ratio R, or its projective concentration ratio C and the integrated solar irradiance S, as given in (7.25), (7.26), and (7.34).

As opposed to the case of the compound parabolic concentrator (CPC), the Fresnel lens concentrator (FLC) is designed by taking both pairs of acceptance half-angles into account. While the concentration ratio of the CPC can be found with the cross-sectional acceptance half-angles θ, the concentration ratio of the FLC also depends on the perpendicular acceptance half-angle ψ. Both pairs of acceptance half-angles have been used in the design of the lens, and determine the amount of irradiance on the concentrator's absorber.

The secondary acceptance half-angle ψ is a cut-off angle, meaning it prevents radiation from the perpendicular direction from being concentrated, if incident at an angle larger than ψ. For the stationary concentrator, ψ should be as large as possible; for the hypothetical case, $\psi = 90°$ and $\sin\psi = 0°$, leading to the elimination of the disadvantage of the linear refractive concentrator over the reflective one.

Although ψ does not positively contribute to a higher geometrical concentration ratio, ψ can prevent a high optical concentration ratio. The linear nonimaging Fresnel lens may be called a 'quasi-3D' concentrator, if one keeps in mind the true nature of the perpendicular acceptance half-angle.

In order to set up the objective function to optimize the 'quasi-3D' concentrator, we resort to [142] and find a concentration ratio $C(\theta, \psi)$ for the Fresnel lens concentrator given by

$$C(\theta, \psi) = \frac{n}{\sin\theta \, \sin\psi} \, . \tag{7.36}$$

This yields an unrealistic concentration ratio for the Fresnel lens concentrator, but serves the purpose of including ψ in the optimization. The value $C(\theta, \psi)$ does not include the optical performance of an actual FLC. It is a value for comparison. We note the effect of ψ on the optical concentration ratio. If $C_{\text{lin}} = \psi/\theta$ (6.27) is too large, the losses increase and (7.36) cannot meaningfully be employed. In the stationary optimization, the angles ψ and θ are not expected to differ enough to give rise to such concern.

The simulation of the Fresnel lens concentrator follows the optimization of the compound parabolic concentrator outlined earlier. As the Fresnel lens design incorporates two pairs of acceptance half-angles, two angles in two planes have to be employed when the edge ray principle of cutting off incidence that exceeds the limits of either acceptance half-angle is employed. Accordingly, the incidence angles are tested with the following equations (see (7.22) and (7.23)):

$$\theta_{\text{p}} = \arctan\left(\frac{\mathbf{c}.v}{\mathbf{c}.w}\right) , \tag{7.37}$$

for the projection of the cross-sectional acceptance half-angle θ onto the vw plane, while the projection of the perpendicular acceptance half-angle ψ onto the uw plane results in

$$\psi_{\text{p}} = \arctan\left(\frac{\mathbf{c}.u}{\mathbf{c}.w}\right) . \tag{7.38}$$

Naturally, the FLC, hindered by the second acceptance half-angle, cannot yield or outnumber the theoretical performance of the CPC. The real performance of both concepts depends on their respective optical concentration ratios.

Following the procedures concerning the definitions of an energy concentration ratio E and that of a concentration ratio C or radiation concentration ratio R, accounting for the two pairs of acceptance half-angles in the Fresnel lens concentrator (7.36), the energy concentration ratio E of the FLC is constructed as

$$\begin{aligned} E &= R + S \\ &= \frac{1}{\sin\theta\ \sin\psi} \sum \text{radiation}\,(w_1, w_2) + \sum \text{radiation}\,(w_1, w_2) . \end{aligned} \tag{7.39}$$

This process is recommended to achieve convergence. See [165] for a similar methodology including the geometrical concentration ratio and the interception factor.

The computer simulation yields the optimum linear Fresnel lens (maximum energy concentration ratio) in the east–west orientation and having a tilt of 35°. The optimum cross-sectional acceptance half-angle θ is calculated to be 25°, while its counterpart, the perpendicular acceptance half-angle ψ, is 36°. The optimum tilt of a 25/35 lens (this notation expresses the acceptance half-angles θ and ψ) can be seen in Fig. 7.16.

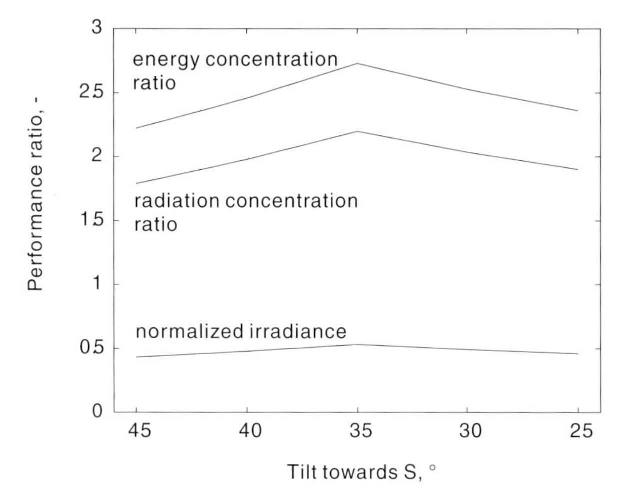

Fig. 7.16. Optimum tilt of a Fresnel Lens Concentrator (FLC) with a cross-sectional acceptance half-angle $\theta = 25°$ and a perpendicular acceptance half-angle $\psi = 35°$. The optimization criterion is the energy concentration ratio E

The optimum values for the acceptance half-angle pairs are given in Fig. 7.17. The initial simulation employs a $5° \times 5°$ grid for finding the optimum lens. The energy concentration ratio incorporates the two counteracting trends of accepting as much radiation as possible through a wide aperture (i.e. wide acceptance half-angles) while trying to keep the very same aperture small in order to achieve a high concentration ratio. The resulting lens with the theoretical optimum 'sees' the best part of the radiation model including the solstice peaks while keeping the acceptance half-angles moderately small.

Figure 7.17 also reveals the sensitivity of the performance of the lens in the east–west orientation on the cross-sectional acceptance half-angle θ. The

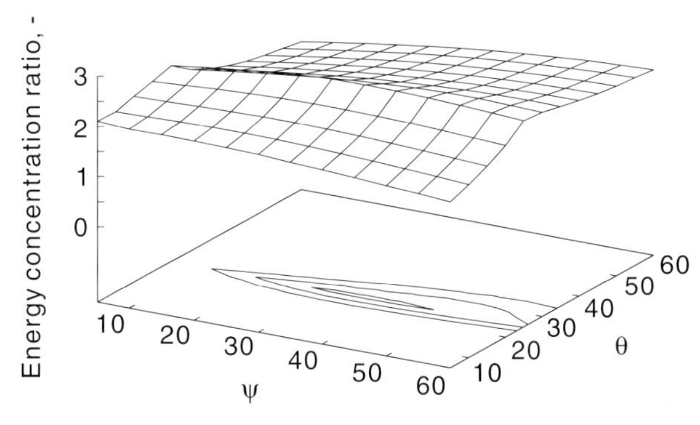

Fig. 7.17. Optimum acceptance half-angles for 35° tilted Fresnel lenses (east–west oriented) in Tokyo at 35.5°. The projections of the contour lines are valued at 2.5, 2.6, and 2.7, counting towards the center

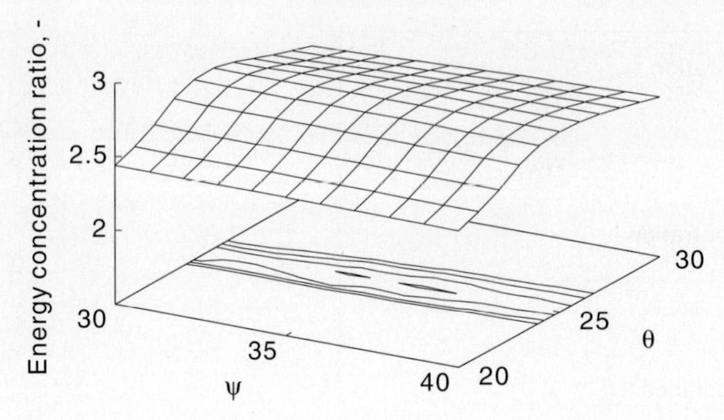

Fig. 7.18. Optimum acceptance half-angles for east–west oriented lenses in Tokyo (35.5°). The maximum energy concentration ratio is rather insensitive to changes in the perpendicular acceptance half-angle ψ. The contour lines peak at 2.73, at an increment of 0.01

Fresnel lens is far more sensitive to changes in the cross-sectional design angle than to changes in the perpendicular design angle. One should remember that the radiation band (the band of possible positions of the sun in the sky) over Tokyo, Japan at 35.5° and anywhere else is stretched from east to west, following the sun from sunrise to sunset. While a small change in the width of the cross-sectional acceptance half-angle θ might include or exclude the solstice chains, a change in the perpendicular acceptance half-angle ψ does not have this sudden consequence, and radiation is more gradually accepted.

Thus, while the lens has optimum performance for the combination of cross-sectional and perpendicular acceptance half-angles of 25° and 36°, respectively, it can be designed for a wide range of values of the perpendicular acceptance half-angle ψ. Losses in the energy concentration ratio are small compared to changes in the design of the cross-sectional acceptance half-angle θ (Fig. 7.18).

The optimum value of the perpendicular acceptance half-angle, $\psi = 36°$, corresponds to an aperture exposure time that allows rays to reach the absorber for an average duration of 4 hours and 45 minutes.

Table 7.4. Results of the simulation of the optimum theoretical compound parabolic concentrator (CPC) and a theoretical Fresnel lens concentrator (FLC) under the radiation of Shimizu, for Tokyo, Japan, at 35.5° northern latitude

	CPC	FLC
Orientation	east–west	east–west
Tilt towards south	34°	35°
Cross-sectional acceptance half-angle θ	27°	25°
Perpendicular acceptance half-angle ψ	-	36°

Table 7.4 compares the main results of the simulation of a compound parabolic concentrator (CPC) and a Fresnel lens concentrator under the simulated radiation of Tokyo, Japan, at 35.5° northern latitude. Both concentrators are theoretical models defined by their acceptance half-angles. Optical losses are not taken into account as they are considered to be more or less constant over the range of results from this optimization.

8 Prototype Design, Manufacturing, and Testing

8.1 Prototypes of Choice

Prototyping describes the practical realization of the theoretical concept of the nonimaging Fresnel lens. The steps that are to be taken in order to get a working prototype of the lens manufactured are as follows:

- Decide what prototype to build. The number of possible combinations of acceptance half-angle pairs for the nonimaging lens is infinite. An educated guess towards the most useful lens in terms of suitability for testing its characteristics and later applications has to be performed.
- Redesign the lens, based on the existing computer simulation, to incorporate manufacturer's constraints. The shaped nonimaging lens is produced as a flat-sheet lens.
- Specify the actual manufacturing process, in order to assess tolerances and possibilities for future mass production.
- Assemble the lens in a solar collector in order to test the characteristics of the obtained prototype.

For the solar Fresnel lens project two different lenses (Fig. 8.1) were manufactured. From the beginning, it was clear that the two designs should be as different as possible, while bearing a resemblance to future lenses to be actually used in an application. The main characteristic of the nonimaging lens is its concentration ratio. Thus, the first difference between the two prototypes should be one of different concentration ratio. Concentration ratios (i.e. the geometrical concentration ratio C) are commonly classified as being 'low' for $C \leq 10$, or 'medium' for $10 < C \leq 100$, or 'high' for $C > 100$.

There is, of course, a thermodynamic limit to solar concentration, the proof of which is given in the previous section on nonimaging concentration on p. 19. For a linear concentrator the ideal limit for the concentration is

$$C_{\text{max, 2D}} = \frac{1}{\sin \theta_{\text{s}}} \approx \frac{1}{\sin(0.275)} \approx 208 \,, \tag{8.1}$$

where θ_{s} describes the radius of the solar disk in degrees when seen from the earth. The result is an approximation due to the seasonal fluctuations of $\pm 1.7\%$ in the sun–earth distance and the diffusion of the sun's edge, which

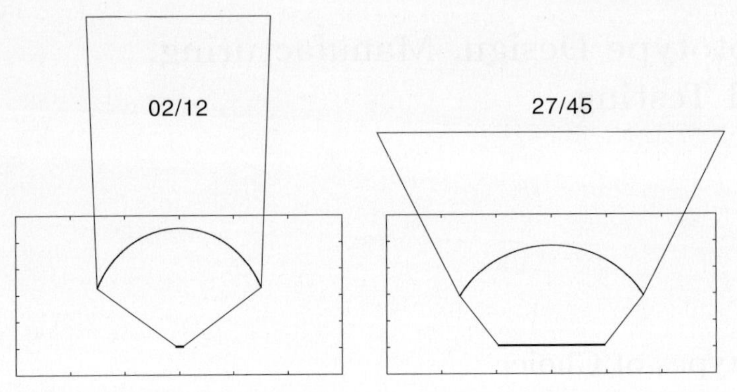

Fig. 8.1. Schematic comparison of tracking and stationary-type nonimaging Fresnel lenses. The acceptance half-angles are 2° in the cross-section (plane of paper) and 12° in the perpendicular plane for a lens with a geometrical concentration ratio $C = 19.1$. For the stationary concentrator of 27°, 45°, the concentration ratio is $C = 1.7$. Projection onto the cross-sectional plane

is gaseous, as well as effects in the terrestrial atmosphere. For ideal concentrators of rotational symmetry

$$C_{\text{max, 3D}} = \frac{1}{\sin^2 \theta_s} \approx \frac{1}{\sin^2 (0.275)} \approx 43\,400 \,. \tag{8.2}$$

The concentration ratios of actual collectors may vary from the ideal by factors of 2–10 for linear designs, and by factors of 5–30 for point-focusing devices ([18, 31]). In the former reference the authors put forward the argument of a maximum practical concentration ratio, based on their experiences with both refractive and reflective concentrators. Interestingly, they put the practically achievable concentration ratio of the shaped Fresnel lens 2.6 times higher than that of a flat Fresnel lens. The primary acceptance half-angle of such a practical lens is given as $\theta = 1°$.

We agree with these findings, and state a rule of thumb for actual linear solar concentrator systems based on Fresnel lenses: imaging lenses have optical concentration ratios of less than 5 (for example [67]), while nonimaging linear lenses yield optical concentration ratios greater than 10. The highest flux achievable in practical linear lens systems (see Chap. 9 for more details) is restricted by the size of the solar disk and the dispersion of the solar spectrum to typically less than 70. Table 8.1 summarizes these numbers.

The simulated optical concentration ratios of the nonimaging Fresnel lenses described here deviate from the ideal by factors of 1.8–2.2 and 2–3, for two-dimensional, and three-dimensional concentrators, respectively. See the examples in Sect. 6.5 for details. In comparing the optical concentration ratios of the actual lenses with the theoretical ideal we must take into account the optical efficiencies of the concentrators and the refractive index of the lens material.

Table 8.1. Optical concentration and flux density properties of typical linear refractive solar concentrators. Rules of thumb found for lenses of similar width

Property	flat imaging	arched nonimaging
Optical concentration	< 5	> 10
Flux density[a]	40	70
Aspect ratio[b]	2	1

[a] Typical values for maximum flux density on the absorber.
[b] Here, the aspect ratio $f/2r$ equals the f/number of the flat imaging lens. The nonimaging lens is arched; f and r refer to its maximum height and its maximum width, respectively.

The required accuracy for tracking the sun's apparent movements is usually the restricting factor in the design of concentrators in the upper medium range of concentration ratios. It is also true that the gap between the ideal concentration ratio and the actual concentration ratio becomes relatively larger when the acceptance half-angle approaches θ_s.

Given the restrictions that are imposed by tracking requirements and solar disk size, a prototype lens of medium concentration is positioned in the lower part of the 'medium' range, with $C \approx 20$.

Having previously found the thermodynamic optimum of a combination of acceptance half-angles θ and ψ for the stationary nonimaging Fresnel lens concentrator to yield a concentrator of rather low concentration, a basic choice concerning the two prototypes' main characteristics, their concentration ratios, is made.

Table 8.2. Main characteristics of the two nonimaging Fresnel lenses under consideration for prototyping. Acceptance half-angles θ/ψ in deg

	2/12	27/45
Orientation	north–south	east–west
Concentration	medium	low
Tracking	one-axis	stationary
Photovoltaic cells	concentrator	plain

The values for the acceptance half-angle pairs of the medium concentration type of nonimaging Fresnel lens are evolving to be in the range of $2°/12°$, by taking into account the points listed in Table 8.2. Should tracking be confined to more simple and less costly one-axis tracking, for a concentrator in north–south orientation (daily tracking mode), then the perpendicular acceptance half-angle ψ defines the seasonal acceptance of radiation. One may tilt the collector twice per year, or one may allow for a peculiarity in the lens design process (the curved shape of the illumination band in the absorber plane, to be explained later), and chose an acceptance half-angle that cuts the maximum declination of the sun at the solstice of $\pm 23.45°$ in half.

Of major concern in selecting the acceptance half-angles of the lens with higher concentration is the suitability of the lens for testing its design parameters and performance. For a lens with comparatively large acceptance half-angles, it will be more difficult to confirm its optical properties, such as geometrical losses or partial illumination due to color dispersion or the refractive influence of the perpendicular acceptance half-angle ψ.

Table 8.3. Possible losses of the nonimaging Fresnel lens concentrator considered during the prototyping decision process. The extent of these errors relates to the suitability of the lens for testing

	2/12	27/45
Geometrical losses	important	important
Size of solar disk	of some concern	of no concern
Color dispersion	important	of little concern
Partial illumination due to influence of ψ	important	of some concern

Table 8.3 lists some points concerning possible losses that were considered during the decision on what acceptance half-angles to employ in the two prototype lenses. Clearly, a lens with higher concentration factor is better for having its optical performance tested. The geometrical losses are expected to be obvious. Absorber misses are strongly defined due to the relatively small size of the absorber. Color dispersion is said to be of greater concern for Fresnel lenses with acceptance half-angles θ smaller than $5°$ [89], and could play a role here.

Measurements of the lens characteristics, such as flux density, are simplified as the absorber width is smaller for a lens with higher concentration ratio, at constant aperture.

While it was clear from the start how to obtain one lens which has the specifications of the thermodynamic optimum stationary concentrator, the final choice of the primary acceptance half-angle θ of the lens of medium concentration was made only after reviewing the available and simple tracking technologies, described in Sect. 8.8.

The prototype lenses are to be truncated at half-height. Previous results had demonstrated the better performance/size ratio for the truncated version of the lenses. Outer or lower prisms do not contribute as much refracted radiation onto the absorber as their central counterparts. Their optical properties are inferior compared to the prisms closer to the optical axis of the lens.

8.2 Prism Size

The dimensions of the prisms have a negligible influence on the performance of the lens. A sensitivity analysis based on prism sizes pictured in Fig. 8.2 yields two main results. First, the smaller the prisms, the better the performance of the ideal lens (in a narrow range), and the smaller the amount of material. Second, the practical lens will show increasing losses for increasing numbers of grooves and prism tips, as these cannot be manufactured the way the ideal lens assumes them to be. Reference [166] gives an experimental proof for facets of different sizes; numbers for groove and tip radii are also given (0.02 mm and 0.008 mm, respectively).

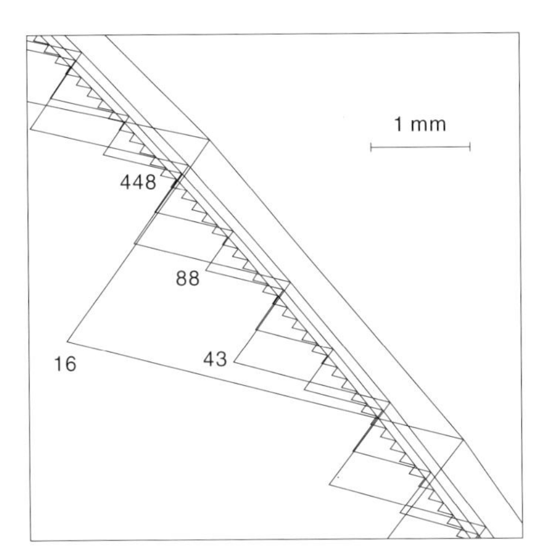

Fig. 8.2. Cross-sectional view of an ideal Fresnel lens with acceptance half-angles $\theta = \pm 30°$ and $\psi = \pm 45°$ for four different prism sizes. Number of prisms per half lens, according to lens width 100 mm

The projective optical concentration ratios for two extreme lenses with 16 and 448 prisms for each half of the lens are shown in Fig. 8.3. With acceptance half-angles $\theta = \pm 30°$, and $\psi = \pm 45°$, the lens is of low concentration ratio. While the overall optical concentration ratio is almost the same for both lenses, as stated by the 1.0-contour line in Fig. 8.3, the plateau is very slightly higher for the lens with a larger number of prisms, when looking at the 1.2-contour line.

The higher performance of smaller prisms is attributed to the fact that during the design process, only the center of each prism is aligned with the

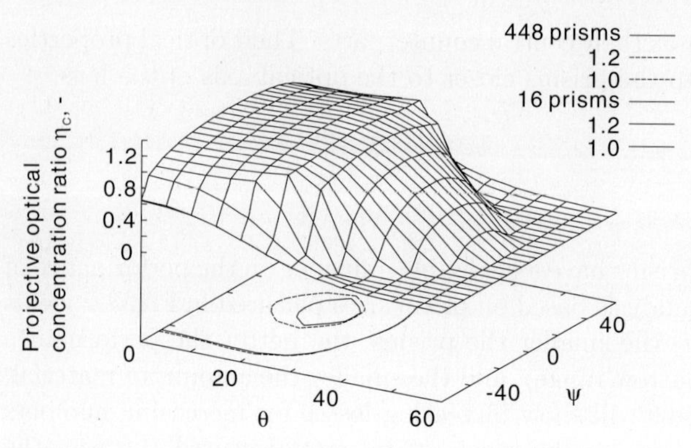

Fig. 8.3. Influence of prism size on projective optical concentration ratio. For an ideal Fresnel lens with acceptance half-angles $\theta = \pm 30°$ and $\psi = \pm 45°$. Number of prisms per half lens

absorber edge. Larger prisms mean larger geometrical losses for rays entering the prisms near their outer reaches at extreme incidence angles (prism surfaces are flat, not curved).

It should be recalled that the number of prisms results from the segment angle ω. Contrary to common imaging designs, the prisms in the lens are not equidistant when assembled horizontally. In the shaped version of the lens, each prism covers a angular segment ω similar to those formed by the spokes of a wheel, but without its circular shape. The number of sixteen prisms per half-lens converts into a segment angle $\omega = 5.0°$; 448 prisms mean $\omega = 0.20°$.

Although the prism size can be freely adjusted, the prisms are not to be chosen too small to limit losses due to manufacturing inaccuracies, such as blunt tips or grooves not clearly formed. Questions like this are best solved by the experiences of the manufacturer, who should be able to find a compromise between manufacturing needs and ideal performance.

The minimum size of the prisms is limited by Fraunhofer diffraction [66], which limits the application of the simple rules of geometrical optics. Fraunhofer diffraction for a single opening and with a light source at infinite distance is

$$b = \frac{m\,\lambda}{\sin\theta_s}\,, \tag{8.3}$$

where $m = 1.0$ for linear slits (linear lenses) and $m = 1.220$ for circular openings (lenses of rotational symmetry), λ is the wavelength of the light passing through the slit, and $\theta_s = 0.275°$ is the half-angle of the sun.

The minimum slit width b is a condition for the occurrence of the first maximum of diffraction. For the typical longer wavelength of the solar spectrum ($\lambda = 0.001$ mm), b (as a measure for minimum prism size) is estimated conservatively for the 02/12-prototype lens to be

$$b_{\text{linear}} = 0.21 \text{ mm}\,, \tag{8.4}$$

$$b_{\text{circular}} = 0.25 \text{ mm} . \tag{8.5}$$

If the opening b is wide enough, the maxima of the diffraction pattern are overlaid by light passing through the slit without diffraction. Diffraction is practically negligible if the opening is large in comparison to the wavelength. Even for small openings diffraction can be negligible for nonimaging lenses, where the size of the absorber exceeds the extent of the first few maxima of the diffraction: for the first, brightest maximum in (8.3), θ equally characterizes the extent of both source and image. One finds

$$d = f \, \frac{m \, \lambda}{b} \, , \tag{8.6}$$

where f is the focal length of the lens (or the distance between lens and absorber where the lens is nonimaging), and d is the distance between the minima of the diffraction pattern, equal to the half-width of the first maximum. For the linear nonimaging Fresnel lens with $\theta = \pm 2°$ and $\psi = \pm 12°$, $d = 0.67$ mm. Thus, d is much smaller than the absorber width.

Diffraction only influences the performance of the lens if either prisms are very small, the focal distance is large, the wavelength of the light is short, or the acceptance half-angle θ of the lens is small. All conditions can be assumed false for typical solar concentrators.

8.3 Lens Redesign

In the initial stages of contacting potential manufacturers for the shaped nonimaging Fresnel lens, it became clear that the lens could not be produced in its final shape. Some of the prism tips create an undercut, i.e. the prism tip is located closer to the optical axis than the prism groove, making it impossible to detach core and mold, or necessitating the use of a collapsible core, an option which actually has been used [110]. Molding or extrusion of the shaped lens was not considered to be a viable option.

When pressing was chosen as the manufacturing method, another constraint arose in what we call the 'centerline' requirement. One prism is chosen to serve as the reference for setting the position of the centerline. Prisms from the reference prism towards the center of the lens are increased in size in order to have their bottom faces (which are almost parallel to their front faces, making the prism very 'thin') cross the centerline at a point where the elevation distances between groove and tip are equal. The outward prisms are decreased in size until the same condition is fulfilled. The grooves and tips are made to displace the same volume of material during the pressing process. The centerline condition facilitates the ease of pressing the lens shape into the acrylic sheet, by parallel molds. The average thickness of the lens over its width remains constant.

Also, the distance between the groove and tip was not allowed to exceed some 0.5 mm, for the same reason of limiting material displacement. This results in a relatively large number of prisms, but limited material requirements. The lens is of almost uniform thickness over its width, enabling it to be mounted with case.

The initial simulation of the lens had to be changed to accommodate the requirements of the prototype production. The following steps are taken, in this order:

1. Design a lens of finite thickness, i.e. the introduction of the possibility to 'backstep', or start a subsequent prism at any point of the face of the previous one.
2. Create a flat lens by moving and rotating the individual prisms until their faces form a smooth horizontal line.
3. Resize the prisms around the centerline.
4. Recreate the shaped lens by rotating and moving the resized prisms back into their original places. Due to the change in the prisms' sizes, a new shape of the lens results, which has to be known to be able to produce a mount into which the bent lens is to be fixed.
5. Simulate the lens width/absorber width relation for a flat lens of width $s = 400\,\mathrm{mm}$, the maximum size offered by the manufacturer.

Finite Thickness

First of all, the lens had to be one of finite thickness. It is helpful to be able to have an additional degree of freedom in the simulation, which is the possibility to backstep, i.e. starting a new prism from any given point on the front face of the previous one, thus preventing the thickness of the lens to be zero at the grooves. Further measures (resizing) have to be taken for prisms very close to the optical axis of the lens. There, the prism angle β is very small.

In the simulation, the starting point A' for a new prism, No. 156, is determined while the current prism, No. 155, is created. Refer to Fig. 8.4 for the nomenclature. The backstep angle ω_b is expressed as a fraction of the pitch angle ω, and can be set at will in the simulation.

The determination of the length w to find the location $F = A'$ for the start of a new prism follows the sine theorem:

$$\frac{w}{\sin\omega_\mathrm{b}} = \frac{g}{\sin\gamma} \,,$$

$$g = \frac{\sin\omega_b}{\sin\gamma} \,, \tag{8.7}$$

where g is given by

$$g = \sqrt{x^2 + y^2} \,, \tag{8.8}$$

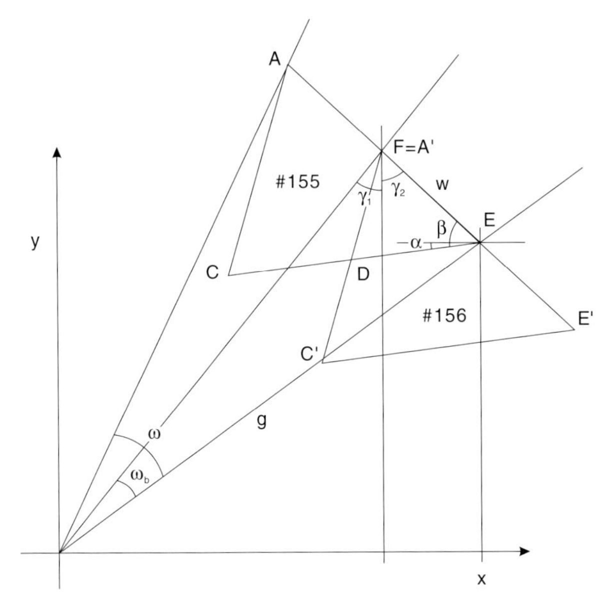

Fig. 8.4. Overlapping prisms to achieve finite thickness of the lens. Not to scale

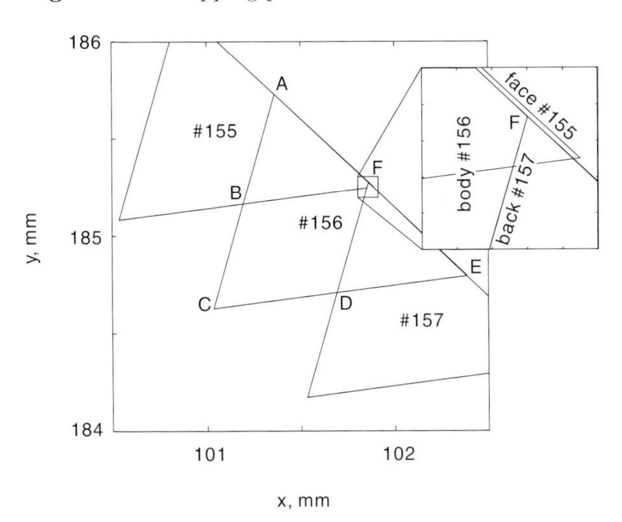

Fig. 8.5. Finite thickness of the 2/12-lens. The overlapping factor is 0.5

and γ is given by

$$\begin{aligned}\gamma &= \gamma_1 + \gamma_2 \\ &= \left(\frac{\pi}{2} - \left(\arctan\left(\frac{y}{x}\right) + \omega_{\mathrm{b}}\right)\right) + \left(\frac{\pi}{2} - (\beta - \alpha)\right) .\end{aligned} \tag{8.9}$$

The coordinates of the point F are found as

$$F_x = x - w \sin \gamma_2 ,$$
$$F_y = y + w \cos \gamma_2 . \qquad (8.10)$$

Figure 8.5 shows the resulting overlap between prisms No. 156 and 157. The inset in the figure is an enlargement of the area around point F to illustrate the difference between the faces of the two prisms.

Flat Lens

The coordinates of the prisms of a flat sheet Fresnel lens are found by moving each prism to a horizontal line, and subsequently rotating each prism around its point A in the counterclockwise direction, so that the prism faces form a smooth line when projected onto the cross-sectional plane. The results of this procedure are shown in Fig. 8.6 as lens (b), originating from the optimum-shaped lens (a).

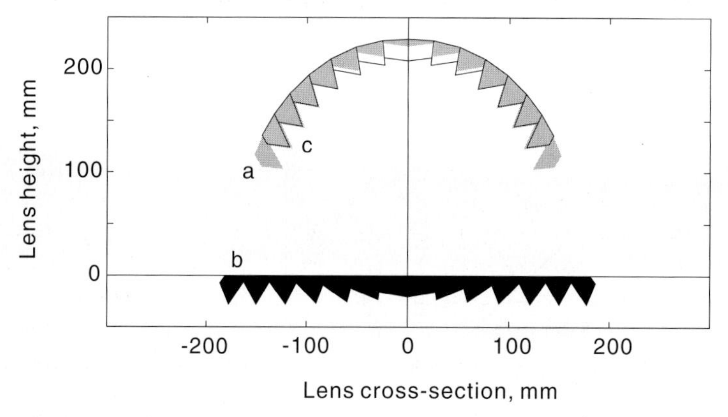

Fig. 8.6. Preparations for manufacturing the Fresnel lens prototype. The prisms of an optimum-shaped lens (**a**) are moved and rotated to form a flat sheet (**b**). Prism tips and grooves are arranged equidistant to a centerline for ease of moulding. Resized prisms are brought back into shape (**c**)

The relocation of each prism's points A, C, D, E, and F (see Fig. 8.4) is completed by simply subtracting their elevation y over the absorber at $y = 0$ for the y coordinate. For the x coordinate the dislocation due to the 'stretching' of the flat lens in comparison to the shaped lens is taken into account by adding the distance between the end point F of the prism just put into the horizontal line, and the new prism's starting point A', $\overline{FA'}$.

A coordinate system like the one in Fig. 7.4 is placed at point A. For each point of the prism, the distance from A to the optical axis at $x = 0$ is subtracted, only to be added after the rotation has been finalized. Rotating the prism in the counterclockwise direction is equivalent to rotating the coordinate system in the clockwise direction by an angle ϕ, which is found from the prism angle β and the prism inclination α as

$$\phi = \beta - \alpha \, . \tag{8.11}$$

The rotation of the coordinate system yields, in analogy to (7.5), the 2×2 matrix for the plane of x' and y':

$$\begin{pmatrix} x' \\ y' \end{pmatrix} = \begin{pmatrix} \cos\phi & -\sin\phi \\ \sin\phi & \cos\phi \end{pmatrix} \begin{pmatrix} x \\ y \end{pmatrix} \, . \tag{8.12}$$

The coordinates of the prism points are found by multiplication with the matrix in (8.12). One problem concerns the prism's groove at point B of the current prism, or point D at the previous one. Due to the flattening of the originally shaped lens, B can no longer be found on the prism's back. A gap between prisms has opened, seen in Fig. 8.7. The solution is, of course, to extend the previous prism's bottom line until it intersects with the current prism's back. Only the points B, C are relevant to the manufacturer of the flat lens.

The procedure of intersecting lines is given in (5.17–5.23).

Centerline and Reshaping

Once the prisms form a smooth flat surface, the second step deals with the changes necessary due to the centerline requirement for manufacturing. Pressing the prisms into a thin sheet of acrylic is eased by displacing at each prism approximately as much material as can be used to fill the groove next to the tip. Moreover, all prisms are located around a centerline, which runs parallel to the smooth surface of the lens. Pressing of a sheet between plane parallel molds is facilitated.

One prism is chosen to serve as reference for setting the centerline. Point G is determined as marking halfway between points C and D. As shown in Fig. 8.7, the prisms to the left of the reference prism are enlarged, and the ones to its right are reduced in size.

Resizing calls for the moving of all points of the prism. Points C and G are simply moved vertically in accordance to the distance Δy between G and the centerline. Points A and E move along the surface of the lens according to the new locations of the lines $\overline{AC}$ and $\overline{CG}$, respectively. These lines do not change their slopes, or the optical properties of the prisms would be affected.

Point B (or point D at the previous prism) is recalculated by intersecting the lines $\overline{CG}$ of the previous prism and $\overline{CA}$ of the current one. Subsequently, point F is found as point A of a new prism.

Finally, in a third step (Fig. 8.6, lens (c)) the flat lens deducted for the manufacturing process has to be reshaped to allow the construction of the fixtures holding the lens. Contrary to the procedure in (8.12), the prisms are

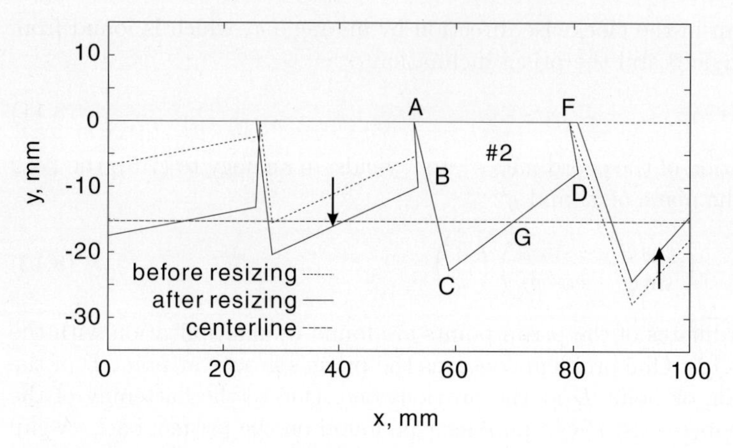

Fig. 8.7. Resizing prisms around the centerline of prism No. 2, in order to facilitate manufacturing of the flat lens

rotated clockwise by the same angle ϕ used previously in the counterclockwise rotation. The matrix changes to

$$\begin{pmatrix} x' \\ y' \end{pmatrix} = \begin{pmatrix} \cos\phi & \sin\phi \\ -\sin\phi & \cos\phi \end{pmatrix} \begin{pmatrix} x \\ y \end{pmatrix} . \tag{8.13}$$

Due to the effects of the relocation of point A, the coordinate system's origin has to be adjusted by the horizontal distance between the new point A after resizing and the old point F before resizing. The turned prisms are then moved into their positions in a reshaped lens. Point B has to be adjusted. This reshaped lens appears slightly modified not only in the size of its prisms, but also in the location of their front faces. The optical properties of the lens have not significantly been changed, as the prism angles and prism inclinations remain constant.

The manufacturer's presses limit the width s of the lens to 400 mm. Since the absorber width $2d$ is set as the initial condition in the simulation of the lens, a relation between absorber and lens width has to be found, to fit a truncated lens onto the sheet without giving away unused space nor having to cut outer prisms. This is found by simple trial and error.

The final specifications of the two prototype lenses are given in Table 8.4. The values for a lens with oversized prisms, which has been used for illustration, are included.

8.4 Lens Manufacturing

The Fresnel lens is to be produced as a flat lens according to the specifications listed in Table 8.4. Figure 8.8 gives an impression of the dimensions and shape of the two prototype lenses.

Table 8.4. Specifications of the prototypes of the nonimaging Fresnel lenses for manufacturing. Lens with oversized prisms used in visualization added for comparison

	02/12-lens prototype	27/45-lens prototype	02/12 lens oversized
Cross-sectional acceptance half-angle θ, deg	2	27	2
Perpendicular acceptance half-angle ψ, deg	12	45	12
Absorber width $2d$, mm	15.8	195.0	16.0
Angular step size, or pitch ω, deg	0.375	0.375	9.9
Backstep (overlap between prisms), –	1/2	1/3	1/3
Refractive index for PMMA, –	1.49	1.49	1.49
Number of prisms for full lens, –	590	480	14
Centerline prism, #	40	100	3
Lens diameter flat, mm	398.8	399.0	373.2
Lens diameter shaped, mm	301.8	332.2	288.4
Geometrical concentration ratio, –	19.1	1.7	18.0

In a simulation, the prisms of the shaped Fresnel lens are rotated and moved until they form a flat sheet lens. Manufacturing makes it necessary to increase or decrease the size of each prism in order to obtain a centerline common to all prisms. The flat lens is then of equal average thickness over its cross-section, and the material displaced by each prism's tip fits into the groove next to it. This centerline is defined by choosing the size of a suitable prism as normal. The vertical distance between prism tips and grooves should not exceed 0.5 mm, in order to allow for pressing the lens contours into a PMMA sheet of 1.0 mm thickness.

Prototypes of both the 02/12 lens and the 27/45 lens were manufactured by Nihon Fresnel Ltd., of Tokyo, Japan. Sheets of $400.0 \times 400.0 \times 1.0\,\mathrm{mm}^3$ dimensions become lenses in the following steps:

- A negative of the lens contours is turned on a numerically controlled lathe using a diamond cutter. Although only the surface of the prism bottom face has to be highly finished, the unused prism back receives the same accurate treatment. The machine is designed for accuracies of 1/1000 mm for lengths, and 1/100° for angles. The prism bottom is defined by the

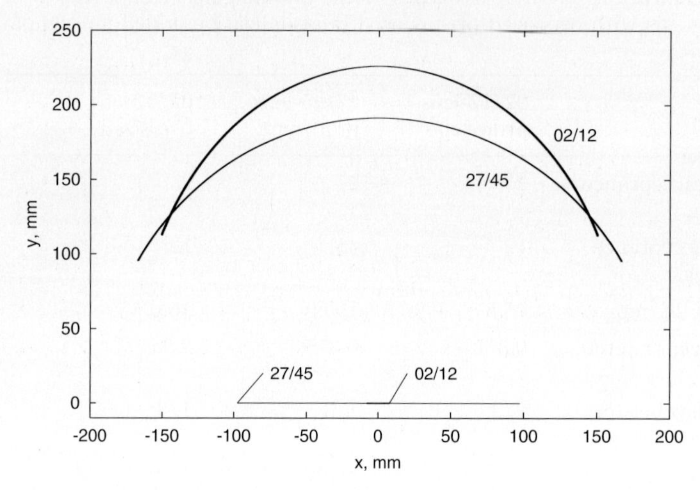

Fig. 8.8. Widths and shapes of the two prototype lenses; and the corresponding widths of the absorbers

angle α in Fig. 8.9 for the enlarged version of the 02/12 lens. The cutting of the lens contours is time intensive, as both the prism's back and bottom are cut for both the right-hand and the left-hand side of the lens.

- A single PMMA sheet is heated for about 210 s at 280°C applying a pressure of 625 kg/m^2 (6.13 kPa). The hot sheet is pressed onto the negative form for another 180 s under similar conditions of pressure and temperature. One negative can be used to produce at least 2000 lenses with sufficient accuracy, if hard spots in the acrylic sheet are manually removed beforehand. Visual control, cleaning of the sheet, removal of hard spots, and the two pressing steps are time-consuming, and rely on manpower.

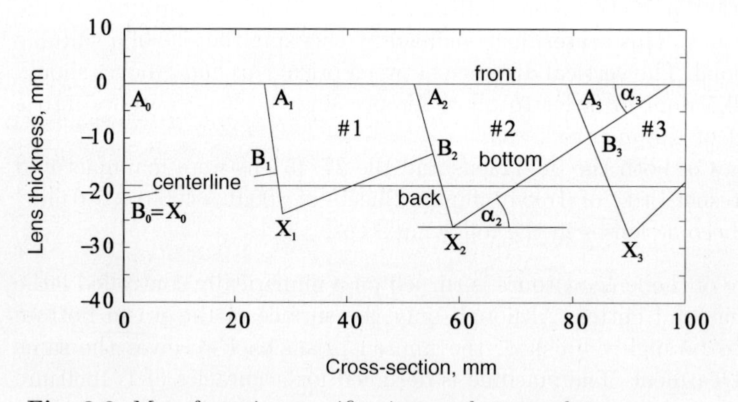

Fig. 8.9. Manufacturing specifications and nomenclature

The manufacturing process described can only be employed if the outer surface of the lens is smooth. A second grooved surface cannot be obtained with satisfactory accuracy with this production method, as both sides would have to be congruent. Furthermore, the prisms have to be small, or pressing is not possible. This leads to a high number of prisms, and high accuracy for a large number of prisms is required in the carving of the metal negative. Also, the size of the lens is limited. Finally, the pressing process is a batch process: output is lower than it would be if extrusion was used.

For the making of the prototypes, the process proved to be highly satisfactory. The accuracies are sufficient for solar energy applications. Once the metal negative is obtained, sheet lenses can be reproduced at low cost. The total cost of manufacturing the two prototypes is in the range of US$ 30 000 (1999).

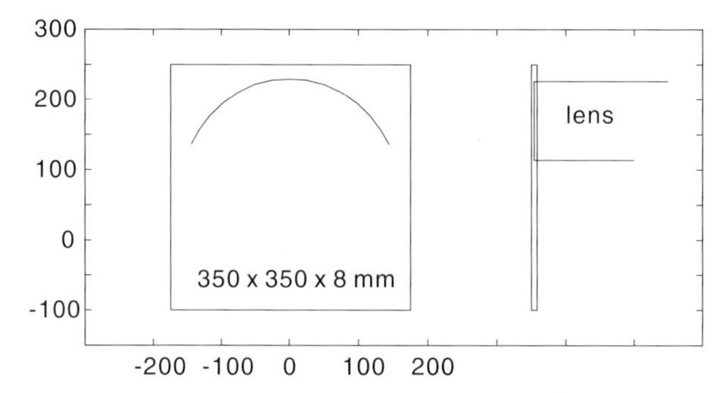

Fig. 8.10. Fixing the lens into an acrylic frame. Schematic

The flat sheet lens is bent into its optimum shape and fixed into a slot in an acrylic frame, as shown in Fig. 8.10. This figure was provided to the manufacturer along with a file containing all points A of the prisms. These points define the outer surface of the lens, and are used to mill the slot for fixing the lens into the acrylic frame. The slot is about 1.5 mm wide, and 3.0 mm deep.

All sides of the lens have to be supported, or fixed to prevent the lens from assuming its original flat shape again. Fixing the two sides not inserted into the slots of a left and right side frame (Fig. 8.10) against their tendency to move upwards pose a problem. For the prototypes, this was solved by using hardwood sticks with a slot running along their lengths. The sticks serve two purposes: connecting the side frames and holding the lens.

The side frames and the bent lens sheet are connected by the mentioned hardwoods, by two metal screws, and by the absorber. Thus, the side frames had to be fitted with some holes. This was done using the machining center of the laboratory.

8.5 Sample

The results of academic work need to be disseminated among the parties interested in new technologies for the collection of solar energy. The aim is not necessarily the commercialization of the nonimaging Fresnel lens, but the communication of results. Also, a sample sent fosters old ties and expresses a token of gratitude. To what extent the novel nonimaging Fresnel lens concentrator deserves attention among the many designs of solar concentrators is left for the experts to decide.

Fig. 8.11. Sample of the 02/12 lens. The shading effect caused by the redirection of beam radiation to the focal area is obvious

In any case, from the surplus of sheet lenses of medium concentration, a number of samples was created by mounting strips of lens into a wooden frame. This sample lens, termed a 'Fresnel lens letterweight', serves as a vehicle of communication.

Figure 8.11 shows the sample. The strip lens is held only at its ends, only approximating the optimum shape. A wooden bar serves to hold the lens, and sports a distancer to allow the focal plane to be on the ground while the lens is held at the correct height. On purpose, the lens does not stand by itself, to prevent the accidential ignition of materials in the focal plane underneath the sample.

The concentrating effect of the lens can clearly be seen in Fig. 8.11: the area underneath the lens apart from the focal area is shaded, since direct radiation is refracted towards the absorber area. The lens in the figure is not held normal to the sun, and the effects on the focal illumination are shown.

8.6 Preliminary Tests

The prototype of the nonimaging Fresnel lens with acceptance half-angle pairs $\theta = 2^\circ$ and $\psi = 12^\circ$ was subject to preliminary testing to confirm the design properties of the lens prior to detailed testing. The lens was mounted on a test rig, and its previously simulated optical design properties could roughly be confirmed. There never had been great concern about the verification of the characteristics of the original optimum-shaped lens, but the lens manufactured as a flat sheet fulfilled all expectations, even though it was bent into shape. It can be stated that the amount of material to be displaced between the prisms when the lens is bent into shape does not influence the optical properties of the lens in any visible way.

Table 8.5. The sun and the moon as light sources for the testing of solar concentrators. Average radii, distances to earth, extension angles on 1 May 1999, 21:32 hours, in Tokyo at 35.5° northern latitude; and relative brightness[a]

	Sun	Moon
Radius, m	$0.695 \cdot 10^9$	$1.73 \cdot 10^6$
Distance to earth, m	$150 \cdot 10^9$	$384 \cdot 10^6$
Solid angle (1 May 1999), °	0.5291	0.4913
Relative brightness	$4 \cdot 10^5$	1

[a] Apparent brightness or magnitude in visible light when the celestial body is in opposition, i.e. on the opposite side of the earth from the sun, usually closest to earth, and best visible. Adopted from [123]. Sun and moon data calculated with Home Planet [178].

In a first experiment, the lens prototype was exposed to the light of the almost full moon (99% full on 1 May 1999, when the experiment was conducted). Utilizing the sunlight reflected from the moon for measuring the optical properties of a solar concentrator offers some advantage over using the rays of the sun directly. The moon appears to have almost the same size as the sun when seen from the earth (Table 8.5). The moon's diameter is 400 times smaller than the sun's, but the moon is on average 400 times closer to earth than the sun. The image of the moon is clearly defined against a dark sky, whereas the sun appears larger than its actual disk size, with an angle of 5.7° per definition filled out by beam radiation. This definition is given for the construction and calibration of pyrheliometers, or devices to measure the direct fraction of sunlight [37].

Moonlight entering the concentrator shows up on the absorber not only bright against the dark background, but can be described as offering a clear threshold between direct rays and darkness. When concentrating sunlight during the day, the region around the sun is of almost the same brightness as the sun itself, making tracing of rays in the concentrator difficult. Bright light and bright absorber are sometimes difficult to distinguish, even when wearing eye protection, which is strongly recommended when working with solar concentrators.

The light reflected from the moon is 'cold' light; its intensity is too low to heat up the absorber. On the other hand, sunlight geometrically concentrated by a factor of 20, and much higher flux in the 'hot spot' region, may damage a dummy absorber or measuring device. Still, the sunlight reflected from the moon at night does have a similar spectral distribution to sunlight coming directly from the sun during the day, since the moon lacks an atmosphere that could interfere with its characterization as a grey body.

A computer-rendered photograph of the moon over the lens is shown in Fig. 8.12. The see-through absorber appears in the lower left corner. The photograph captures a time series of ten exposures over a period of fifteen minutes. The moon appears over the longitude of the same location on earth

Fig. 8.12. Testing the Fresnel lens under the 99% full moon on 1 May 1999, 21:20–21:35, Tokyo, Japan, at 35.5°N. The lens is seen from its lower right side. The see-through absorber, seen from underneath, and almost filled out with light, appears in the lower left corner of the computer-rendered picture. The two thin lines of illumination flanking the bright spot on the luminescent absorber are formed by light exiting this secondary concentrator

not every 24 hours like the sun (not taking into account analemma) but about forty minutes later. Thus, the movement of the moon over a point on earth is slightly slower than that of the sun. Not 15° of angle are covered every hour, but only approximately 14.5°.

During the 15 minutes of photographic exposure in Fig. 8.12 the moon changed its position relative to the observer by an angle of some 3.5 degrees. The absorber was almost fully illuminated from one side to the other with light concentrated by the lens, which has an acceptance half-angle of $\theta = \pm 2.0°$, as expected. Furthermore, the lens was inclined towards the south, or in the direction perpendicular to its cross-section. The tilt angle amounts to the sum of the elevation angle of the moon and the design acceptance half-angle ψ (which can be positive or negative). As will be shown later, aberrations of the refractive behavior of the lens are smallest in this position at the perpendicular design angle $\psi = \pm 12°$.

A similar test was conducted under the sun at clear skies a few days after the moon experiment (Fig. 1.1). With the increased brightness, it is difficult to distinguish the exact focal area from its immediate surroundings, but the optical properties of the lens were confirmed.

In contrast to imaging lenses under similar conditions, we could not observe color aberrations at the nonimaging lens, which are common near the focal point of imaging devices. Of course, any prism refracts rays depending on wavelength, but the nonimaging lens mixes those rays in the absorber plane, and no color lines accompanying the focal area are observed.

8.7 Partial Absorber Illumination

Before we describe the pattern of wavelength-dependent refraction at the nonimaging Fresnel lens, we shall calculate how the receiver is illuminated when light is incident from different directions. All light entering the lens aperture within the acceptance half-angle pairs is refracted towards the absorber. This is true for each of the prisms constituting the lens: one may imagine an upside-down pyramid of light on any given point on the prisms' surfaces contributing to the light reaching the absorber.

Not all light reaches the absorber, even after the losses on and inside the lens body are accounted for. Some rays miss, because of a combination of design principle and refractive laws. The design angles for the lens are the maximum combinations of the acceptance half-angles, $+\theta/\psi$, and $-\theta/\psi$. The perpendicular angle ψ is symmetrical along the 2D lens, but some rays entering at $\psi_{\mathrm{in}} < \psi$, and maximum θ, miss the absorber because refraction does happen in the perpendicular plane as well as in the cross-sectional plane of the lens (see Fig. 5.8).

Rays entering the lens, their refraction at the front and back faces of the prism and their intersection with the plane of the absorber are shown in Fig. 8.13. In this top view of a typical prism of the 02/12 lens, it is clearly

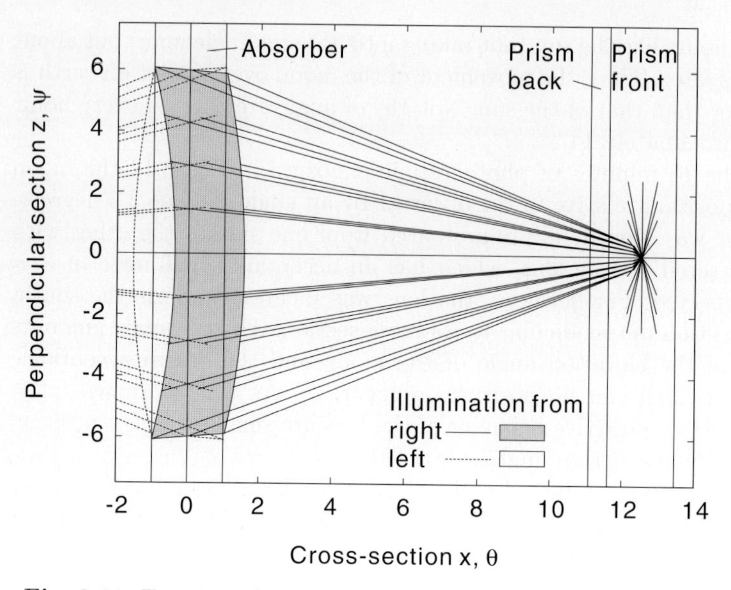

Fig. 8.13. Top view of a typical prism and absorber of the nonimaging Fresnel lens with acceptance half-angle pairs $\theta = \pm 2°$ and $\psi = \pm 12°$. The incident light fills an upside-down pyramid on any point of the lens, is refracted twice at the prism's faces, and is shown intersecting the absorber level, where rays form a curved band of light due to perpendicular refraction. Oversized prism, for yellow light, and refractive index $n = 1.49$

shown that the band of light incident on the absorber level is curved, and some rays incident at perpendicular angles smaller that the perpendicular design angle miss the receiver. Fig. 8.13 has been drawn using data for yellow light, with the refractive index of 1.49 used throughout the design of the lens.

The curvature of the concentration band becomes stronger with increasing design acceptance half-angle pair ψ. This is the reason why ψ cannot be increased beyond values of C_{lin} that are reasonably comparable to the cross-sectional acceptance half-angle θ (see (6.27)). A large ψ is desirable from the viewpoint of capturing solar rays during one-axis tracking, where ψ should theoretically be $\pm 23.45°$. In the light of the previous discussion of the lens' concentration ratio $C = 1/\sin\theta$, the ability of ψ to restrict the ability to concentrate in this 2D collector becomes evident. So does the need for ray tracing when determining the actual optical concentration ratio.

The incompleteness of illumination of the absorber can be overcome by employing a secondary concentrator, e.g. a compound parabolic concentrator. The acceptance half-angle of this concentrator depends on the position of the outermost prism of the lens $P\,(x, y)$, and the extent of the absorber d, enlarged by the maximum extent of the curved band in the cross-sectional x direction, or even for bands formed by light of extreme wavelengths, if desired. The absorber of the truncated lens becomes the first aperture of the secondary concentrator, whose acceptance half-angle must exceed

$$\theta_{\text{secondary}} > \arctan\left(\frac{x + d_{\max}}{y}\right) . \tag{8.14}$$

For the truncated prototype of the 02/12 lens, this procedure results in a secondary concentrator of a minimum acceptance half-angle pair of around 55°, which in turn yields a geometrical concentration ratio of 1.2. This illustrates that additional concentration after the optimum lens is hardly possible, although the incomplete and nonuniform illumination of the absorber can be corrected (see Sect. 9.5). The nontruncated, nonimaging Fresnel lens approaches the ideal concentrator for $\psi \to 0$, neglecting losses. Further concentration is impossible with secondary concentrators having the same refractive index as the lens. The condition of identical refractive indices follows from the findings in Sect. 6.5.

8.8 Tracking

The sun's apparent movements can be grouped into seasonal movements and daily movements. Daily, or azimuth movements describe the path of the sun over the sky from the east in the morning to west in the evening. Seasonal movements concern the sun's elevation over the horizon; the sun stands directly over the tropics of Cancer (+23.45°) and Capricorn (−23.45°) during summer and winter solstices in the northern hemisphere. Seasonal movements are fastest around the equinox days.

Three-dimensional solar collectors are usually designed for high concentration ratios. Full tracking of the sun around two axes is required. Linear focusing collectors, such as the nonimaging Fresnel lens, may be designed as concentrators of low and medium concentration ratio.

Low concentration can be achieved with stationary concentrators; no tracking is performed. The concentrator is oriented in the east–west direction to allow the primary acceptance half-angle θ to cover seasonal differences in the sun's altitude. Seasonal movements require θ to be greater than 23.45°, while the sun's position in the east–west direction changes at a rate of 15° per hour. Although calculations have to take into account the tilt of the collector, towards the south in the northern hemisphere at an angle equal to latitude; as a rule of thumb the perpendicular acceptance half-angle should exceed $\psi \geq 45°$ to collect six hours of sunshine per day.

Concentrators of medium concentration ratio have to be equipped with tracking around at least one axis. The concentrator is best oriented in the north–south direction to allow for a small θ and, again, a relatively wider ψ. The more sensitive primary acceptance half-angle θ can be as small as the tracking accuracy allows, as long as it is larger than the sun's half-angle of 0.275°, and in a reasonable relation to the secondary acceptance half-angle ψ.

The acceptance half-angle pairs of the prototype concentrator lens of medium concentration were chosen to be $\theta = \pm 2°$ and $\psi = \pm 12°$. This decision was based on the assumption to employ the lens in a north–south direction, with azimuthal tracking.

The most simple one-axis tracking devices achieve tracking accuracies of around ±2° [81, 140]. This is typically the case for sunny days, i.e. a high fraction of direct radiation. Two tracking concepts can be differentiated: photovoltaic systems and thermohydraulic ones. The working principle of the latter is based on the expansion of a liquid of low boiling point. When exposed to the sun, the liquid starts boiling, and the system moves in on the sun, until a position normal to the sun is reached, where the liquid is shaded or otherwise protected from boiling. The advantages are that the machine contains no electronics and few moving parts; the thermohydraulic tracker can be operated without using organic refrigerants. These media have been banned for their ozone-depleting potential.

Trackers based on photovoltaic modules use a variety of geometries to expose one of two counteracting photovoltaic cells to the sun. Figures 8.14a–c show some of them. The cells are connected electronically in antiparallel mode (Fig. 8.14d), and drive a reversible DC motor via a gearbox. The tracking

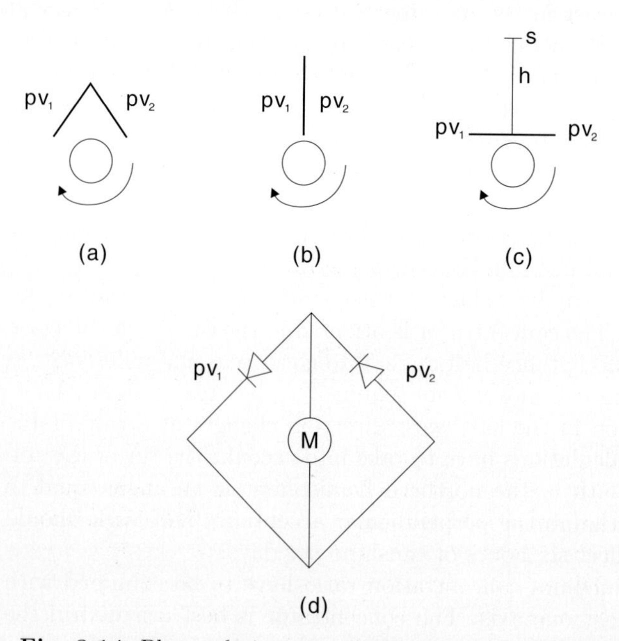

Fig. 8.14. Photovoltaic-powered one-axis tracking systems. Three tracker geometries are shown in (**a**)–(**c**). These are connected electronically in antiparallel mode in (**d**)

accuracy does not depend only on the amount of direct radiation, but also on the size of the panels, the rotational speed of the motor and its response, and on the geometry of the panels.

The tracking system in Fig. 8.14c is the most accurate of the three listed. A configuration of the solar cells as shown for tracker (a) will serve only systems with low tracking accuracy requirement. The height of the shading fin h and the width of the top fin s allow for the design of higher tracking accuracies. For locations with high and constant direct solar radiation, trackers (b) and (c) are employed, often with a forward tilt of the bifacial cells or the top fin. This asymmetrical feature allows for coordinated movements of the tracker and the sun, instead of following only. The feature is not suitable for tracking backwards, which can be desirable in locations with less constant availability of direct solar radiation.

The nonimaging 02/12 lens was mounted on a simple tracker of the type shown in Fig. 8.14c. An array of nine flux sensors covers part of the absorber plane, in a cross-sectional way. Three of the sensors next to each other are just wide enough to cover the width of the absorber.

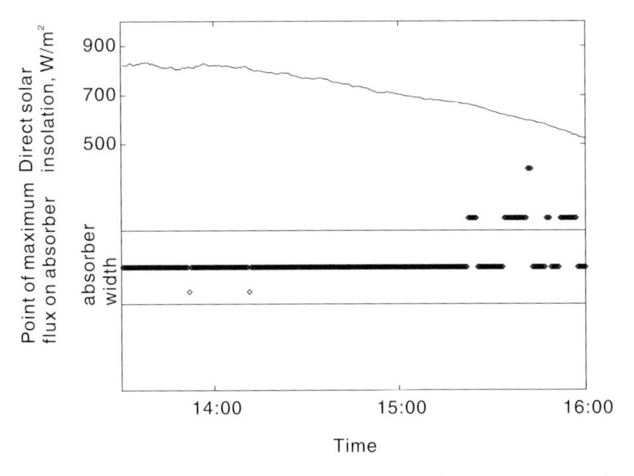

Fig. 8.15. Performance of one-axis tracking system with simple tracker. The tracker moves too slowly to follow the sun when direct radiation intensity on the attached photovoltaic cells drops below 600 W/m^2 (Tokyo, 11 October 1999)

The results of this experiment are shown in Fig. 8.15. Solar direct radiation is relatively high on this fine afternoon of October 11, 1999, in Tokyo. The points in Fig 8.15 indicate the center of the sensor with the highest response, showing that it receives the highest amount of flux. For most of the two hours, the tracker follows the sun exactly. Only when the direct radiation intensity on the attached photovoltaic cells drops below 600 W/m^2, and the tracking mechanism reacts too slowly, does it trail the sun.

The reason for the slow reaction at lower insolation is the small size and low efficiency of the photovoltaic panels used. Larger panels, used in a second experiment, are powerful enough to over-react, i.e. rotate the system too fast too far. An extra gear was added in the gearbox to optimize the tracking.

In conclusion, it can be stated that the nonimaging Fresnel lens of medium concentration is well suited for tracking systems in climates rich in direct solar radiation. Simple one-axis systems for azimuth tracking keep the tracking accuracy well within the required $\pm 2°$, as long the design is carefully optimized and direct radiation is available.

More sophisticated two-axis trackers (see Sect. 9.6) are either electronically controlled, or use a double pair of sensors assembled on the sides of a pyramid. Larger trackers may carry higher loads than the simple one-axis trackers. Accuracies are usually well below one degree for any direction. Two-axis trackers, if installed within latitudes of $\pm 23.45°$, may have to be rotated by 180° on equinox days to facilitate elevation tracking for solar positions both on the north side and on the south side of the tracker.

9 Concentrated Sunlight and Photovoltaic Conversion

9.1 Flux Density

The subject of this chapter is the flux characterization of the nonimaging Fresnel lens designed by the authors in terms of the irradiation uniformity and homogeneous color distribution required for optimum performance of multijunction photovoltaic cells.

The simulation presented here ultimately aims at the clarification of flux density issues in nonimaging Fresnel lens concentrators for use with multijunction solar cells. In contrast to the performance of crystalline silicon concentrator cells, the performance of multijunction devices is sensitive to both nonuniform flux and nonuniform color distribution.

In the following, flux densities are described as dependent on the optical concentration ratio of the lens, the solar disk size and the related brightness distribution, and the spectral dispersion of the incident sunlight. Flux uniformity over the area of the multijunction device and color flux homogeneity penetrating the cell layers are distinguished as 'parallel' and 'series connection' types. Both types of intensities are crucial for the optimum performance of the multijunction device.

Grilikhes [49] has presented a generalized model of sunlight concentration. Differing from highly mathematical models, and bearing in mind the demands of the simulation procedure of the Fresnel lens designed here, the following approach is developed with the help of ray tracing. Note that flux densities are very difficult to calculate analytically, in particular for nonimaging concentrators of difficult shapes under a realistic degraded sun, as incidence from various angles representing different amounts of energy must be accounted for.

Flux density is given here as dimensionless, as it is of great advantage for the model to use normalized values for all parameters. Normalization makes it possible to check the results, because all sums in the model must be unity. For example, the integrated direct irradiance over the solar disk and the circumsolar region is unity. The same is true for the energy transmitted in the solar spectrum. The most convenient absorber size (width or half-width for linear systems, radius or diameter for three-dimensional concentrators), again, is one.

Table 9.1. Total solar spectral irradiances[a] in W/m^2

Spectrum	Designation[b]	Irradiance
Terrestrial global-tilt	AM2G	691.2
Terrestrial global-tilt	AM1.5G	962.5
Terrestrial direct-normal	AM1.5D	767.2
Extraterrestrial	AM0	1367

[a] Source: [137], except AM2G [45].
[b] AM is the air mass the sunlight passes on its way through the earth's atmosphere; see (7.12).

If a unit is required, one may multiply the flux density ratio given by the irradiance used in the model (see Sect. 9.3). The terrestrial direct-normal spectral irradiance is given in [137] as 767.2 W/m^2, while for the extraterrestrial case, the irradiance is 1367 W/m^2 (Table 9.1). Furthermore, the model presented here allows for spectral 'windows' representing the spectral responses range of photovoltaic devices, typically 75–90% of total spectral irradiance. Care has to be taken when interpreting the flux density ratios given here. Flux density is defined as the radiant energy crossing an area element [160]. The flux density ratio is for the absorber what the optical concentration ratio is for the lens. But while it may be sensible to average the optical concentration ratio of the lens over the whole lens and still obtain a meaningful figure, the flux density should always be a distribution, with some maxima and an extension.

In the nonimaging 2D lens, the absorber width is usually greater than the cross-section of the area filled by refracted light originally incident at a combination of rays θ_{in}/ψ_{in}. The size of the fraction filled and its location on the absorber must be known in order to evaluate the suitability of the lens for photovoltaic applications, where homogeneous illumination of the absorber in terms of both flux density and color spectrum is essential for the performance of the system.

The flux density on the absorber of the linear nonimaging Fresnel lens can be calculated by tracing incident edge rays from each prism of the lens to the absorber. The edge (outermost) rays for any combination of angles of incidence are traced, and their intersections with the absorber plane are found in a cross-sectional projection, resulting in a part of the absorber plane ΔD being illuminated. The flux density $\delta\xi$ on any part of the absorber plane is a function of the angles of incidence representing the relative position of the sun over the concentrator, and is given by

$$\delta\xi_{\delta \mathrm{d}}\left(\theta_{in}, \psi_{in}\right) = \sum_{-i}^{i} w_{e,i}\, \tau_i\, s_i \ , \tag{9.1}$$

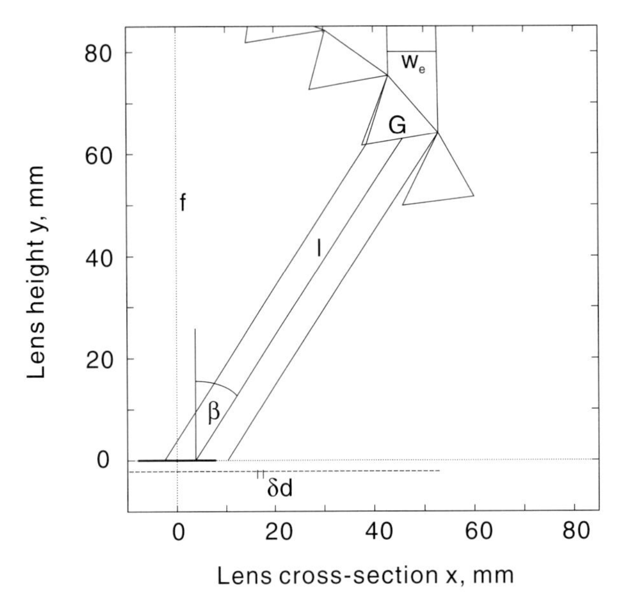

Fig. 9.1. Flux density factor. Nomenclature for calculations, here shown for an ideal thin lens of acceptance half-angle pairs $\theta = \pm 5°$ and $\psi = \pm 10°$. Incidence at $\theta_{in} = 1°$ and $\psi_{in} = 6°$

where the geometrical losses (tip and blocking losses, but not absorber misses) are discounted, resulting in an effective width $w_{e,i}$ of the prism i (Fig. 9.1). Transmittance losses τ accounting for first-order reflections are deducted. Depending on the distance of the prism from the absorber, a factor s describing the intensity of the refracted beam on the absorber is calculated as

$$s = \Delta D \cos \beta \, l \, . \tag{9.2}$$

The prism's height over the absorber plane defines the cosine losses of the beam when hitting the absorber at an angle β other than normal. Closer distance means higher flux density. A factor l is introduced to describe this distance, normalized in respect to the lens height f.

In the simulation, the lens size is individually normalized in relation to the absorber size, as are power densities related to solar disk size and its brightness distribution, and power densities attributed to each ray related to spectral irradiance. This results in the comparability of lenses of different acceptance half-angle pairs, under set conditions. Differences between the flux densities of the lenses show only in the shape of the graph of the flux density factor. Thanks to the normalizations, the area under the graph is unity. This, however, is not always desirable.

Different acceptance half-angles lead to different concentration ratios or higher flux densities as lenses with higher concentration ratios become wider with constant absorber width, accepting more radiation. To visualize this effect, the flux densities are multiplied by the average concentration ratio of the lens. While this average concentration ratio accounts for reflection losses and geometrical losses at the prisms, it does not consider absorber misses, because these are to be shown in the flux distribution. Thus, resulting flux densities may be exaggerated by a factor estimated at less than 10 percent, depending on the properties of the lens, mainly on the relation $C_{\text{lin}} = \psi/\theta$.

A test to confirm the simulation results of flux density was conducted in Koganei, Tokyo at latitude 35.5°N and longitude 138.3°E, and at GMT +9 hours. The 02/12 lens is mounted on a test rig, shown in Fig. 9.2. The lens can be seen in the center of the photograph; a transducer to measure the flux density is situated under the lens, in the absorber plane, on a sledge which can be manually moved in the cross-section of the system.

The orientation of the system towards the sun is measured with a sundial-type element (seen in the top right of Fig. 9.2), consisting of a pointer sticking out by $h = 224\,\text{mm}$ above a plate marked with concentric rings denoting the movements of the sun with a distance r given by

$$r = h \tan \alpha \,, \tag{9.3}$$

where α marks the sun's apparent movement in elevation and azimuth. The orientation of the test rig was set with the help of calculated positions of the sun for the day of an early experiment, shown in Fig. 9.3.

The calculations [178] illustrated in Fig. 9.3 can also be used to determine the best orientation for the concentrator. For the sake of the experiment, it is important to minimize the relative movements of sun and concentrator for the duration of the manual measurements. It can be seen in the figure that the sun's elevation is almost constant just before noon, while the azimuth angle increases quite strongly at all times.

It was decided to test the lens with its axis oriented east–west, tilting the system by the solar elevation angle. The small, and sensitive, cross-sectional acceptance half-angle θ now corresponds to the slow changes of the solar elevation, while the secondary acceptance half-angle ψ is subject to changes in the solar azimuth angle. The best conditions for the least angular changes for the experiment are found around 11:50 a.m., local time.

The transducer used in the experiment to measure the heat flux is a calibrated device made by Medtherm Co., Model 64-5SB2mm-20, with a sapphire window of diameter 2 mm. Its output is linearly rising with $5.33\,\text{mV}/5\,\text{W}\,\text{cm}^{-2}$. The device was directly connected to a data logger, which converted the input voltage into flux density automatically.

The weather on the day of the experiment was exceptionally clear; there was little of the haze usually present during many summer days in the humid subtropical climate of Tokyo. The fraction of direct solar radiation to the total radiation was about $750\,\text{W/m}^2$ to $910\,\text{W/m}^2$, or approximately 82%. This makes any disturbing influence of diffuse radiation in the measurements

Fig. 9.2. Test rig for flux density experiments with the prototype lens of acceptance half-angle pairs $\theta = \pm 2°$, and $\psi = \pm 12°$. Transducer on sledge in the absorber plane. Orientation towards the sun measured with the sundial seen in the top right. Cross-sectional view

of the transducer unlikely. The acceptance of the lens for diffuse radiation would be too small.

The experiment was conducted by manually turning the wheel in Fig. 9.2 by 1/16 rotation, moving the sledge with the transducer by approximately 1.5 mm. In the pause between movements the flux density values from the transducer were logged. This procedure was repeated for different combinations of angles of incidence on the lens.

The resulting data was normalized, converting it into flux density factors, independent of the value of the direct solar irradiance on the lens outer surface. A comparison for simulated and measured flux density factors is given

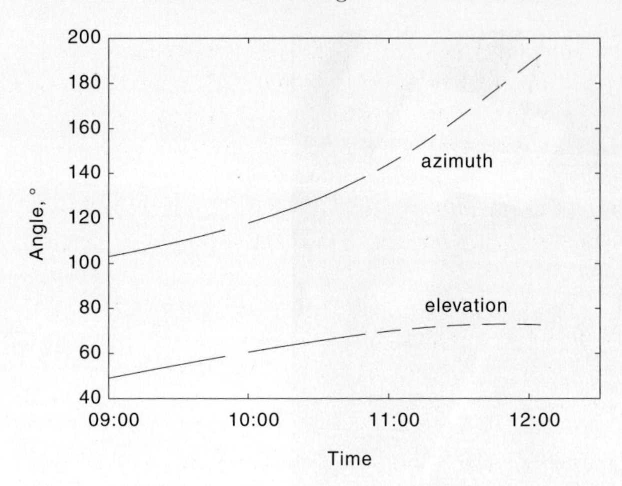

Fig. 9.3. Solar elevation and azimuth angles on 30 July 1999, for Koganei, Tokyo at latitude 35.5°N and longitude 138.3°E. Calculated with [178]

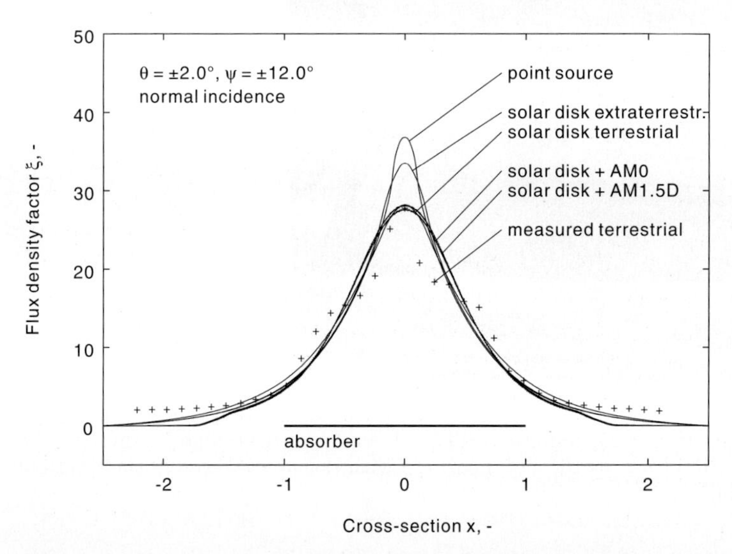

Fig. 9.4. Flux density factors for the nonimaging Fresnel lens of acceptance half-angle pairs $\theta = \pm 2°$ and $\psi = \pm 12°$. Measured terrestrial data, comparison with simulated values for point source of radiation, extraterrestrial and terrestrial cases incorporating solar disk size. The effects of the terrestrial direct normal solar spectrum (AM1.5D) and the extraterrestrial solar spectrum (AM0) have been added. Normal incidence

in Fig. 9.4. The figure is used also for verification of the simulation. The simulated graphs in the figure are the subject of the next sections.

Recapitulating, the problems encountered during the test procedure were:

- orientation towards the sun: for best results, the lens must be oriented normal to the sun. This is difficult to achieve by hand, only with the help of the sundial casting its shadow. The accuracy of the procedure is estimated to be about $\pm 1°$. An electronic tracker is indispensable for accurate experiments;
- step size: the movements of the sledge with the mounted transducer have to be well controlled. Smaller steps may allow closer following of the flux density maximum;
- concentrator symmetry: the flux density ratio is very sensitive to changes in the distance between lens and the absorber;
- transducer symmetry: the exact symmetry of the flux meter is not known. In particular the unknown distance of sensor and window must have some influence on output;
- pyrheliometer accuracy: make sure to readjust the tracking path of the pyrheliometer used in the experiment.

9.2 Solar Disk Size and Brightness

The radius of the solar disk is, corresponding to the solar disk half-angle, $\theta_s = 0.275°$. This limits the concentration ratio of ideal solar concentrators, which for linear devices is $1/\sin\theta_s \approx 208$. The influence of solar disk size on the flux density at the absorber of a concentrator can be simulated by ray tracing. Proper simulations for imaging concentrators and nonimaging ones with acceptance half-angles approaching θ_s incorporate the brightness distribution of the solar disk, plotted in Fig. 9.5, from Rabl [143]. The diffusion over the edge of the solar disk is caused by absorption and scattering of radiation in the photosphere of the sun and the atmosphere of the earth.

In the simulation, the brightness distribution is normalized, and a corresponding power density factor attributed to every ray traced through the prisms of the lens to the absorber plane. For terrestrial applications, the solar disk and circumsolar radiation are considered, while for the extraterrestrial case, only radiation originating from the solar disk is used to simulate the flux density in the absorber plane of a nonimaging Fresnel lens.

An example showing the insensitivity to solar disk size effects for a lens where the acceptance half-angle is larger than the half-angle of the sun ($\theta > \theta_s$) is given in Fig. 9.4 for a lens of acceptance half-angle pairs $\theta = \pm 2°$ and $\psi = \pm 12°$.

The 'sunshape', or brightness distribution, of the solar disk can be approximated by a fit of the solar and circumsolar regions. These sunshape calculations [143, 177] are based on measurements carried out in 1976–81 with a circumsolar telescope. For comprehensive data, see [124].

A possible fit of the sunshape is given in [177]. For the solar disk itself:

$$L = L_s \left(1 - \left(0.5051\frac{\gamma}{\alpha}\right)^2 - \left(0.9499\frac{\gamma}{\alpha}\right)^8\right) , \qquad \gamma \leq \alpha , \tag{9.4}$$

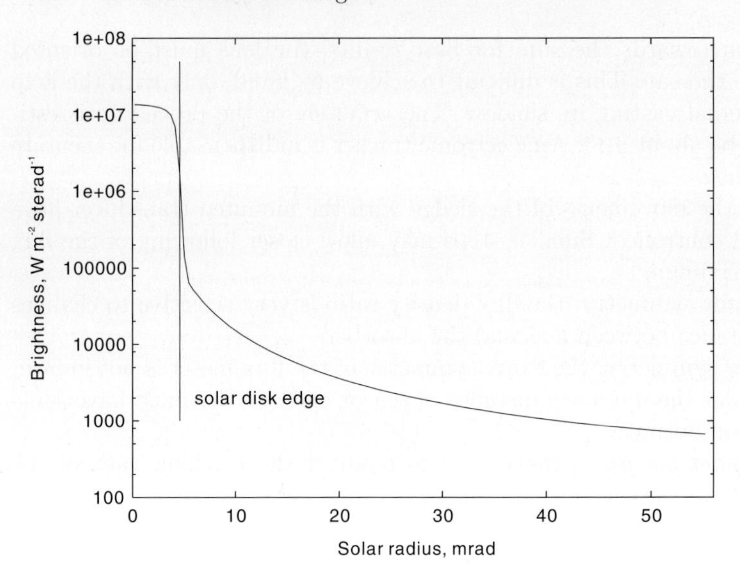

Fig. 9.5. Brightness distribution of the solar disk, using data from [143].

where $\alpha = 4.653\,\text{mrad}$ is the angular radius of the sun. The angle γ is measured between a vector pointing to a particular surface element on the solar disk and the vector pointing to the sun's center from the earth. The central radiance $L_\text{s} = 13.639 \cdot 10^6\,\text{W/m}^2\text{sterad}$ is taken from the standard circumsolar scan.

For the circumsolar region, the radiance of the aureole is

$$L = L_\text{a}\sqrt{\frac{\gamma}{\alpha}}\,, \qquad \alpha < \gamma < 55.85\,\text{mrad}\,, \tag{9.5}$$

where $L_\text{a} = 72\,200\,\text{W/m}^2\text{sterad}$. Integrations of (9.4) and (9.5) yield

$$\begin{aligned} E_\text{s} &= \int_\text{sun} L\,\mathrm{d}\Omega = 0.7402\,L_\text{s}\pi\alpha^2\,, \\ E_\text{a} &= \int_\text{aureole} L\,\mathrm{d}\Omega = L_\text{a}\pi\alpha^2 \cdot 2\ln 12\,. \end{aligned} \tag{9.6}$$

The total energy from the sun in this case is

$$E = E_\text{s} + E_\text{a}\,, \tag{9.7}$$

or $711.1\,\text{W/m}^2$, which corresponds to the irradiance given in Table 9.1 for the AM1.5D spectrum. If other irradiances are required, new values for the radiances L_s and L_a can be found from (9.6) and (9.7).

In our simulation, each ray thought to be incident on a prism from a point on or near the solar disk carries a normalized power density given by the solar brightness at the point of origin, and a normalized power density determined by the color of the respective solar spectrum. All rays of all origins carry all colors once; in other words, the solar spectrum is thought to originate equally from all over the solar disk (and circumsolar radiation). This is an assumption only; ultraviolet radiation is more strongly attenuated in the sun's photosphere than infrared, thus the rim of the solar disk appears darker and redder than its center. Giving an example, Grilikhes [49] describes the radiance of the solar disk to decrease towards the edge by 50% in the wavelength range 450–500 nm, and by 30% in the wavelength range 700–1000 nm.

The wavelength-specific description of the sunshape is not reported to have any significant influence on the performance of solar concentrators; the spectral brightness distribution induces errors which are small in comparison to optical distortions. Schubnell [156] cites the design of solar-pumped lasers as one area where spectral limb darkening of the sun should be considered. Parameters and calculations are given there.

9.3 Spectral Color Dispersion

In refractive solar concentrators, color dispersion will have an influence on total flux intensity. Ray bundles may be refracted in such a way that they miss the absorber, or concentrate on a 'hot spot', leading to partial illumination of the absorber, e.g. a photovoltaic cell.

Color behavior, or dispersive power, depends on the refractive index n of the lens material, here polymethylmethacrylate (PMMA) with $n = 1.49$ for yellow light. The refractive index is wavelength-dependent, for PMMA over the range of 1.515 to 1.470 from blue to red light. The refractive index can be calculated from a number of formulas. One of them is the Hartmann formula (3.43). Dispersion is temperature- and humidity-dependent; these effects have been considered earlier. For the flux model, we assume the influences of temperature and humidity on the refractive index of PMMA to be small.

The flux density simulation uses standard solar spectral irradiances ([137], Fig. 9.6) for the extraterrestrial (AM0) and the terrestrial (AM1.5D) case to incorporate dispersion effects at the rays' refractions. Most solar concentrators can convert only direct solar radiation, a fact that can prohibit the efficient use of solar concentrators in some locations on earth. In space, there is no diffuse solar radiation.

The standard spectra do not guarantee an accurate representation of the spectral irradiance. For the simulation of the performance of concentrating solar collectors in the south-west of the United States, for example, a total irradiance of 850 W/m^2 is suggested [101]. Moreover, the authors found that the global-tilt spectral distribution more accurately represents the actual spectral irradiance incident on the concentrator than the direct-normal spectrum.

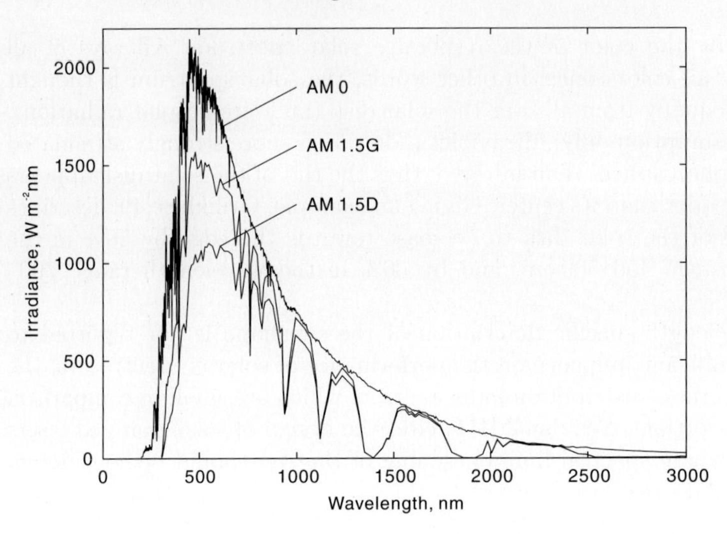

Fig. 9.6. Terrestrial direct-normal (AM1.5D), terrestrial global-tilt (AM1.5G), and extraterrestrial (AM0) standard solar spectral irradiances. From data in [137]

Table 9.2. Typical solar spectral response ranges of semiconductor materials[a]

Semiconductor	Response range[b], nm	Response range, eV
Crystalline silicon	300–1200	1.00–3.44
Amorphous silicon	360–900	1.38–3.44
InGaP	360–670	1.85–3.44
GaAs	360–871	1.42–3.44
Ge	360–1850	0.67–3.44

[a] Sources: [70, 149].
[b] The limits of the response ranges for longer wavelengths are found by the respective minimum band gap energies E_g of the semiconductors; shorter wavelengths are limited by the ultraviolet end of the solar spectrum or the transmittance limit of the concentrating lens material, and are set accordingly in this simulation.

The limits of the solar spectrum for this part of the simulation have been determined by the window of the response range of crystalline silicon, 300–1200 nm. Response ranges for other semiconductors are given in Table 9.2. The photon energy in the radiation follows the relation [149]

$$h\nu = h\frac{c}{\lambda} = \frac{1.24}{\lambda\,[\mu\mathrm{m}]}\,[\mathrm{eV}]\ , \qquad (9.8)$$

where c is the velocity of light, $h\nu$ the energy of one photon, $h = 6.6260755 \cdot 10^{-34}\,\mathrm{Ws}^2$ is the Planck constant, ν is the frequency of a spectral component of the radiation, and λ is the wavelength. Therefore, the longest wavelength

λ_g, below which photons can contribute to the photovoltaic effect in a semiconductor material of band gap E_g is

$$\lambda_g = \frac{1.24}{E_g} . \tag{9.9}$$

Longer wavelengths are not absorbed as they do not carry enough energy to produce carriers in the semiconductor. Shorter wavelengths, or photons of higher kinetic energy than required to produce electron–hole pairs in the semiconductor, will lose the excess kinetic energy as heat in the lattice. Thus, the conversion efficiency of photons of wavelengths shorter than λ_g must be reduced by a factor λ_g/λ.

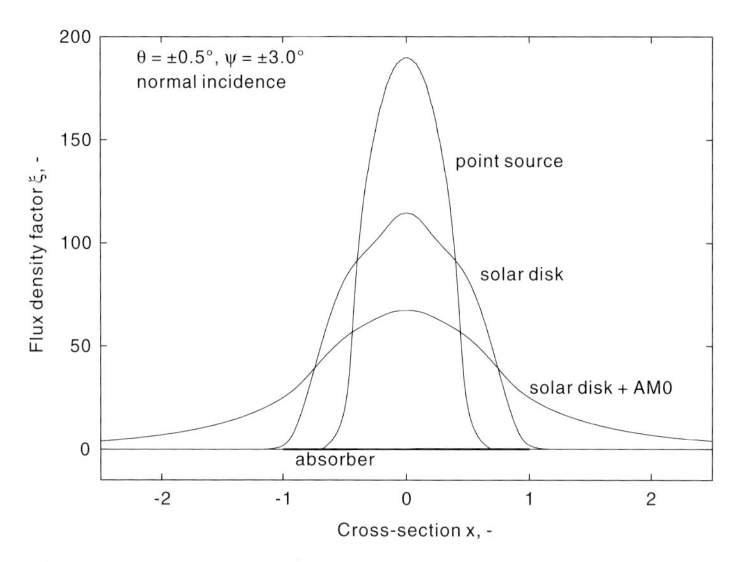

Fig. 9.7. Flux density factors for the nonimaging Fresnel lens of acceptance half-angle pairs $\theta = \pm 0.5°$, and $\psi = \pm 3.0°$ in space. Comparison of simulated values for point source of radiation (no sunshape, yellow light only), extraterrestrial solar disk size and brightness distribution (no circumsolar radiation, yellow light only), and dispersive effects of extraterrestrial solar spectral irradiance, including extraterrestrial sunshape. Normal incidence

Color dispersion changes the resulting flux densities quite significantly, once the acceptance half-angle of the concentrator gets close to the radius of the solar disk. The flux densities of a lens of acceptance half-angle pairs $\theta = \pm 0.5°$ and $\psi = \pm 3.0°$ have been analyzed in the described way. In Fig. 9.7, flux densities are shown for the following three extraterrestrial cases:

- point source of radiation: solar disk size $\theta_s = 0$, yellow light only;
- solar disk size with extraterrestrial brightness distribution: the sunshape has been normalized for radiation from the solar disk only; circumsolar radiation is omitted. For yellow light only;

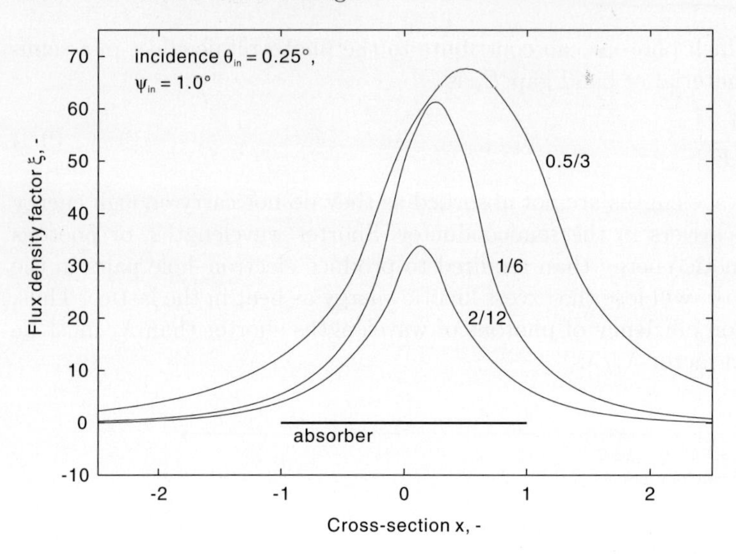

Fig. 9.8. Flux density factors for nonimaging Fresnel lenses. Incidence at $\theta_{\text{in}} = 0.25°$ in the cross-section of the linear lens (here, from the left in the plane of the paper), and at $\psi_{\text{in}} = 1.0°$

- solar disk plus dispersion due to the AM0 spectral irradiance: extraterrestrial sunshape and spectral dispersion effects.

The effects shown for normal incidence are similar, and are more pronounced for incidence off normal. A modest tracking error of $\theta_{\text{in}} = 0.25°$ in the cross-section of the linear lens and of $\psi_{\text{in}} = 1.0°$ in the perpendicular direction leads to flux densities as shown in Fig. 9.8. For small acceptance half-angles and high concentration ratios, dispersion dominates the flux distribution, setting a practical limit of solar concentration for linear Fresnel lenses (even for nonimaging lenses built from minimum deviation prisms) to about 70.

Flux density, describing the distribution on the absorber, on the one hand, and the optical concentration ratio of the lens, being an efficiency indicator of the concentrator, on the other hand, are different concepts. Mentioning both terms in one sentence as in the paragraph above is only justified in general statements. Conditional adjustments like *evenly distributed over the width of the absorber* for the flux, or *at normal incidence* for the optical concentration ratio are knowingly omitted. At this stage, it should be clear that the *geometrical* concentration ratio is quite useless for the technical description of solar concentrators.

For comparison, the flux density distribution on the absorber of a nonimaging Fresnel lens of $\theta = \theta_{\text{s}} = \pm 0.275°$ and $\psi = \psi_{\text{s}} = \pm 0.275°$ is shown for different cases in Fig. 9.9. This lens represents the highest possible concentration for a real linear refractive nonimaging solar concentrator, except for it

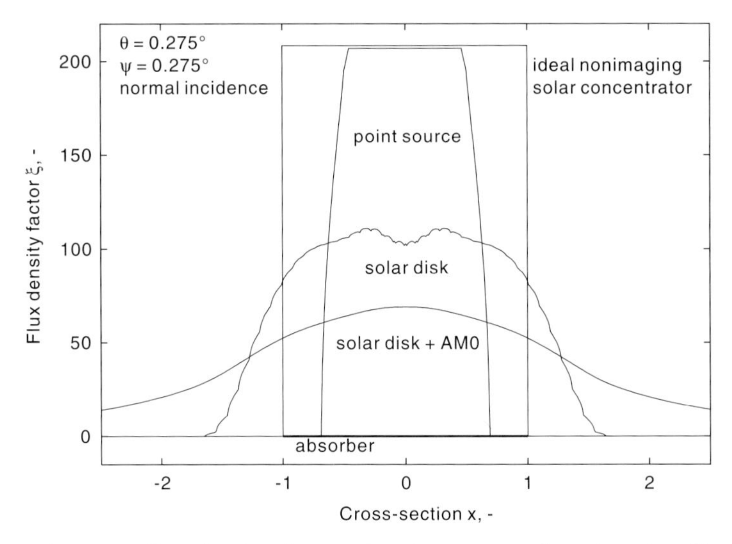

Fig. 9.9. Flux density factors for the truncated nonimaging Fresnel lens of acceptance half-angle pairs $\theta = \theta_s = \pm 0.275°$, and $\psi = \psi_s = \pm 0.275°$. Comparison of simulated values for point source of radiation, extraterrestrial solar disk size and brightness distribution, and effects of extraterrestrial solar spectral irradiance on dispersion with flux density distribution of the ideal nonimaging solar 2D concentrator. Normal incidence

being truncated. The lens' lower reaches have been truncated at half-height over the absorber, and the optical concentration ratio used in the multiplication with the normalized flux is thus less than ideal. This explains part of the mismatch between the real and ideal ares for the extended light sources in Fig. 9.9.

A second reason for this mismatch is the refractive index of the practical lens, which is much smaller than the ideal refractive index of infinity. Furthermore, the point source of radiation cannot deliver an ideal flux density, as this contradicts the finding that in ideal nonimaging concentrators the first aperture in its limits of acceptance half-angles has to be filled completely with uniform radiation, in order for the concentrator to qualify as a Lambertian radiator.

Clearly, the concentration is limited by the thermodynamic maximum of (2.8), which is one way to assess the correctness of the simulation. Once solar disk size and spectral dispersion are considered, the flux density of the real lens and the flux density of the nonimaging ideal are very different indeed. Of course, this effect is most pronounced for lenses of high concentration ratios, where the design with yellow light, and the evaluation for white light is least compatible.

9.4 Concentrator Cells

The concentration factors of solar concentrators are commonly classified according to the optical geometrical concentration ratio, denoting the ratio of entry aperture width to absorber width. A possible classification of photovoltaic cells does not necessarily correspond with these optical categories (Table 9.3). Solar cells have to be manufactured according to the flux they will receive under a concentrator. The ability of solar cells to convert concentrated sunlight is limited by their base series resistance, by the contact resistance between the grids and the semiconductor material, and by the emitter resistance caused by the spaces between the grid lines.

Two rules of thumb related to the performance of solar cells under concentrated radiation can be given as:

- The product of concentration ratio and cell series resistance is a constant. Additionally, the resistance increases proportionally to cell width. Small concentrator cells are desirable.
- Solar cells under concentration potentially see their efficiency increased by a factor $\ln(C)$.

Concentrator cells are of higher quality than one-sun cells, both in manufacturing and the ability to convert sunlight into electricity at a higher efficiency. Surface reflections are kept low; light is trapped in the chemically or mechanically grooved surface. Surface recombination of the carriers can be minimized by local doping of the semiconductor [159].

Table 9.3. Classification of solar concentrators according to optical concentration ratio. Solar cells are classified according to their suitability under X suns of irradiance

Classification	Geometrical concentration ratio C	Class of solar cell X
Low	$1 < C \leq 10$	$1 < X \leq 2$
Medium	$10 < C \leq 100$	$2 < X \leq 50$
High	$100 < C \leq C_{max}$ [a]	$50 < X \leq 300 \cdots 1000$ [b]

[a] C_{max} is the thermodynamic limit of concentration.
[b] Respective limits for silicon Si and gallium arsenide GaAs cells [18]

In Table 9.3, C_{max} is the thermodynamic limit of concentration: $C_{2D\ max} \approx 208$ for line focusing concentrators, $C_{3D\ max} \approx 43\,400$ for point-focusing devices. Concentration ratios of actual collectors may vary from the ideal by factors of 2–10 for linear designs and by factors of 5–30 for 3D concentrators. In the previous Sect. 6.5, the actual performance of the arched Fresnel lens was found to be superior to the values listed above.

Resistance and Concentration

Boes and Luque [18] give an example for the base series resistance R_s and its limiting role for the concentration under which the semiconductor may be used. The maximum efficiency of a solar cell is reached when

$$I_{sc} R_s = V_{th} , \tag{9.10}$$

where I_{sc} is the short-circuit current of the cell and V_{th} its thermal voltage

$$V_{th} = \frac{q}{\sigma T} , \tag{9.11}$$

with the electron charge $q = 1.602177 \cdot 10^{-19}$ As, the Stefan–Boltzmann constant $\sigma = 5.67051 \cdot 10^{-8}\,\mathrm{Wm^{-2}K^{-4}}$, and the semiconductor temperature T. A typical value for silicon is $V_{th} = 0.02\,\mathrm{V}$. In this example, the thermal voltage is taken to be constant. The series resistance R_s of the cell depends on its base resistivity, its size, and its thickness. The short-circuit current can be calculated from (9.10) as $I_{sc} = V_{th}/R_s$.

Since the short-circuit current is directly proportional to the irradiance E, a concentration ratio of

$$C = \frac{I_{sc}}{I_{sc_0}} \tag{9.12}$$

can be defined, which expresses the maximum concentration C for the irradiance E at one sun (1X, or 1000 W/m^2, and standard spectrum), represented by the short-circuit I_{sc_0} for this case. The first rule of thumb holds true: for a particular semiconductor, the product of a given series resistance R_s and the concentration ratio is a constant.

Concentrator Cell Efficiency

The conversion efficiency of a solar cell under nonconcentrated radiation can be expressed as

$$\eta = \frac{I_{sc} V_{oc} FF}{P_{in}/A} , \tag{9.13}$$

where A is the area of the cell. The fill factor FF describes the quality of the I–V characteristics of the solar cell. The more rectangular the curve for current and voltage appears, the is higher FF, which is given by

$$FF = \frac{I_{mpp} V_{mpp}}{I_{sc} V_{oc}} , \tag{9.14}$$

where the subscript mpp corresponds to the maximum power point, or the point for which the area of the rectangle $P = I\ V$ is a maximum. The fill factor describes the 'squareness' of the graphs in Fig. 9.10. Common values for FF for GaAs cells under one sun are around 0.87; they decrease rapidly with

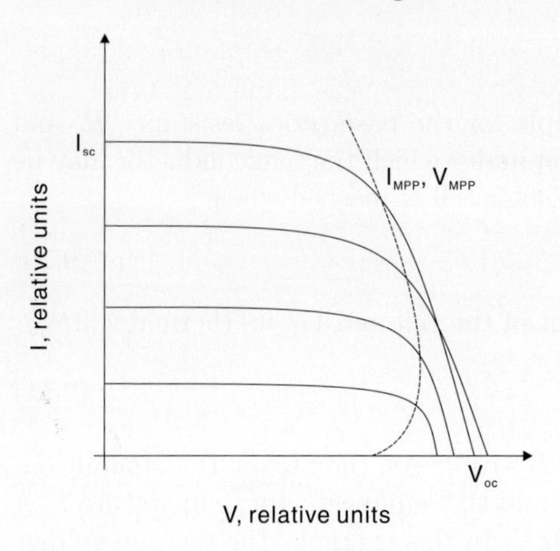

Fig. 9.10. I–V characteristics of a solar cell under practical conditions. Increasing insolation (higher concentration) leads to a higher short-circuit current I_{sc}, and a degradation of the fill factor FF, coupled with a rising operating temperature and its negative influence on the open-circuit voltage V_{oc}. Schematic

increasing ohmic losses of higher series resistance due to the concentration [3, 18].

Under concentration, the current increases linearly with irradiance (or optical input power P_{in}, as the latter is often termed) as

$$I_{\mathrm{sc}} = C I_{\mathrm{sc}_0} \, . \tag{9.15}$$

The open-circuit voltage V_{oc} increases logarithmically as

$$V_{\mathrm{oc}} = V_{\mathrm{oc}_0} + n\, V_{\mathrm{th}} \ln (C) \ . \tag{9.16}$$

The zero in the subscript denotes the case of one-sun irradiation. The factor $n\ V_{\mathrm{th}}$ expresses the temperature dependency of the efficiency of the cell (n is the diode quality factor of the cell: $n = 1$ for good cells, $n = 2$ for bad ones).

For a photovoltaic cell under concentrated light, kept at constant temperature, (9.13) can be expressed as a function of concentration by

$$\eta (C) = \frac{C\, I_{\mathrm{sc}} \cdot V_{\mathrm{oc}} (C) \cdot FF (C)}{C\, P_{\mathrm{in}_0} / A} \, , \tag{9.17}$$

where the fill factor FF is a strong function of the concentration ratio C, as given by

$$\begin{aligned} FF (C) &= FF_0 (V_{\mathrm{mpp}}) \ , \\ V_{\mathrm{mpp}} &= V_{\mathrm{mpp}_0} + n\, V_{\mathrm{th}} \ln (C) - R_{\mathrm{s}} C I_{\mathrm{mpp}} \, . \end{aligned} \tag{9.18}$$

In the dominant factor $R_{\mathrm{s}} C I_{\mathrm{mpp}}$, I_{mpp} may be approximated by I_{sc}. The expression for V_{mpp} has to be inserted in (9.14) to obtain a suitable value for FF in (9.17).

The efficiency of a solar cell under concentrated light has the potential to increase by a factor $\ln(C)$. In reality, some other influences have to be considered. As with the concentration, one observes not only the positive effects noted above, but also the following effects:

- increasing resistance R_s: resistance can be reduced by closer arrangement of the top contacts, but this results in the trade-off of blocking incoming radiation. Secondary concentrators [4], such as prism sheets, can be used to partially solve this problem by refracting light past the fingers, but optical losses as well as the complexity of the concentrator system increase;
- temperature dependency: the influence of the temperature on the efficiency of the concentrator cell is severe (9.16). Common Si cells suffer a degradation of about 0.45% per increase of 1 K [15]. For GaAs cells the degradation is in the range of 0.15–0.2% per 1 K [82]. The temperature limit for operation of high-quality cells is in the region of 120℃.
  The temperature coefficient decreases with increasing concentration. Other small effects related to an increasing temperature include a rise of I_{sc} for increasing temperatures, and a degradation of the fill factor FF;
- distributed diode effects: under concentration these degrade the fill factor [8, 80];
- effects of inhomogeneous illumination, both in terms of flux and color. These are discussed in the following section.

The physical conversion efficiency of solar cells increases for high-quality concentrator cells, until even their low series resistance R_s creates losses, which cause the total efficiency to drop sharply.

9.5 Multijunction Devices

Multijunction photovoltaic cells are used with medium and highly concentrated sunlight. The conversion efficiency of concentrator cells increases with the concentration ratio, but the fill factor is degraded by increasing resistance losses. For state-of-the-art multijunction cells in concentrator systems, the flux characteristics of the concentrated sunlight must be accurately controlled to ensure optimum performance. The irradiation over the cell area should be kept constant.

Color effects, or the spectral mismatch between the cells in the multijunction device, can be expressed by a rule of thumb: if the spectral mismatch between cells in a multijunction device amounts to 20%, the total performance of the device is cut by 10% [80].

Model of Nonuniform Illumination

Two forms of nonuniform illumination can be distinguished for the purpose of evaluating their influence on the performance of a multijunction cell:

- inhomogeneous illumination over the width of the cell. The flux density is not uniform for all points on the surface of the cell. In the model, this will be called 'parallel connection' of the cell. The effect does not depend on color, and can be observed in any concentrator cell.
  Inhomogeneous illumination of this type is caused by any practical concentrator optics, due to optical distortions, sunshape, or tracking errors;
- nonuniform color distribution over the width of the cell. Photovoltaic cells are designed to respond to the solar spectrum; the band gap energy of the semiconductor is matched to a fraction of the solar spectral irradiance. In single junction cells, like silicon concentrator cells, nonuniform color flux is generally perceived by the cell as nonuniform illumination, since the color differences are usually still within the relatively wide window of response wavelengths, or band gap energies.
  Multijunction devices consist of cells which individually have been designed to respond to a more narrow window of response wavelengths. The band gaps of the stacked cells are successively ordered. Each cell's thickness corresponds to its task of converting a specific wavelength into electrical energy. The stacked cells are electrically connected in line, and, within limits, the worst cell's output determines the output of the whole multijunction device. Thus, it is necessary to reproduce the design spectrum over the depth of all cells at any point of the multijunction device. The problem is termed 'series connection' type in our model.
  Nonuniform color distribution is primarily of concern with refractive concentrators, where dispersion makes an accurate image of the solar spectrum unlikely.

The nonuniformly illuminated multijunction device can be approximated as a parallel connection of cells, where the total output current is the sum of cell components [8]. Partial illumination with medium and high concentration ratios causes distributed diode effects manifested in a degradation of the fill factor due to rounding of the I–V characteristics near the maximum power point [7]. For a typical multijunction cell, pictured schematically in Fig. 9.11, this type of inhomogeneous flux density is referred to as the 'parallel connection' type.

Multijunction cells additionally require color uniformity within area segments. Uniform color is defined here as the proportionality of band gaps, for which the cells have been designed, usually according to the direct normal solar spectrum. The thickness of each layer is designed so that each junction produces the same photocurrent [79]. Should the design spectrum change, as in the case of nonuniform color flux, one or more junctions will operate suboptimally, and limit the output of the cell, which is understood as being formed by junctions connected in series (Fig. 9.11). This type of inhomogeneous color flux shall be called the 'series connection' type.

Other authors described the problem of color nonuniformity. Early descriptions are found in [60] and [111]. Papers including a quantification of

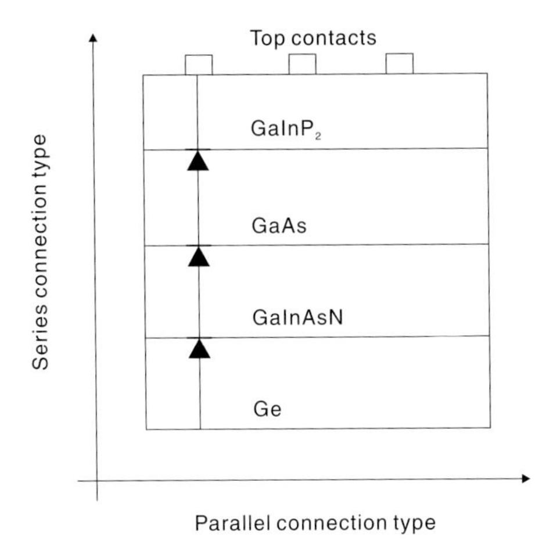

Fig. 9.11. Typical multijunction photovoltaic concentrator device. For an explanation of the axis see text. Not to scale

the chromatic aberration losses for multijunction cells are [59] (with imaging lenses), and [78, 80] (with O'Neill's lens of medium concentration). The consensus seems to be that losses due to color dispersion are smaller than other losses, and that they may be controlled by design measures such as resistance and diode quality. Still, for testing and operating high-performance cells, the color behavior of the optical system must be well understood.

Kurtz and O'Neill [80] mention 4% losses due to chromatic aberrations, if the system was aligned optimally. Our simulations show that spectral changes in the image are very sensitive and will be of concern for practical systems, more so than for laboratory tests. Testing of multijunction cells is performed in controlled surroundings, where the image is larger than the cell. Such overdimensioning of the concentrator system leads to more uniform flux over the area of a small cell, but to losses unacceptable in practical applications.

The two types, parallel and series, should be separately treated in their effects on the cells, if the concentrator happens to be a Fresnel lens, or other refractive or dispersive concentrator. For reflective concentrators (mirrors), there is virtually no color dispersion, and the 'series connection' type ideally does not exist. The model in Fig. 9.12 illustrates the refractive case. The 'parallel connection' type problem is caused by the design of the lens; imaging Fresnel lenses create a focal point of infinite concentration, or 'hot spot', whereas ideal nonimaging lenses produce uniform flux on the absorber, if the entry aperture of the system (the lens itself) is uniformly illuminated. This shifts our attention to the solar disk size and brightness distribution, and also to issues of ideal concentration (Sect. 6.5).

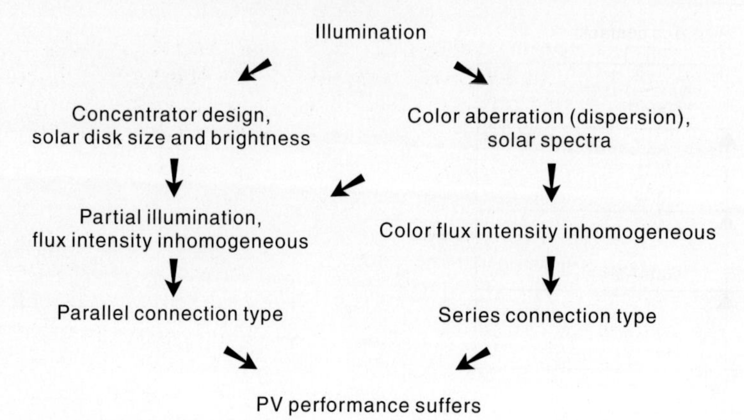

Fig. 9.12. Inhomogeneous illumination: flux and color flux nonuniformities. Influences of design and radiation source and their effects on the multijunction photovoltaic cell, described as 'parallel' or 'series connection' type

Color Behavior of the Nonimaging Fresnel Lens

Color flux nonuniformity is caused by the dispersive power and geometry of the optical system. The nonimaging lens examined here is designed of minimum deviation prisms, which are minimum dispersion prisms. The arched shape of the lens and its symmetry minimize color aberrations.

Fresnel lenses have been the concentrators of choice for photovoltaic power generation. Until now, imaging Fresnel lenses have been used for solar concentration. The focal point of imaging lenses is relatively sensitive to changes in the angle between the optical axis and nonparaxial incidence, and makes high precision tracking imperative (see Fig. 4.7). Imaging Fresnel lenses produce inhomogeneous color flux in and near the focus. To minimize the effects, the absorber is usually placed at the circle of least confusion (CLC), which is located closer to the lens than the actual focus. Dispersion becomes progressively greater in optical materials like glass or polymethylmethacrylate (PMMA) towards wavelengths shorter than that of yellow light (600 nm) used for the design of the system.

An indication of the color problems of imaging lenses are the rainbow-colored edges of the focal area. Nonimaging Fresnel lens concentrators can be designed with higher tolerance for tracking errors, accounting for effects due to the size of the solar disk, and with better color behavior than conventional imaging lenses.

Adding color to the considerations of partial illumination allows for more accurate descriptions of the refraction at the prism. Following the data used in Fig. 3.6, values for refractive index, solar irradiance, transmittance, and cumulative energy corresponding to ultraviolet, yellow, and infrared radiation can be found.

The infrared part of the solar spectrum is of no great concern as PMMA remains transmitting, and the refractive index is almost constant at long wavelengths (not so in the far infrared). The ultraviolet part of the solar irradiance is, while not being absorbed in PMMA with UV-enhanced transmittance (dotted line in Fig. 3.6 (middle)), refracted increasingly stronger than visible wavelengths.

In the three-dimensional Fig. 9.13 rays incident at combinations of $\theta = -2^\circ$ and $\psi = -12, 0, +12^\circ$ are drawn with refractions as well as intersections with the absorber level. Edge rays (those incident at design angles, and calculated with design refractive index) hit the edge of the absorber, while ultraviolet or infrared rays are generally more likely to miss the absorber, although for the case of $\theta = -2^\circ$ and $\psi = 0^\circ$ from the left-hand prism, the strong refraction of the ultraviolet ray lets it reach the receiver while the design ray (yellow light) misses.

The wavelength-dependent refractive power of a prism reaches a minimum for minimum deviation prisms. The angle of deviation is defined as the angle between the ray incident on the prism and the ray exiting the prism after two refractions. Although a prism can have only one angle of minimum deviation, the idea of 'reversible' prisms used in the novel nonimaging Fresnel lens creates conditions for reduced dispersion.

Emphasis must be put on the effects of color aberration, and the behavior of the 2D lens, where rays are incident within a pair of cross-sectional acceptance half-angles $\pm\theta$ from both sides of the symmetrical lens, strongly influenced by the perpendicular acceptance half-angle ψ, as was seen in

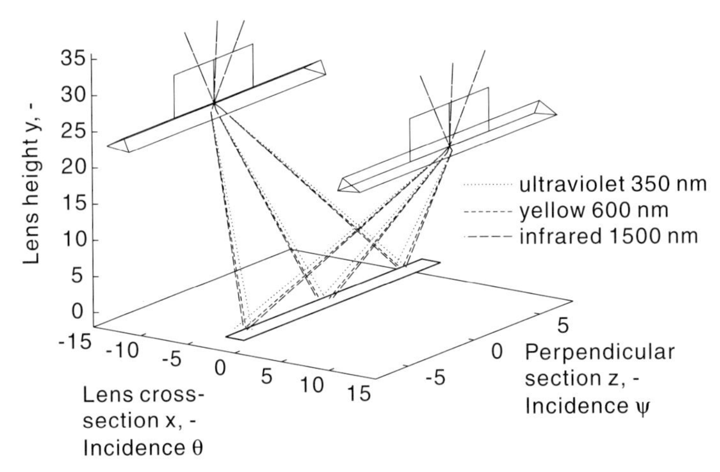

Fig. 9.13. 02/12 lens. Combinations of rays incident at $\theta = -2^\circ$ and $\psi = -12, 0, +12^\circ$ are drawn with refractions at symmetrical prisms, and intersections with the absorber level. Ultraviolet, yellow, and infrared rays roughly describing the extent of the solar spectrum, with corresponding refractive indices of polymethylmethacrylate

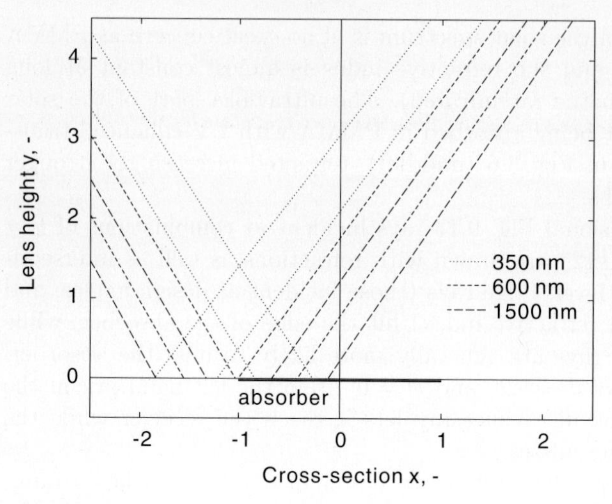

Fig. 9.14. 02/12 lens. Combinations of extreme rays incident at $\theta = -2°$ and $\psi = -12, 0, +12°$ are drawn with refractions at symmetrical prisms, and intersections with the absorber level. Ultraviolet, yellow, and infrared rays mix for incidence other than that at design angles. Cross-sectional projection

Figs. 8.13, and 9.13. Presenting the rays in the latter in a cross-sectional projection, Fig. 9.14 is obtained. The yellow rays from both sides hit the edge of the receiver only when $\psi_{\mathrm{in}} = \psi_{\mathrm{design}}$. If the perpendicular incidence is not equal to the design angle, the colors mix on the absorber.

Since the usual case of operation of the nonimaging Fresnel lens is collecting solar rays incident anywhere within the acceptance half-angle pairs, mixing of refracted and color-separated rays can be assumed. In fact, the concentrated sunlight on the absorber appears white in an experiment. It lacks the colors lining the focus characteristic of imaging Fresnel lenses that may be observed in a similar experiment conducted with conventional lenses, where the color aberrations increase at the rate at which the incidence deviates from the paraxial centerline of the optical system.

Spectral Mismatches in the Multijunction Device

Solar concentrators must display uniform color flux over all areas of the photovoltaic device. This uniformity may be called vertical, as useful band gap energies of the solar spectrum must be reproduced when the radiation penetrates the multijunction device to be converted into electricity.

The thickness of each cell in a multijunction device is matched to the fraction of the solar spectrum this cell can convert in such a way that the photocurrent of all cells is the same. The band gap range of the solar spectrum for each cell carries different energies for each cell. Table 9.4 lists values for the $GaInP_2$/GaAs/GaInAsN/Ge cell used as the example in the subsequent calculations.

Table 9.4. Band gap characterization of a multijunction device under the nonimaging Fresnel lens solar concentrator. Wavelengths λ in nm, band gap[a] E_g in eV, cumulated solar spectral energy (AM2D)[b] $f_{0-\lambda}$ in %, refractive index n of polymethylmethacrylate (PMMA, calculated with the Hartmann formula)

Semiconductor	λ	E_g	$f_{0-\lambda}$	n_{PMMA}
	360	3.44	2	1.508
$GaInP_2$	670	1.85	35	1.485
GaAs	871	1.42	58	1.481
GaInAsN	1200	1.00	80	1.477
Ge	1850	0.67	93	1.474

[a] Source: [79].
[b] Source: [184].

The multijunction device intercepts wavelengths accounting for 91% of the energy carried by the terrestrial solar spectrum. For the extraterrestrial spectrum, this value is slightly lower at approximately 88% [37], because the extraterrestrial sunlight extends further into the blue range of wavelengths.

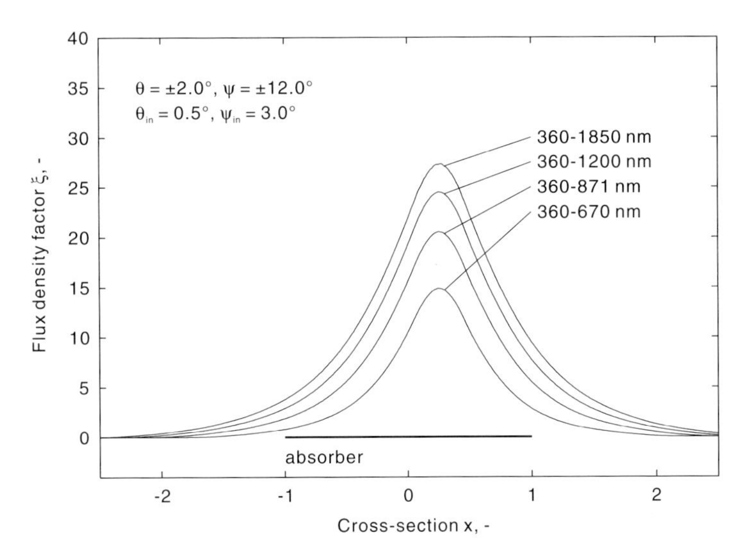

Fig. 9.15. Color flux density factors and solar spectral reproduction for the nonimaging Fresnel lens of acceptance half-angle pairs $\theta = \pm 2°$ and $\psi = \pm 12°$. Incidence away from the normal by $\theta_{\mathrm{in}} = 0.5°$ and $\psi_{\mathrm{in}} = 3.0°$

The spectral distribution in terms of band gaps for each cell can be simulated, and is shown in Fig. 9.15. The calculations are for light incident at $\theta_{\mathrm{in}} = 0.5°$ and $\psi_{\mathrm{in}} = 3.0°$ from the normal in the cross-sectional and perpen-

Table 9.5. Solar spectral band gap reproduction of the nonimaging lens of $\theta = \pm 2°$ and $\psi = \pm 12°$ for four combinations of incidence angles, calculated in % of total spectral energy (360–1850 nm, values cumulative) at four equidistant locations from the center of the absorber to its right-hand edge. Spectral energy for AM2D spectrum[a]; simulation of flux densities ξ uses AM1.5D spectrum[b]

Incidence θ_{in}/ψ_{in}		360–670	360–871	360–1200	flux ξ_{max}
0.0/0.0 at	0.00	54	75	90	26
	0.33	45	69	87	19
	0.67	34	61	83	11
	1.00	27	55	80	6
0.5/3.0 at	0.00	48	71	88	22
	0.33	54	75	89	27
	0.67	41	67	86	17
	1.00	30	58	82	9
0.5/24.0 at	0.00	34	59	81	9
	0.33	33	59	81	9
	0.67	34	59	81	9
	1.00	35	61	82	7
2.0/12.0 at	0.00	14	21	52	1
	0.33	13	29	64	4
	0.67	21	51	79	17
	1.00	53	78	91	66
spectral energy		36	62	86	1

[a] Source: [184].
[b] Source: [137].

dicular direction, respectively. The colors in Fig. 9.15 seem well distributed. No range of wavelengths is clearly dominant in any segment over the absorber.

The color flux density roughly corresponds to the energy delivered by the color range, but depends on the location on the absorber and the angles of incidence. Table 9.5 lists results of a simulation using the lens of acceptance half-angles $\theta = \pm 2°$ and $\psi = \pm 12°$ with four combinations of angles of incidence. Color fractions corresponding to the band gaps listed in Table 9.4 are calculated for four positions on the absorber (center at 0.0, 0.33, 0.67, and edge at 1.00). When the flux density is high, the area of the multijunction device receiving maximum flux may operate inefficiently due to the presence of a high proportion of blue light. For low flux densities near areas of high flux, infrared light may be dominant, and the photovoltaic cells may work suboptimally.

The closest fit to the solar spectra used here is achieved when the flux densities are well below the concentration ratio of the lens. The term 'spectral energy' in Table 9.5 refers to an AM2D spectrum of [184], while the simulation of flux density uses the AM1.5D spectrum of [137], but no significant differences are expected in terms of relative color distribution. Good color reproduction over the width of the absorber is found for a combination of incidence angles $\theta_{\rm in} = 0.5°$ and $\psi_{\rm in} = 24.0°$, which in the latter case means exceeding the perpendicular acceptance half-angle by a factor of two, with resulting lower flux density.

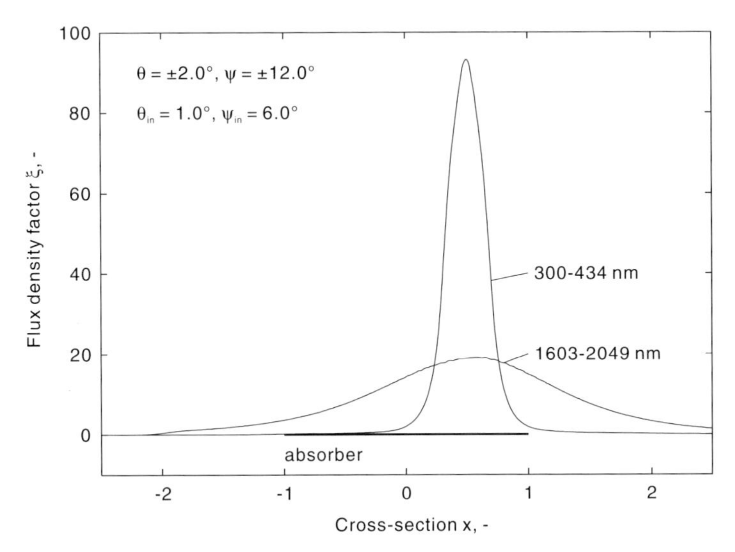

Fig. 9.16. Extreme ultraviolet and infrared color flux density factors for the nonimaging Fresnel lens of acceptance half-angle pairs $\theta = \pm 2°$ and $\psi = \pm 12°$. Equal increments of energies of 5% (normalized graph, each fraction represents 50% of irradiance). Incidence combination: $\theta_{\rm in} = 1.0°$ and $\psi_{\rm in} = 6.0°$

One may pick extreme ultraviolet (300–434 nm) and infrared (1603–2049 nm) wavelengths [184], each representing an equal increment of energy of 5% for the terrestrial direct normal spectrum, and compare the flux densities on the absorber, as in Fig. 9.16. The graphs show the limits of color mixing of the nonimaging lens. According to the 'series connection' type, the multijunction cell does no longer operate efficiently, because the distribution of photons of different band gap energies over the cell area is not uniform.

If the band gaps are designed to be very small (Fig. 9.16), as required in the ideal case of the multijunction device with n cells with $n \to \infty$, color mixing becomes very difficult under concentration. Realistic cases, like the four-junction device with band gaps corresponding to those shown in Fig. 9.15, should allow the cells to operate near their 'series' design point.

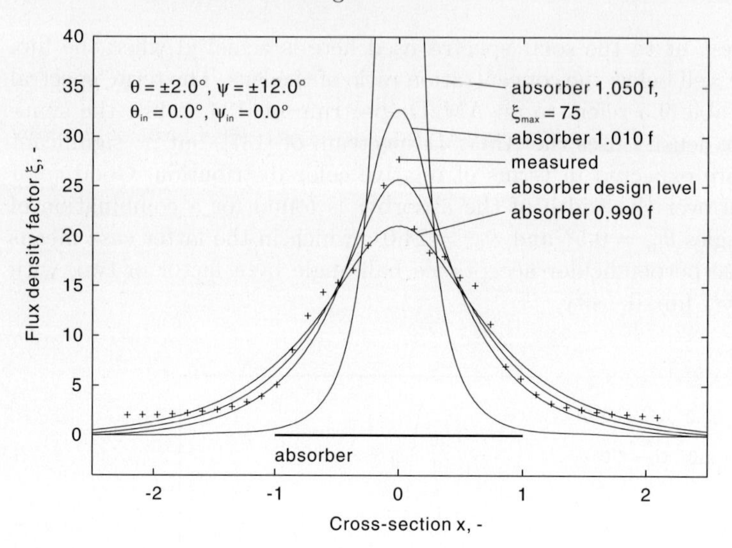

Fig. 9.17. Flux densities on absorber moved towards and away from the lens by 1% of the distance f between absorber and lens. Normal incidence

One should remember that the terrestrial solar spectrum changes naturally over the time of day, and with the weather conditions.

Solar spectral reproduction can be problematic for the optimum performance of the multijunction device according to the 'series connection' theory. Possible solutions include the following:

- The redesign of the nonimaging lens: in theory, the nonimaging lens can be designed to produce uniform flux. The argument is that the degree of freedom remaining between the ideal theoretical lens with its uniform flux and the nonideal practical lens could be used to create a lens with prescribed flux distribution. Related work is under way by several researchers, but definite results are not available yet.
- The movement of the absorber closer to the lens or further away from it: Fig. 9.17 shows the flux density factors for absorbers moved by $\pm 1\%$ with respect to the original distances between absorber and lens. A closer absorber performs marginally better in terms of color reproduction, but the absorber misses increase. In the case of the absorber being further away from the lens, less absorber misses are recorded, but worse color reproduction is observed. For an absorber located at $1.05f$, color reproduction at the center of the flux is quite good, while the system becomes more sensitive to incidence angles off normal, and flux uniformity over the absorber is strongly reduced, showing a strong peak (see Table 9.6 for details).
- The placement of a reflective or refractive secondary concentrator in place of the original receiver: the secondary concentrator is called a homogenizer if the aim is to control the flux adding little or no geometrical concentration.

Table 9.6. Solar spectral band gap reproduction of the nonimaging lens of $\theta = \pm 2°$ and $\psi = \pm 12°$ for normal incidence, calculated in % of total spectral energy (360–1850 nm, values cumulative) at four equidistant locations from the center of the absorber to its edge. Absorber moved towards or away from the lens by multiples of the design distance f between absorber and lens. Spectral energy for AM2D spectrum[a]; simulation of flux densities ξ uses AM1.5D spectrum[b]

Absorber position		360–670	360–871	360–1200	flux ξ_{max}
$0.99f$ at	0.00	51	73	88	21
	0.33	46	70	87	18
	0.67	38	64	85	12
	1.00	32	59	82	7
$1.01f$ at	0.00	57	77	90	33
	0.33	42	68	86	20
	0.67	28	57	92	10
	1.00	20	49	77	5
$1.05f$ at	0.00	41	72	90	75
	0.33	34	48	75	15
	0.67	38	43	64	3
	1.00	49	61	70	1
spectral energy		36	62	86	1

[a] Source: [184].
[b] Source: [137].

The homogenizer can be an option to redirect rays and make the color flux more uniform, but first-order reflection losses at the secondary are substantial, and may exceed the losses caused by inhomogeneous flux. The prism sheet secondary [131, 150] is used to refract rays away from the grid on the surface of a photovoltaic cell to avoid shading losses.
A kaleidoscope-based secondary homogenizer [32, 146] for the linear lens can be constructed from two parallel mirrors forming a trough of absorber width under the original absorber level. In the example given in Fig. 9.18 (inspired by J. Gordon), we use a secondary with a depth of $12.6\,d$, where d is the absorber half-width. Clearly, the flux uniformity and the reproduction of the solar spectrum improve (Table 9.7), when compared to the results presented in Tables 9.5 and 9.6. Additional reflection losses are calculated with (6.12) to 9.8% using a reflectivity of the mirror $\rho_{\mathrm{m}} = 0.95$ and the average number of reflections $n = 2.01$.

- The design of photovoltaic cells: instead of the redesign of the concentrator, the layers of the multijunction device could be designed and assembled with varying thickness according to the spectral fraction incident at that point.

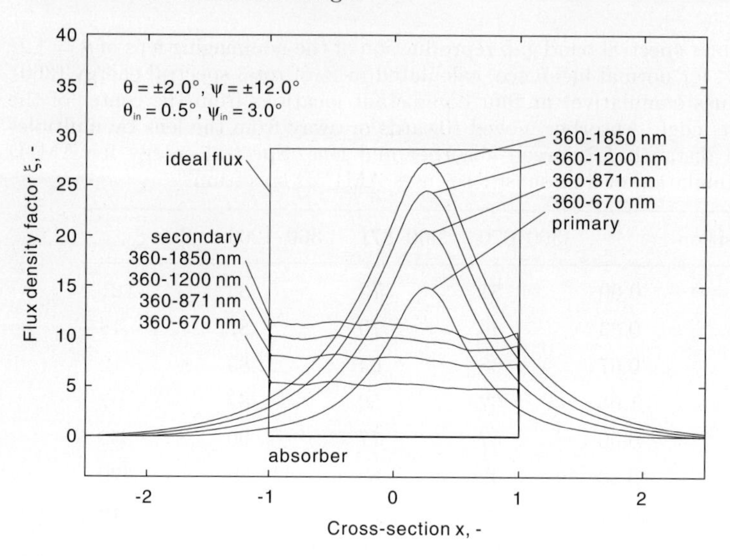

Fig. 9.18. Color flux density factors and solar spectral reproduction for the nonimaging Fresnel lens of acceptance half-angle pairs $\theta = \pm 2°$ and $\psi = \pm 12°$ with and without secondary kaleidoskope-based homogenizer. Incidence away from the normal by $\theta_{\text{in}} = 0.5°$ and $\psi_{\text{in}} = 3.0°$. Depth of the secondary $12.6\,d$

This may not be practical, since the angle of incidence of the radiation on the concentrator has a strong influence on the flux distribution at the absorber. Still, multijunction devices which are to be used with refractive solar concentrators must be designed for the flux and spectral reproduction the lens produces, i.e. the spectral transmittance of the lens material is a design parameter.

- The design of cells for systems intended for use in space: the solar spectrum is a constant, and efforts to correct for color-induced losses could be taken. Furthermore, extraterrestrial systems may prevent losses using a lower sheet resistance in the emitter layer of the multijunction device [78]. Modern multijunction devices with low sheet resistance may be priced suitably for terrestrial application in the near future, too.

These measures not only complicate the design of a photovoltaic concentrator, but also carry the risk of lower performance at incidence angles off normal. Careful analysis and design are necessary.

It has been shown that flux density factors in nonimaging Fresnel lens solar concentrators can be strongly influenced by the solar disk size, and to some minor extent by the brightness distribution over the solar disk and circumsolar radiation (nonparaxial radiation), as well as by solar spectral irradiance (color dispersion). These factors determine the performance of a multijunction device in terms of the 'parallel connection' type.

Table 9.7. Solar spectral band gap reproduction of the nonimaging lens of $\theta = \pm 2°$ and $\psi = \pm 12°$ with kaleidoskope-based secondary homogenizer for incidence away from the normal by $\theta_{\mathrm{in}} = 0.5°$ and $\psi_{\mathrm{in}} = 3.0°$, calculated in % of total spectral energy (360–1850 nm, values cumulative) at four equidistant locations from the center of the absorber to its right-hand edge. Spectral energy for AM2D spectrum[a]; simulation of flux densities ξ uses AM1.5D spectrum[b]

Incidence $\theta_{\mathrm{in}}/\psi_{\mathrm{in}}$		360–670	360–871	360–1200	flux ξ_{max}
0.0/0.0 at	0.00	48	71	88	11
	0.33	48	72	89	12
	0.67	47	71	88	12
	1.00	47	71	88	12
0.5/3.0 at	0.00	48	71	88	11
	0.33	47	70	87	11
	0.67	49	73	89	10
	1.00	46	70	88	10
spectral energy		36	62	86	1

[a] Source: [184].
[b] Source: [137].

The absorber area is not always fully illuminated, and usually some peak of illumination exists. The closer is the acceptance half-angle to the extension of the solar disk, the greater are the influences of both solar disk size and color dispersion; the latter dominates for lenses of small acceptance half-angle pairs, and practically limits the concentration ratio of all lenses. Optimum truncated linear Fresnel lenses may reach a concentration of 70 extraterrestrial suns, roughly one-third of the theoretical limit. This ratio is more favorable for lenses of lower concentration; the losses progressively increase for higher concentration ratios.

The differences between the terrestrial case and the extraterrestrial case (different spectra, no circumsolar radiation for the latter) are not very strongly emphasized by the results of a simulation, which uses normalized values of brightness distributions and solar spectral irradiances and calculates the performance of novel linear nonimaging Fresnel lens concentrators.

Multijunction devices require uniform color distribution over each point of the cell aperture to operate efficiently. For band gaps of practical width, the nonimaging lens provides a spectral distribution close to that of the solar spectrum over the illuminated area of the cell for some (lower) flux densities and combinations of incidence angles. Other cases may require the redesign of the cell, and of the optics of the concentrator. If the band gaps are designed to be very small, color representation could be a factor limiting the performance of the multijunction device in practice.

The concept of dividing flux issues into two types, the 'parallel connection' type and the 'series connection' type, provides useful information for further optimization of multijunction devices. General flux density and color flux density each influence the performance of the multijunction device in different ways.

9.6 Photovoltaic System Performance

A test of a photovoltaic concentrator system was conducted in Koganei, Tokyo at northern latitude 35.5° and eastern longitude 138.3°. The test took place around noon on a clear winter day on 17 January 2001. Direct radiation amounted to $730\,\mathrm{W/m^2}$. A schematic of the concentrator system is shown in Fig. 9.19.

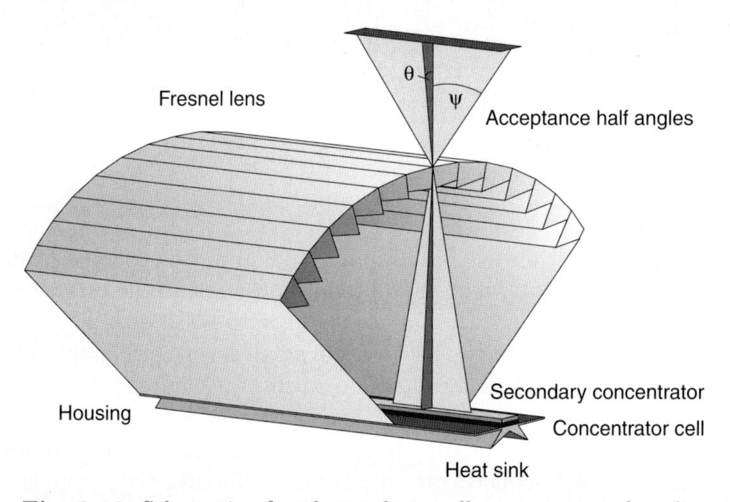

Fig. 9.19. Schematic of a photovoltaic collector system of medium concentration, incorporating the nonimaging Fresnel lens. Not to scale

The 02/12 nonimaging Fresnel lens is mounted on a two-axis tracker (Fig. 9.20). The orientation of the system towards the sun is controlled by a remote optical sensing system for tracking. The optical sensor constitutes of four photo-diodes, arranged on the four sides of a plastic rod with a cone tip. The rod is about 10 cm long, the diodes are inserted close to the rod's base, at an angle near 45°, directed normal to a point between horizon and zenith. One pair of photodiodes faces east–west, the other pair faces north–south. The controller seeks to equalize the sunlight received by opposing sensors for each axis. The tracker follows the sun in two axis, achieved by an azimuthal rotation of the tracker around its vertical axis, and by inclination of the table

Fig. 9.20. Two-axis tracking system [9] with test equipment mounted on the tracking table. The optical sensor (resembling an obelisk) is visible in the lower left corner

according to the declination of the sun. Tracking accuracy is about 1/20 of a degree.

While the nonimaging Fresnel lens of medium concentration is designed for one-axis tracking, full tracking allows for accurate testing. Most importantly, normal incidence for cell testing can be maintained over extended periods of time.

Sunlight at normal incidence is concentrated by the nonimaging Fresnel lens onto a solar concentrator cell. The cell is designed by K. Araki and M. Yamaguchi of the Toyota Technological Institute (TTI), Nagoya, Japan, as a crystalline silicon cell [6]. The device is called, in reference to its production process, the Single Photolithography Fine Grid, or SPFG cell. The designed concentration ratio is 20X. The basic structure of the cell is quite similar to commercially available crystalline silicon cells for flat plat photovoltaic applications, except for fine grid electrodes. There is no anti-reflection (AR) surface texture. Commercial crystalline silicon cells have a resistance R =1–5 $\Omega\mathrm{cm}^2$. The best concentrator cell resistance reaches 0.1 $\Omega\mathrm{cm}^2$; the TTI cell has a resistance of 0.15–0.20 $\Omega\mathrm{cm}^2$. The concentrator cell's characteristics are shown in Table 9.8. A similar crystalline silicone cell had previously been obtained from the group of A. Blakers at the Australia National University, Canberra, but the width of the TTI cell is better suited for our prototype concentrator.

Table 9.8. Designed and experimentally found characteristics of a crystalline silicon concentrator cell[a]. Temperature of the cell T, short-circuit current I_{sc}, open-circuit voltage V_{oc}, fill factor FF, and concentrator cell conversion efficiency $\eta(C)$. Optical concentration in suns X

Condition	T	I_{sc}	V_{oc}	FF	$\eta(C)$
	℃	mA/cm^2	V	%	%
1X, TTI	25	28.1	0.60	78	13.2
Without lens[b]	23	13	0.60	75	8.0
10X, 02/12 lens[c]	30	110	0.66	73	14.9

[a] Designed by K. Araki and M. Yamaguchi, Toyota Technological Institute, Nagoya, Japan. Cell size 10 × 48 mm^2.
[b] 730 W/m^2 direct insolation.
[c] Optical concentration ratio of the lens $\eta_C = 13.47$ multiplied with the direct insolation, normalized for standard conditions of 1X.

The solar concentrator cell under the nonimaging lens is pictured in Fig. 9.21. The TTI cell's I–V characteristics are measured by the four-probe method, not using an I–V curve tracer. This method of measurement is accurate enough for practical use. The solar concentrator cell is bonded with its back to the structural aluminum plate using a silicone grease loaded with diamond particles to increase the thermal conductivity and electrical resistance. In this experiment, the temperature of the concentrator cell's back surface reaches almost 30℃ under a concentration of about 10X; the temperature of the cell without concentration is 23℃.

The TTI solar concentrator cell is less wide than the absorber for which the nonimaging Fresnel lens is designed: half-width $d_{abs} = 7.9$ mm, $d_{cell} = 5.0$ mm. The I–V characteristics of the TTI cell resulting from the experiment are given in Fig. 9.22.

The open-circuit voltage V_{oc} increases under concentration as predicted by (9.16), and the fill factor FF is reasonably good, while not reaching state-of-the-art values for silicon cells of 0.80. The conversion efficiency of the cell is clearly higher under concentration. In this experiment, the short-circuit current I_{sc} of the cell under concentration is 8.8 times higher than the short-circuit current of the cell without any concentration. The short-circuit current was expected to be higher (9.15), considering the nonimaging Fresnel lens's optical concentration ratio of 13.47. The gap can be explained by mismatches between the cell and the nonimaging Fresnel lens, and by some measurement problems.The main problems are

- the cell is less wide than the absorber, i.e. not all direct insolation incident at the concentrator aperture reaches the surface of the cell. Therefore, the short-circuit current I_{sc} is lower than expected;

Fig. 9.21. Crystalline silicon concentrator cell (Toyota Technological Institute) under a concentration of about 10X, achieved with the nonimaging Fresnel lens of acceptance half-angle pairs $\theta = \pm 2°$ and $\psi = \pm 12°$ at normal incidence

- the nonimaging Fresnel lens delivers an optical concentration ratio of 13.47 for a direct radiation of 73% of the standard value for 1X, resulting in a flux of 10X. The concentrator cell is optimized for 20X, but this reduction of the cell's conversion efficiency is of minor importance;
- inaccurate measurement, due to not using an I–V curve tracer;
- there are some possibilities of electrical problems, e.g. related to solder and back surface field.

If the solar concentrator cell had a more suitable size matching the absorber width, and if the cell design targeted the flux delivered by the nonimaging Fresnel lens, better results could be expected. This section has shown that there are significant obstacles to tackle when matching a laboratory cell to outdoor conditions under a real concentrator.

9.7 Concentration and Cost

Solar photovoltaic generation of electricity can satisfy the demand in niche markets today and in a restructured power sector tomorrow. The cost of photovoltaic cells is relatively high. The use of concentrating solar collectors is one way to reduce costs, as long as the concentrator is less expensive than the substituted solar cell. This study compares flat plate photovoltaic systems

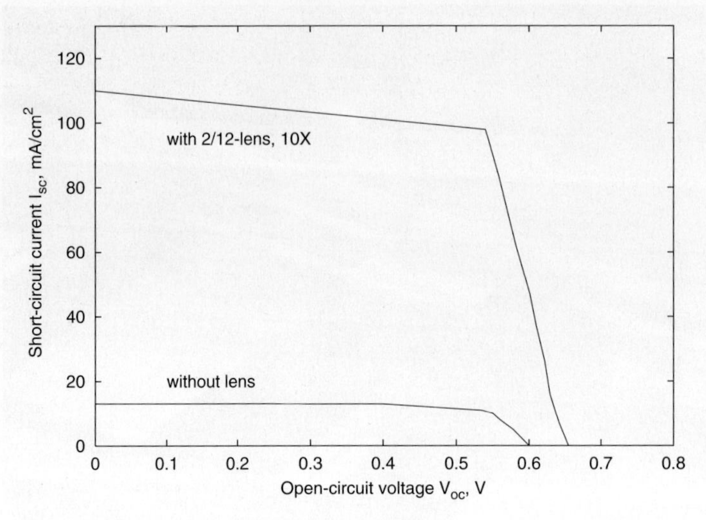

Fig. 9.22. I–V characteristics of TTI's concentrator cell under the 02/12 nonimaging Fresnel lens, and without the concentrator. Normal incidence, suboptimal operation

(1X) with those of concentration factors 2X, 20X, 1000X. Fresnel lenses are used to achieve significant cost reductions; systems of medium concentration are most promising. Still, it is found that risk-averse market players are not likely to soon develop and introduce concentrator systems.

Triggered by the apparent unavailability of concentrator cells on the photovoltaic market in recent years, this section examines the cost of photovoltaic systems of various concentration factors, and the willingness of manufacturers to expand the corresponding market segments. The technology representing solar concentration is the use of Fresnel lenses of low, medium and high concentration for photovoltaic systems.

Since the concentrator technology is essentially the same for all ranges of concentration, the influences of the factors accompanying the choice of concentration ratio, such as, complexity of tracking, radiation utilizability, and concentrator cell efficiency, on the system cost can be assessed. This section is about the actions of market players concerning the development and market introduction of concentrating systems. The optical performance of two concentrators of low and medium concentration factor, respectively, is presented after the limits of concentrator cells are clarified. The costs of concentrating and nonconcentrating photovoltaic systems are compared.

The development of concentrators for photovoltaic systems faces a dilemma. The higher the geometrical concentration ratio of the concentrator, the smaller is the solar cell needed to generate a fixed amount of electricity, but the higher is the system complexity, mainly in terms of the tracking accuracy and the related mechanics and controls.

Although concentrator cells become smaller, they are more expensive, due to lower production volume and higher cell complexity. Shipments of solar cells from the United States have been listed in Table 9.9; three-quarters of these are exported, more than half of those to Germany and Japan. The world's largest solar cell manufacturers are located in America; they manufacture about one-third of the global pv production. The production volume of concentrator cells is very small, and points to batch-type manufacturing of concentrator cells specifically for larger projects.

Concentrator cells cost about two to ten times as much for the same area as flat plate crystalline silicon cells. Concentration factors may be in the range of 20–50, in some extreme cases up to 1000. The costs for the photovoltaic cell have a potential for being lower for higher concentration, but the system cost may partially offset this effect due to higher complexity.

Table 9.9. Shipments from the USA of solar cells in kW_p over the period 1993–1999 by type[a]

Year	c-Si[b]	t-Si[b]	conc-Si[b]	other[b]	total
1993	20 146	782	21	2	20 951
1994	24 785	1061	231	0	26 077
1995	29 740	1266	53	0	31 059
1996	33 996	1445	23	0	35 464
1997	44 677	1886	154	0	46 354
1998	47 186	3318	58	0	50 562
1999	73 461	3269	57	0	76 787

[a] Source: [38].
[b] c-Si: crystalline silicon; t-Si: thin-film silicon; conc-Si: concentrator cells silicon; other: gallium arsenide etc. Cells for space/satellite applications are not included.

Estimating the cost of a new solar concentrator is, as with most products, a difficult undertaking. Not only is little known about the actual performance and application of the novel nonimaging Fresnel lens, but cost data for comparable systems are difficult to obtain, and, due to the systems' limited dissemination in an often subsidized (national) market, unreliable. The technology of Fresnel lenses for solar applications is mature, but the markets are not.

A condensed cost study for solar concentrators, including Fresnel lenses for photovoltaics, has been published [18]. The following considerations (Table 9.10) are based on the referred meta analysis, and the data are backed up by traceable numbers circulated on the internet [163]. Similar numbers can be found in [71], and in the work of other authors in the same volume.

Table 9.10. Cost estimates for photovoltaic systems of concentration (in X suns[a])

	Unit	1X	2X	20X	1000X
Fresnel lens concentrator	$\$/m^2$	–	30	30	30
Photovoltaic cells[b]	$\$/m^2$	320	160	25	75
Module[c]	$\$/m^2$	**400**	**190**	**140**	**200**
Array structures/tracking[d]	$\$/m^2$	80	80	155	230
Power conditioning	$\$/m^2$	20	20	40	40
Land	$\$/m^2$	4	4	4	4
Direct cost	$\$/m^2$	**524**	**294**	**384**	**544**
Indirect cost[e]	$\$/m^2$	175	98	113	158
Total cost	$\$/m^2$	**699**	**392**	**452**	**632**
Optical efficiency η	–	0.90	0.75	0.75	0.80
Cell, power efficiency η_c[f]	–	0.15	0.15	0.20	0.25
Radiation utilizability η_r[g]	–	1.0	0.9	0.8	0.8
Output[h]	W_p	135	101	120	160
Turn-key cost	$\$/W_p$	**5.18**	**3.88**	**3.76**	**3.95**

[a] Geometrical concentration ratio, calculated for irradiance of $1000\,W/m^2$ and standard spectrum, as stipulated by the definition of 'sun'.
[b] 1X, 2X: crystalline Si cells ($0.032\$/cm^2$), stationary array; 20X: nonimaging 2D Fresnel lens, concentrator cells ($0.05\$/cm^2$), one-axis tracking; 1000X: imaging or nonimaging 3D Fresnel lens, space cells ($7.50\$/cm^2$), full tracking.
[c] Including assembly, housing, others [18].
[d] Cost of single-axis tracking $75\$/m^2$ [141].
[e] Assumed to be 33% of direct cost [192].
[f] System efficiency, including inverter and excluding lens; slightly optimistic values.
[g] Flat plate with latitude tilt; one-axis, N–S tracking collector with latitude tilt; full tracking. Values for Puerto Rico or Guam (1.0/0.7/0.7); Hawaii (Honululu only: 1.0/1.0/1.0); Barstow, CA (1.0/1.0/1.2) [120].
[h] At 1 sun: output $= 1000\,W \cdot \eta \cdot \eta_c \cdot \eta_r$.

The idea behind solar concentration for photovoltaics is, of course, to reduce costly area of semiconductor cells. While concentrator cells are more expensive than conventional ones, and cost is added for the concentrator and a tracking mechanism, the specific turn-key cost of a project tends to decrease when concentration technology is used. The following factors related to the considerations in Table 9.10 should be pointed out:

- concentration factor: 2X is the practical limit for stationary concentrators, 20X the limit for concentrators tracking in one-axis, and 1000X is the limit imposed by the GaAs cell on concentration;
- availability and technical maturity of concentrator technologies: full tracking requires high accuracies, if a concentration of 1000X is to be employed. The

reliability of both the optical system and potential trackers may be uncertain for some applications, and only suitable for systems in areas where the infrastructure allows constant supervision;

- radiation: concentrators can collect only direct solar radiation. While they track the sun, which results in a potentially higher yield due to reduced cosine losses, concentrating systems may be unsuitable for tropical climates where the diffuse fraction of sunlight is larger. The values given for Guam, and Barstow, CA, in the notes to Table 9.10 illustrate this. Diffuse radiation is assumed to be isentropic, and concentrators with wide acceptance half-angles do intercept part of the scattered rays;
- flat plate developments: emerging manufacturing technologies and new materials for flat plate photovoltaic systems are reducing the cost of these considerably. Crystalline silicon cells account for roughly the same cost as concentrator technology. It remains to be seen whether the potential for cost reductions is greater for cells or for concentrating technology;
- time horizon: current cost for c-Si system, future costs for concentrator modules for 10 MW_p production volume. While 20X-systems are commercially available, 1000X-systems are not. Significant cost reductions for polycrystalline thin film [192] or amorphous cells [29] are envisaged, reducing module cost by a factor of 8–10, while conversion efficiency will drop to around $\eta_c = 0.10$, leaving a real cost reduction of more than 50%. All costs are conservative estimates, often based on pre-1992 data, see [18, 183].

Single items in Table 9.10 have a strong influence on the overall cost of the concentrator. This is particularly true since the alternatives in this case are characterized by costs in the same range. More difficult than the determination of changing costs is the assessment of the practicability of the alternative for photovoltaic power generation, from the engineering point of view.

The nonimaging Fresnel lens is an emerging concentrator technology. A comparable system [131] is in its market penetration phase, with apparently good results and great expectations of possible cost reductions.

Manufacturers of photovoltaic systems are risk averse, as all market players are. Most manufacturers can adjust their product portfolio to contain any photovoltaic technology, concentrating or not. The decision-making process of whether to invest in the development and marketing of a pv-system of concentration has to take into account imponderabilities of a technical and economic nature. Tracking systems may be complex, expensive, not readily available, and prone to erroneous functioning in harsh environments. Badly designed concentrators may lower the efficiency of the concentrator cell. Concentration systems may find only a small niche market, and may be difficult to position, because photovoltaic markets are largest in the industrial, residential, and commercial sectors of industrialized nations.

The direct fraction of solar radiation in developed nations is rather unfavorable for the use of concentrator systems that cannot collect diffuse solar radiation. The environment in developing countries, though often blessed with high direct solar radiation, may be considered less suitable for complex systems of renewable energy conversion, since financing and operating conditions are often difficult and not clearly predictable.

Most importantly, the costs of flat plate crystalline silicon and its competitors (polycrystalline, amorphous, and thin film technologies) are expected to continue to fall [98]. Thus, concentrating technologies have to offset the anticipated cost reduction of conventional photovoltaic systems in addition to overcoming the problems related to complexity.

Solar photovoltaic systems of low and medium concentration factor cost less than 75% of conventional flat plate systems. Medium concentration shows the lowest achievable cost and the technology is mature. Special cost reduction potentials for higher concentration systems could not be identified. While all concentrator systems should be located in areas with a comparatively high fraction of direct solar radiation, systems of low concentration use conventional cells, and may therefore operate (with weak performance) even under diffuse radiation, reducing the risk of complete unavailability.

Cell manufacturers are often module and system manufacturers, while the core business remains cell production. From their viewpoint, the value of cells relative to the complete product should be as high as possible. Existing mass production for conventional c-Si cells should be used optimally.

In a risky environment, the decision of major manufacturers to focus on conventional photovoltaic systems makes sense economically. Little future development in the concentrator market segment is to be awaited from major manufacturers. This means bad news for the competitiveness of the renewable energy industry as a whole.

10 Solar Thermal Concentrator Systems

10.1 Solar Resources

Nonimaging Fresnel lenses can only concentrate the direct fraction of solar radiation. Solar resource assessments are necessary to evaluate the suitability of locations for the installation of solar concentrating collectors and to estimate their potential for energy conversion.

When planning to install concentrators it is important to design the system according to a minimum standard in terms of the solar fraction of the planned total energy supply. This will determine the size and quality of the solar system to be built. The backup system has to be sized simultaneously.

Asia and Oceania have long been neglected in solar radiation measurements. The database of the World Radiation Data Center (WRDC) [189] is a comprehensive tool for evaluating both global and beam radiation at locations in Asia and Oceania. Quality factors vary, but data was verified by comparison with a recently released database [100]. To facilitate the evaluation of the full potential of solar energy solutions in Asia and other developing parts of the world, quality-controlled measurements of both diffuse and direct solar fractions must be conducted at more locations.

Available data is given in [37, 189] as daily irradiation (averaged over one month) on a horizontal plate at a certain latitude ϕ. Assessment of differences between global and beam radiation and suitable collector types requires tracking of the sun. Monthly direct radiation values $\overline{H_{\mathrm{b}}^{\mathrm{m}}}$ must be corrected for cosine losses by a factor

$$\overline{H_{\perp}^{\mathrm{m}}} = \frac{\overline{H_{\mathrm{b}}^{\mathrm{m}}}}{\cos\left(\phi - \delta^{\mathrm{m}}\right)} , \tag{10.1}$$

where δ^{m} denotes the monthly declination of the sun (values are given in Table 7.1). Diffuse radiation is assumed to be isentropic.

The design insolation is defined as the radiation on a tracking plate for the month with the lowest value. The type and quality of the solar collector to be used and the size of the solar field can thus be roughly calculated for a given application at a given location.

The radiation for Asian and Pacific locations is shown in Figs. 10.1 and 10.2 for global and direct radiation, respectively. The yearly average daily

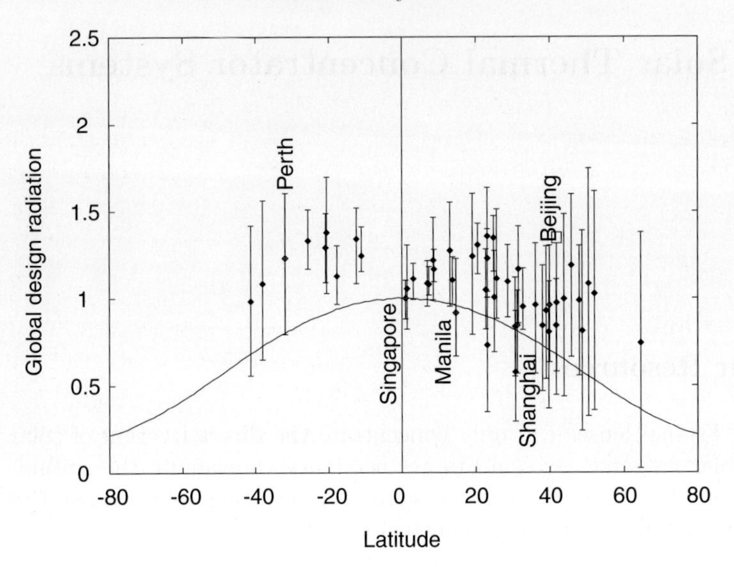

Fig. 10.1. Global monthly normalized solar radiation on a tracking plate. Comparison for Asia and the Pacific

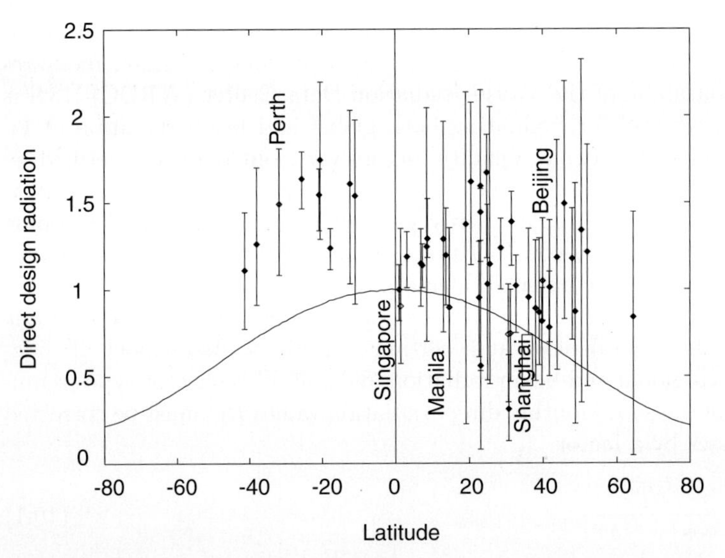

Fig. 10.2. Direct monthly normalized solar radiation on a tracking plate. Comparison for Asia and the Pacific

irradiance value is shown for the best month and the design month, i.e. the months with the highest and the lowest average daily insolation value. All data have been normalized in respect to the values of Singapore. Singapore receives a daily average of global irradiation on a tracking surface of $17.5\,\mathrm{MJ/m^2}$, with a beam fraction of $8\,\mathrm{MJ/m^2}$. The choice of Singapore as the reference point has been influenced not only by its equatorial location, but also by the fact that the large solar water-heating plant at its Changi Airport uses concentrating CPC collectors with satisfactory operating results [28].

When comparing the global radiation in Fig. 10.1 with the beam radiation values in Fig. 10.2 the greater deviation of the direct maxima and minima (compared to the distance of global maxima and minima) from the average is obvious. The difference between the global maxima and minima is not as large as in the case of the direct radiation, meaning that the design of flat plate collectors for most locations is less prone to failure due to large differences between monthly values than the design of concentrator systems. The latter should not be designed without a close look at the design insolation of the worst month, as this value might be very different from the yearly average. Moreover, this deviation cannot be captured by a 'rule of thumb' but must be found for each location, as seasonal influences on direct radiation can be very strong.

It is clearly visible that most locations in the earth's sunbelts (20–35°) in both hemispheres receive global and direct radiation above average, while most places near the equator and further north, or south, receive comparatively less solar radiation. The direct design insolation is worst in places affected by the monsoon and in south-eastern China. It is best in Australia, some Pacific islands, and northern China. Western India and places from Pakistan westward are also offering excellent direct insolation. The insolation increases for higher elevations, such as the Himalayas, where direct radiation can be extreme.

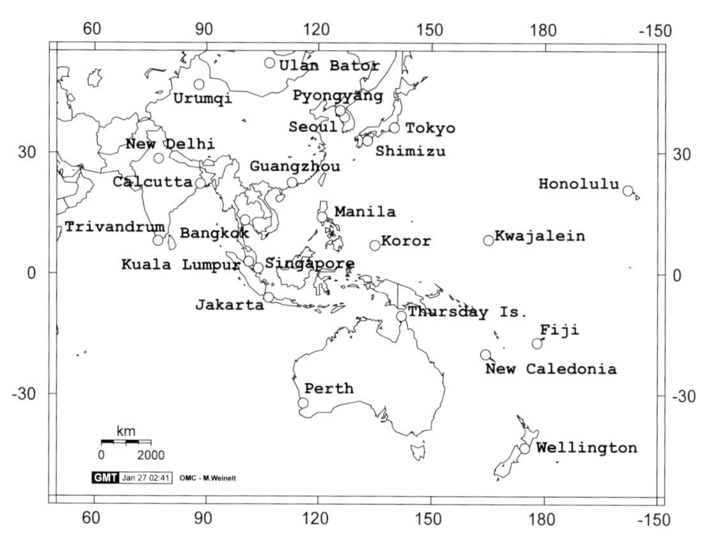

Fig. 10.3. Solar resources in Australasia: some locations corresponding to data given in Tables 10.1 and 10.2

The locations and values used in plotting the Figures 10.1 and 10.2 are given below. These values serve as a rough guideline of what to expect when designing a solar concentrator system. Values can equally be used for thermal and photovoltaic designs. The example of Australasia is chosen because this world region is rarely presented in other publications.

Figure 10.3 shows some of the locations for which the design insolation $\overline{H}$ is given in Tables 10.1 and 10.2. The design insolation comprises yearly average daily insolation data on a tracking plate.

Table 10.1. Solar resources in Australasia. Yearly average daily global $\overline{H}_{\mathrm{g}}$ and direct insolation $\overline{H}_{\mathrm{b}}$ on a tracking plate. Daily average data for best (index h) and design (index l) months in MJ/m^2 for locations at latitude ϕ in deg

Location	ϕ	$\overline{H}_{\mathrm{g}}$	$\overline{H}_{\mathrm{b}}$	$\overline{H}_{\mathrm{g,h}}$	$\overline{H}_{\mathrm{g,l}}$	$\overline{H}_{\mathrm{b,h}}$	$\overline{H}_{\mathrm{b,l}}$
[a]Ahmadabad, India	23.1	22.1	14.1	26.6	16.4	19.0	4.7
[b]Akita, Japan	39.7	13.1	7.2	17.9	6.0	8.9	4.0
[b]Aspendale, Australia	-38.0	17.6	11.2	25.4	10.6	15.0	8.1
[b]Bangkok, Thailand	13.7	18.0	10.6	20.1	15.7	13.0	8.0
[a]Beijing, China	39.9	15.6	9.3	21.2	10.3	12.5	7.3
[a]Bombay, India	19.1	20.2	12.1	26.0	12.9	18.9	2.0
[a]Calcutta, India	22.7	17.0	8.4	20.9	13.4	11.3	5.4
[a]Chengdu, China	30.7	9.4	2.7	15.1	4.8	5.5	1.1
[b]Chita, Russia	52.0	16.7	10.7	26.3	5.8	16.2	3.9
[b]Colombo, Sri Lanka	6.9	17.7	10.2	19.8	15.5	11.9	8.0
[b]Darwin, Australia	-12.4	21.8	14.2	24.7	17.6	18.0	9.1
[b]Fiji	-17.8	18.3	10.9	21.2	15.7	11.9	9.9
[a]Guangzhou, China	23.1	11.9	4.9	16.9	5.7	10.2	0.7
[a]Haeju, Korea	38.0	13.7	7.9	19.4	7.6	11.3	4.2
[a]Harbin, China	45.8	19.4	13.2	26.0	10.8	19.4	7.3
[b]Honolulu, Hawaii, USA	23.1	20.0	12.8	22.8	16.3	14.0	11.3
[b]Karachi, Pakistan	24.8	21.9	14.8	24.0	18.8	16.7	10.4
[b]Koror, USA	7.3	17.6	10.1	19.2	16.6	11.2	8.9
[b]Kuala Lumpur, Malaysia	3.1	18.1	10.5	19.5	16.2	11.8	8.9
[a]Kunming, China	25.0	16.4	9.1	20.7	11.3	13.2	4.2
[b]Kwajalein, USA	8.7	19.0	11.4	20.8	17.8	13.5	10.5
[a]Kyongsong, Korea	41.7	13.7	6.9	18.1	7.3	9.7	4.4
[b]Lahore, Pakistan	31.5	19.0	12.3	23.0	14.0	13.8	10.0
[a]Lanzhou, China	36.1	15.6	8.4	21.4	9.3	11.9	4.8

Continued in Table 10.2.

Table 10.2. Continued from Table 10.1. Solar resources in Australasia. Yearly average daily global $\overline{H}_g$ and direct insolation $\overline{H}_b$ on a tracking plate. Daily average data for best (index h) and design (index l) months in MJ/m^2 for locations at latitude ϕ in deg

Location	ϕ	$\overline{H}_g$	$\overline{H}_b$	$\overline{H}_{g,h}$	$\overline{H}_{g,l}$	$\overline{H}_{b,h}$	$\overline{H}_{b,l}$
[b]Madras, India	13.0	20.7	11.4	25.2	15.5	17.4	6.7
[a]Manila, Philippines	14.6	14.9	7.9	20.0	10.9	12.0	4.3
[a]Nancy, France	48.7	13.2	7.7	22.6	3.9	14.2	2.0
[a]New Caledonia	-20.6	22.4	15.4	27.6	16.7	19.4	11.4
[b]New Delhi, India	28.6	17.8	10.9	21.3	14.5	12.4	8.8
[b]Nome, Alaska, USA	64.5	12.1	7.4	22.4	0.4	12.8	0.4
[b]Perth, Australia	-31.9	20.0	13.2	26.1	12.9	16.0	9.6
[c]Pinares De Mayari, Cuba	20.5	21.3	14.3	23.2	18.6	18.3	10.1
[b]Pretoria, South Africa	-25.7	21.7	14.5	24.5	19.0	15.9	12.9
[a]Pyongyang, Korea	39.0	15.1	7.7	21.1	8.8	11.4	5.0
[a]Semipalatinsk, Russia	50.4	17.6	11.8	28.4	5.3	20.5	3.1
[a]Shanghai, China	31.2	13.7	6.6	17.0	9.3	8.8	4.2
[a]Shenyang, China	41.8	15.8	8.9	22.0	9.4	12.4	6.2
[a]Shillong, India	25.6	18.1	10.1	24.6	14.4	18.1	4.0
[b]Shimizu, Japan	32.7	15.5	9.0	19.1	13.3	10.6	6.6
[a]Singapore	1.4	17.2	8.0	21.3	13.4	11.9	5.1
[b]Singapore	1.0	16.3	8.8	17.9	14.3	10.1	7.3
[a]St. Denis, Reunion	-20.9	21.0	13.7	24.2	17.7	15.0	11.8
[a]Thursday Is., Australia	-11.0	20.2	13.6	23.0	18.8	17.5	8.1
[a]Trivandrum, India	8.5	19.8	11.0	23.8	16.1	18.1	6.2
[a]Ulan Bator, Mongolia	47.9	16.1	10.4	21.2	9.1	13.0	6.3
[a]Urumqi, China	43.8	16.2	10.5	24.1	7.1	16.4	4.7
[a]Wellington, New Zealand	-41.3	16.0	9.8	23.1	9.1	12.8	6.8
[a]Wuhan, China	30.6	13.6	6.6	18.2	9.0	9.1	3.1

[a] Source: [189].
[b] Source: [37].
[c] Source: authors.

Since this work deals with solar concentrators, the direct solar fraction is of prime importance, and has been given for the month with the highest daily average insolation and for the design month. Most locations are within Asia and the Pacific region; some locations elsewhere have been added for comparison.

10.2 Solar Sorption Air Conditioning

The nonimaging Fresnel lens concentrator may be used in solar thermal energy conversion. A prime field of applications comprises solar-powered sorption air conditioning.

This book focuses on photovoltaic applications of the lens, but some work is presented in this section in order to account for the significance of solar air conditioning in our field. It is impossible here to give a complete technical introduction to absorption and adsorption thermodynamics (just as it was deemed unnecessary to explain the production process of photovoltaic concentrator cells in detail). A good introduction to solar heating and cooling is found in [154] and [86]; both books are collections, the latter being quite comprehensive. Another useful overview of the topic is Chap. 6 in [45].

The following section combines the solar resource assessment of Sect. 10.1 with psychrometric findings. A methodology is developed to evaluate air conditioning needs in relation to local solar conditions, which may supply the driving power source.

The demand for air conditioning in industry and for comfort cooling is growing rapidly in Asia. Air conditioning unit production in Japan has been rising from 2500 billion Yen in 1987 to 3500 billion Yen in 1996 [62]. Japan is alongside with the USA home to the world's largest HVAC&R industry. The acronym stands for heating, ventilation, air conditioning, and refrigeration.

In industrially developed Japan, air conditioning accounts for one-third of the electricity demand in the yearly average, and for one-half on some summer days. Asian countries will only be able to meet greenhouse gas reduction targets when reducing the electricity needs for commonly employed vapor compression air conditioning. Absorption air conditioning is environmentally friendly as the driving source can be solar thermal power or waste heat, and the working fluids (H_2O–LiBr, NH_3–H_2O) do not contribute to ozone layer destruction. Non-electricity driven cycles can become an element of demand-side management. Sorption cycles can be solar driven, and solar energy supply is generally highest when air conditioning demand is greatest. Desiccant-enhanced cycles can be energetically superior to conventional ones, as will be shown.

In both Europe and the United States encouraging expectations regarding the use of desiccant wheels in air conditioning have been expressed. Experimental plants are being built in Europe [14], while the southern United States of America has seen a considerable number of commercial systems installed, typically 'ventilation cycles' set up in supermarkets. In America, desiccant wheels have been used since the early 1980s and performance standards are being developed now [162].

Conventional electrically driven vapor compression cycles supercool air below its dew-point temperature to remove moisture (latent heat). The air is then reheated to satisfy the conditions of the 'comfort zone'. The process is shown schematically by the left arrow in Fig. 10.4. Supercooling and reheating is energetically inefficient. More sound air conditioning systems have been proposed, e.g. the design of a hybrid combination of a solar-powered absorption chiller for sensible heat removal and a desiccant wheel for humidity control, or the combination of an adsorption-bed chiller with an absorption machine. These systems are environmentally friendly and energy efficient as they neither require supercooling nor superdrying.

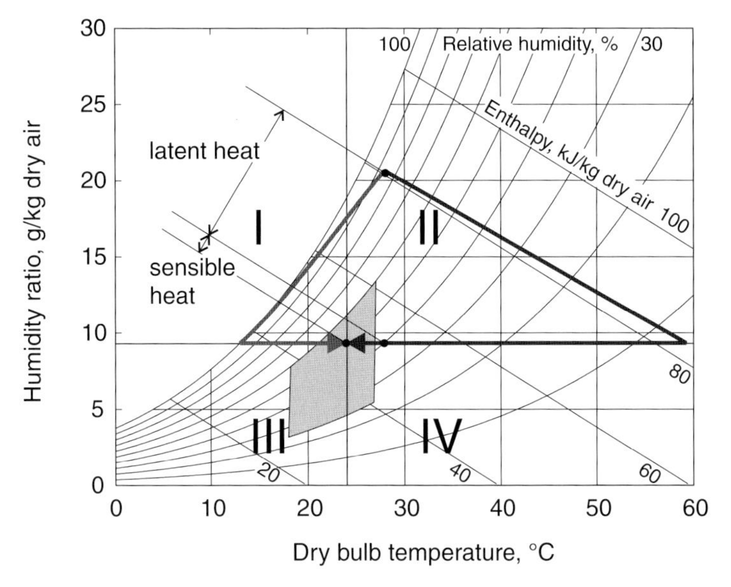

Fig. 10.4. Psychrometric chart. Schematic of vapor compression air conditioning (left arrow), and desiccant wheel-enhanced absorption air conditioning (right arrow). Also shown are iso-enthalpy lines, latent (vertically) and sensible (horizontally) heats, and the comfort zone

Desiccants are hygroscopic materials showing a high affinity for water vapor in air. The performance of desiccant wheels has recently been greatly improved [55], and the desorption temperature for regeneration can be as low as 60℃. In climates with very high absolute humidity ratios (> 15 g/kg dry air), wheels featuring desorption temperatures of 80℃ are to be used. Desiccant wheels are inflammable, nontoxic, and deodorizing.

Combining solar-driven absorption and adsorption (desiccant) cycles results in an environment friendly system that is energy efficient since supercooling can be avoided. 'Asian' outdoor air often is at 25–30℃ and above 80% relative humidity, while air delivered to the indoor space should be at 24℃ at 50% relative humidity (the appropriate value may vary according to the designer of the system), which is within the comfort zone (see Fig. 10.4).

While it is possible to reach the comfort zone by sensible heating or cooling from conditions at the same horizontal level, hot and humid climates

Table 10.3. Cases of sensible and latent loads for air conditioning. For an explanation see text

	Latent heat removal	Latent heat supply
Sensible heating	I	III
Sensible cooling	II	IV

need dehumidification (removal of latent heat) and sensible cooling. Four combinations can occur; they are listed in Table 10.3.

Of prime significance is case II, which requires the removal of both sensible and latent heat. In case I, most often only moisture is to be removed. Reheating is necessary if the moisture is condensed by supercooling the air below its dew point using vapor compression air conditioning. If desiccant wheels are to be used in a ventilation cycle [87], superdrying of the process air is necessary to facilitate evaporative cooling. When desiccant wheels are combined with an absorption cooling cycle, the process indicated by the right arrow in Fig. 10.4 takes place.

The desiccant-enhanced absorption air conditioning cycle may be constructed as in Fig. 10.5. Air is first dehumidified; in a second step, its sensible heat is removed by an absorption cooling unit.

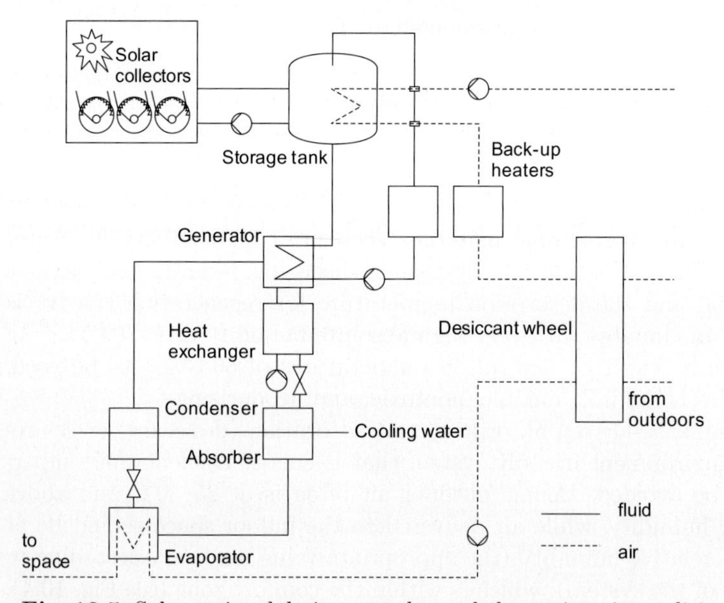

Fig. 10.5. Solar-assisted desiccant-enhanced absorption air conditioning unit. Schematic

The absorption unit can be constructed as a double-stage machine, raising the potential coefficient of performance (COP) from 0.7 to around 1.3. Also, it is possible to imagine the absorption chiller being replaced by a conventional vapor compression unit. This allows for retrofitting existing air conditioning units, but the disadvantage is the reliance on electricity (which could be generated by photovoltaic cells, if the sun was the only energy source available).

Furthermore, the cycle in Fig. 10.5 can be run in the opposite direction, i.e. the air stream may be cooled first and dehumidified afterwards. In the psychrometric chart of Fig. 10.4 this would result in a cycle that moves along the saturation line towards the left and down, until a desiccant wheel can be used to dehumidify and heat the process air, reaching the desired point from the saturation line by a sloped line along an iso-enthalpy line. Energetically, this cycle is of superior efficiency, when compared with the other two cycles mentioned here, which are based on process enthalpy.

In the cycle indicated by the right arrow in Fig. 10.4, and in Fig. 10.5, an air stream enters the desiccant wheel, where its moisture is removed to a value corresponding to the final goal of 50% relative humidity at 24℃ air temperature. The air becomes warmer (60℃) due to the heat of adsorption and desorption at 80℃. The enthalpy remains almost constant. The decisive parameter for the performance of the desiccant wheel is the velocity of the incoming air stream. In a second step the process air is sensibly cooled down to the final desired temperature by an absorption cycle. Both the desiccant wheel and the absorption cycle perform better with increased driving-source temperature.

Two alternatives are to be considered: if the latent heat to be removed exceeds the sensible heat to be extracted, the solar heat should be primarily delivered to the desiccant wheel, and only secondarily should it drive the generator of the absorption unit, and vice versa. Separation of driving sources after the solar storage offers this degree of freedom, and solves the problem of having to deliver hot air for desorbing the desiccant wheel.

Cases III and IV are of some technical significance since latent heat has often to be added to an air stream that is sensibly heated. Examples of simultaneous adding of both latent and sensible heat (in almost equal parts) are all the locations in central Japan. Case IV is represented by desert climates.

Latent and sensible heat can be calculated as enthalpy, with the unit $kJ/kg_{dry\ air}$. Iso-enthalpy lines have been incorporated in Fig. 10.4. The vertical part of the path from a given condition to the desired point is the latent fraction, while the horizontal part stands for the sensible heat. It has to be noted that the enthalpy (energy) required by the air conditioning process is most often different from the thermodynamic enthalpy. The air conditioning process may consist of two or more steps, whose enthalpies have to be added. This total (plus some allowance for system efficiencies) is the process enthalpy. The thermodynamic enthalpy for the same case may be zero, and

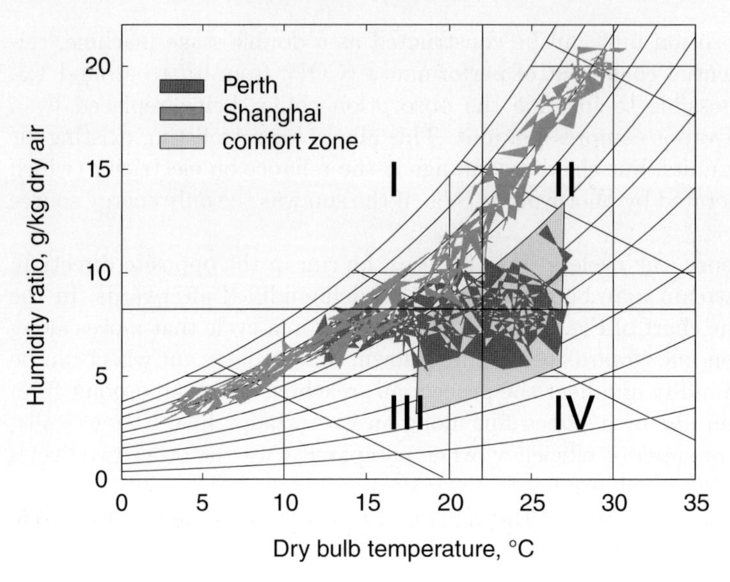

Fig. 10.6. Psychrometric chart. Daily mean data for Perth and Singapore

can best be understood when imagining it to be zero (i.e. the given and the desired conditions are on the same iso-enthalpy line).

A psychrometric chart for Perth, Australia and Shanghai, China, has been plotted in Fig. 10.6, using daily mean data over four years from 1994 to 1997, based on the Federal Climate Complex data sets [119]. Data is read into a simulation program and checked for mistakes and missing sets, the process enthalpies are calculated, and the averaged results are written to file.

Absorption cycles are reversible, and if properly designed can operate both in heating and cooling mode. The main advantage of evacuated tube-type solar collectors over flat plate collectors is their comparatively low heat loss coefficient. When ambient temperatures are low and the required process temperatures are high, evacuated tubes are more efficient. This is the case for areas in Asia where heating is required during the winter season.

It has been pointed out that hot and humid locations generally receive direct radiation levels below the average of locations at the same latitude. Ambient temperatures are high, heating in winter is not required, and the use of flat plate collectors providing the driving source for single-stage absorption cycles might well be sufficient. Stationary (non-tracking) concentrating collectors are able to provide a higher driving source temperature for double-stage/effect cycles, but when the direct fraction in the total radiation drops, higher concentration ratios are necessary. A collector with a higher concentration ratio has a smaller acceptance half-angle, i.e. it 'sees' a smaller part of the sky, and the time during which solar rays are actually concentrated be-

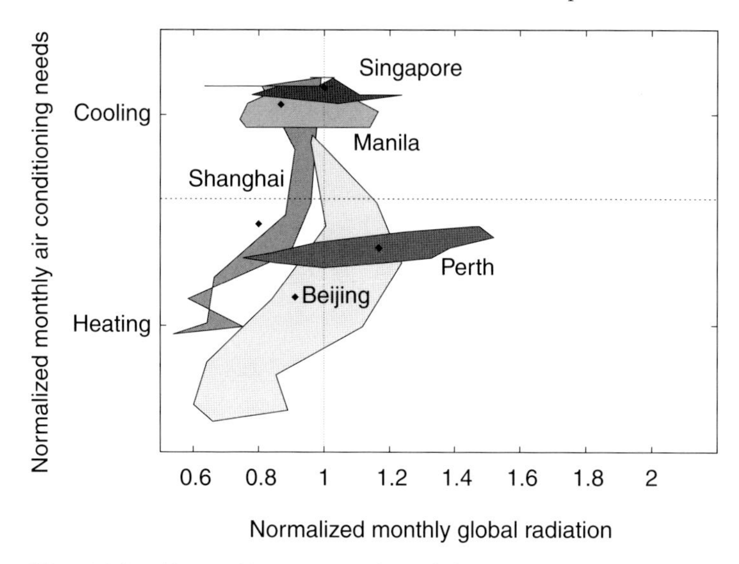

Fig. 10.7. Air conditioning needs and direct monthly normalized solar radiation on a tracking plate. Note that the region around the threshold between cooling and heating can be covered by passive means of cooling and heating buildings

comes shorter unless tracking is employed. Solving this optimization problem is an individual task for every location.

Figures 10.7 and 10.8 show the relation between air conditioning needs and global or direct insolation for monthly data for five locations in Asia. A representative indoor climate of 22°C at a relative humidity of 50% [10] is taken as designer's choice for the threshold between heating and cooling for the locations of Shanghai and Perth. The air conditioning needs incorporate both sensible cooling/heating and latent dehumidification/humidification based on an enthalpy calculation.

While equatorial Singapore is in for active cooling (and dehumidification), Perth has a balanced climate with high solar direct insolation. If passive means to climatize a building have been taken, there is little need for either cooling or heating. Shanghai's monthly climate values stretch over a heating period in winter into a cooling and dehumidification period in summer (Fig. 10.9), where active air conditioning has to be performed.

What kind of solar collector should be used where, and in combination with which air conditioning technology?

While some information on the most suitable solar collector can be extracted from Fig. 10.7, the visualization of the direct solar fraction in Fig. 10.8 allows for identification of areas where concentration collectors are of potential use in air conditioning. Locations with low global or direct solar radiation fractions still offer themselves for successful solar installations, but at the price of higher system complexity.

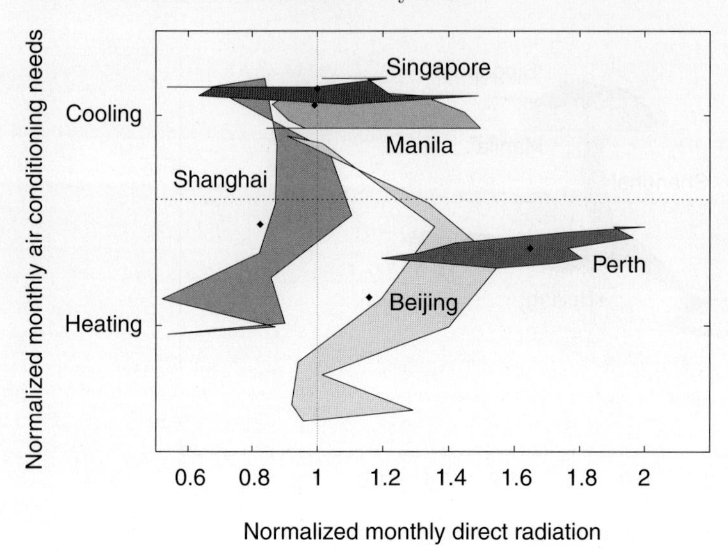

Fig. 10.8. Air conditioning needs and direct monthly normalized solar radiation on a tracking plate

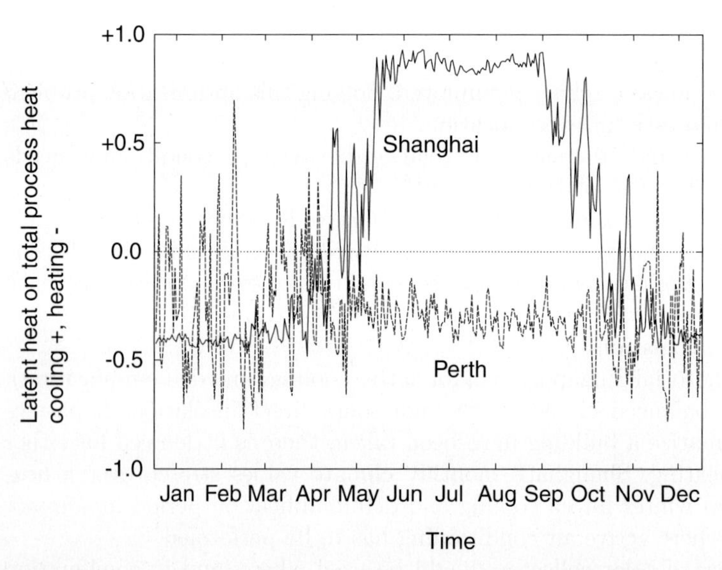

Fig. 10.9. Latent heat fraction on total air conditioning process enthalpy for Perth and Shanghai

The analysis of the ratio of latent heat to total process enthalpy in Fig. 10.9 suggests the use of desiccant dehumidification during summer in Shanghai, while the latent and sensible fractions are very balanced in Perth. Concentrating collectors are needed to provide the necessary 80℃ temperature for desorption of the wheel under given radiation conditions. In winter, the evacuated tubes could be used for heating. Most locations in north–east Asia (Japan, Korea, eastern China) call for this kind of solution.

In Fig. 10.4 four cases have been differentiated. To decide which air conditioning technology suits any case best, the relation of the latent and sensible enthalpies to the total process heat has to be known. Desiccant wheel-enhanced cycles may be superior in cases I and II if the latent heat fraction is larger than the sensible heat. Existing cycles should be enlarged by the addition of an adsorption system, if an exact analysis shows the suitability of an available desiccant wheel or adsorption bed. Sensible heating/cooling can be performed by absorption (or vapor compression) units, and by stand-alone solutions in cases III and IV, as well as in combination with solid sorption in the other two cases. Ventilation cycles are suitable where water for evaporative cooling is abundant.

The choice of solar collectors for Perth is straightforward. Anything warmer than water for a shower should be heated by concentrator-type solar collectors to make use of the high direct radiation fraction. In the case of Shanghai, the choice of collector type is far from clear. Although heating is required in winter, temperatures then are moderate, and direct radiation values are low. One might tend to use advanced flat plate collectors with good insulation, but if funds are available one might as well use evacuated tube-type collectors with small concentration. The latter case could allow for increased performance of a chiller driven by higher temperatures, for necessary dehumidification during summer.

10.3 Energy and Exergy

Exergy, also called available energy, is the part of the energy that can be reversibly converted into potential or kinetic energy. Irreversibilities are exergy losses, as entropy is destroyed exergy. An entropy analysis is usually called a second-law analysis.

The concepts behind analyzing exergy and entropy changes in energy transformation processes are both part of the second-law analysis. The second law of thermodynamics can be expressed as

$$\Delta S = \Delta S_{\mathrm{in/out}} + \Delta S_{\mathrm{irreversible}} \; . \tag{10.2}$$

The entropy ΔS of a closed system can change if entropy is transported over the system boundaries or when entropy is generated within the system. Ideal reversible processes do not show any entropy generation.

The unit of the exergy is the Joule, and the unit of the entropy is the Joule per Kelvin. While the entropy allows an engineer to optimize a part of a system with the temperature levels already set, the exergy concept incorporates temperature levels, thus expressing the quality of a system. The annihilation of exergy ΔE and the generation of entropy ΔS within a system are closely connected by the ambient temperature T_{a}. The equation

$$\Delta E = T_{\mathrm{a}} \Delta S_{\mathrm{irreversible}} \tag{10.3}$$

is known as Guoy–Stodola's theorem.

For the energy form of heat, the threshold between available and non-available energy, or exergy and anergy, is defined with the help of the ambient temperature. The energy of air at ambient temperature and at a pressure of 1 bar consists of 100% anergy and 0% exergy. Concentration differences, chemical energy differences, and everything that contributes to imbalances in a defined ambient balance are further contributors to exergy.

The exergy analysis provides a tool to enhance the classic energy analysis. Energy balances, with mass and heat equations, define the system that can be evaluated by means of an exergy analysis.

Exergy Concept Evolution

Exergy analysis today is a well-founded concept based on extensive research in many disciplines. The first studies dealing with exergy were published more than one hundred years ago. The previously mentioned theorem of Guoy–Stodola describing exergy loss for irreversible processes honors a French and a German pioneer in the field. Other early researchers are mentioned in [42]. Gibbs was the first to express the concept of available work in 1873 [179]:

> We will first observe that an expression of the form
> $-e + Ts - Pv + M_1 m_1 + M_2 m_2 + ... + M_{\mathrm{n}} m_{\mathrm{n}}$
> denotes the work obtainable by the formation (by a reversible process) of a body of which e, s, v, m_1, m_2, m_{n} are the energy, entropy, volume, and the quantities of the components, within a medium having the pressure P, the temperature T, and the potentials M_1, M_2, M_{n}. (The medium is supposed so large that its properties are not sensibly altered in any part by the formation of the body.)

The word *exergy* was coined by Rant in 1956 when he created it from the Greek *erg(on)*, meaning work or force, and *ex*, from from or out of. With the concluding *ie* the German *Exergie* was formed.

Published handbooks on exergy include [23, 42]. Books on general thermodynamics most often include a discussion of exergetic fundamentals. Articles on exergy have appeared in most technical journals. General reading on exergy includes the references [2, 13, 97], and the early reference [69]. The exergy analysis of solar driven absorption cycles is clearly described in [22].

A Fundamental Example

Heat at a temperature above ambient can be transformed into work. This is why exergy may be referred to as available work. The exergy of heat is the part of its energy that can be reversibly transformed into another form of energy, e.g. mechanical energy.

Exergy can never become smaller than zero, and no exergy is available below ambient temperature (for the case of heat), unless one uses ambient temperature to drive the process. This can be proven mathematically (see for example [42]). Exergy of heat can be described by its ability to be transformed into work. Energy of heat is

$$\delta q = c\,\delta T\,, \tag{10.4}$$

or the product of specific heat capacity of the material and its temperature change. Work, or exergy is defined as

$$\delta w = \delta e = c_\mathrm{p}\left(1 - \frac{T_\mathrm{a}}{T}\right)\delta T\,. \tag{10.5}$$

The Carnot factor $1 - T_\mathrm{a}/T$ corresponds to the maximum work that can be extracted from a reversible isothermal process, where the lower bound of the process is equal to ambient temperature. The integration of (10.5) leads to the amount of exergy in the process

$$E = \int_{\mathrm{T_a}}^{\mathrm{T}} c\left(1 - \frac{T_\mathrm{a}}{T}\right)dT\,, \tag{10.6}$$

resulting in

$$E = c\,(T - T_\mathrm{a})\left(1 - T_\mathrm{a}\frac{\ln\left(T/T_\mathrm{a}\right)}{T - T_\mathrm{a}}\right)\,. \tag{10.7}$$

The mean logarithmic temperature T_m, or exergetic temperature

$$\tau_\mathrm{e} = 1 - \frac{T_\mathrm{a}}{T} = \frac{T - T_\mathrm{a}}{T} \tag{10.8}$$

plays a vital role in describing exergies. It is also known from calculating heat exchangers.

An exergy never becomes negative. Equation (10.7) has been plotted over a range of driving temperatures in Fig. 10.10 to visualize this finding. The graph shows the exergy level only qualitatively, as changing values of the specific heat capacity c_p were not considered. For temperatures $T > T_\mathrm{a}$ exergy can be extracted from the driving heat source of the process. For temperatures $T < T_\mathrm{a}$, exergy is extracted from ambient heat. If $T \to 0$, then $c_\mathrm{p} \to 0$ and the amount of exergy extractable from the environment approaches a finite value.

Exergy Balances

A systematic approach in describing the exergy of a system as a black box allows us to distinguish and describe the different processes responsible for exergy changes. Exergy can enter the system, it can exit the system, and it

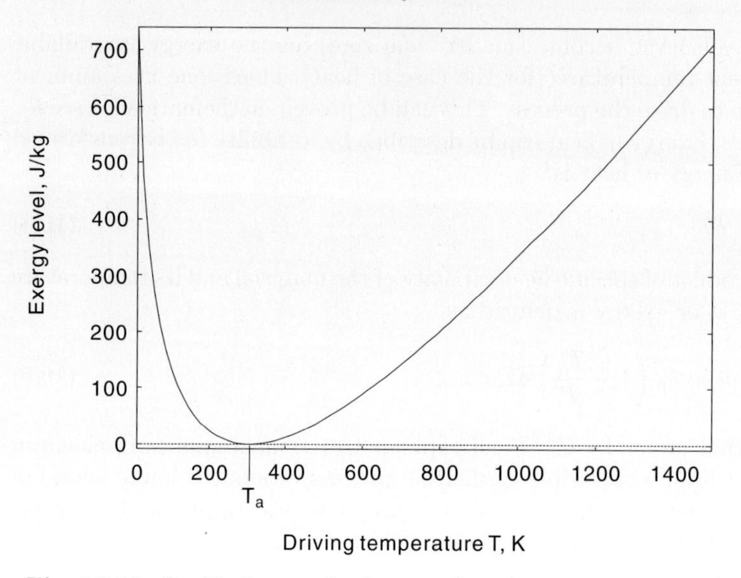

Fig. 10.10. Qualitative graph of exergy from heat over a range of temperatures

can be destroyed within the system due to irreversibilities. Exergy changes can happen with or without mass flow across the borders of the system.

Figure 10.11 introduces the concept of the exergy balance of a system. As in the classical thermodynamic energy balance equations, the sum of the exergy fractions entering and leaving the system must balance. Hence the exergy annihilation, exergy leak, and exergy storage in the system are equal to the difference of E_{out} and E_{in}, i.e.

$$E_{\text{out}} = E_{\text{in}} - E_{\text{leak}} - E_{\text{annihilation}} - E_{\text{storage}} \, . \tag{10.9}$$

Exergy describes thermal and chemical energy in a comparable way. Thus, the inputs and outputs in this model can be directly added. The exergy

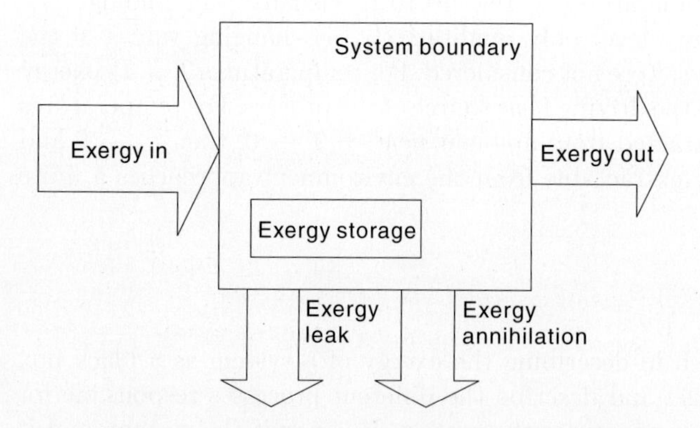

Fig. 10.11. Schematic of the exergy balance of a system

balance equations are explained in more detail in the application for a solar collector described in Sect. 10.4.

The Exergy Valley: Exergy of an Ideal Gas

Exergies can be defined for any process. The example above dealt with the available work from heat. The following section describes the exergy of a system which differs in temperature and pressure from its surroundings. Based on the exergy of a pressurized system as given by

$$E = \int_{\mathrm{V_a}}^{V} P\mathrm{d}V = -\int_{\mathrm{T}}^{\mathrm{T_a}} \mathrm{d}U \,, \tag{10.10}$$

a formulae describing exergy of an ideal gas as function of temperature and pressure differences with the environment can be found according to [136]:

$$E = nRT_{\mathrm{a}} \left(\frac{\Theta - 1}{\Gamma - 1} - \frac{\Gamma}{\Gamma - 1} \ln \Theta + \ln \Pi + \left(\frac{\Theta}{\Pi} - 1 \right) \right) , \tag{10.11}$$

where the ratios of temperature Θ, of the specific heat capacities Γ at constant pressure and volume, as well as the ratio Π of the pressures involved are given by

$$\Theta = \frac{T}{T_{\mathrm{a}}} \,, \tag{10.12a}$$

$$\Gamma = \frac{c_{\mathrm{p}}}{c_{\mathrm{v}}} \,, \tag{10.12b}$$

$$\Pi = \frac{P}{P_{\mathrm{a}}} \,. \tag{10.12c}$$

The ideal gas constant R is defined as

$$R = c_{\mathrm{p}} - c_{\mathrm{v}} = 8.315 \, \frac{\mathrm{J}}{\mathrm{mol\,K}} \,. \tag{10.13}$$

Often an ideal gas represents a good approximation for practical processes, although binary mixtures like steam crossing phases in a Rankine cycle require complex equations for each phase and composition [42]. Based on (10.11) the exergy of an ideal gas can be calculated. Unfortunately, this equation cannot be solved analytically, and instead a numerical way of solving it has to be employed. Newton's method (see p. 83) can be used to approximate the solution of a function.

First, (10.11) can be rewritten in a suitable form as a function of the pressure P, identifying all constants, combined in A, B, and C. Of course, the equation could also be written as a function of the temperature T, but for P rewriting is easier:

$$f\,(p) = -E - A + AB + \frac{C}{P} + A\,\ln \frac{P}{P_{\mathrm{a}}} \,. \tag{10.14}$$

Newton's method is expected to find the solutions for P for various temperatures T, at an exergy level of E. The derivative of $f(p)$ is found to be

$$f'(p) = -\frac{C}{P^2} + \frac{A}{P}. \tag{10.15}$$

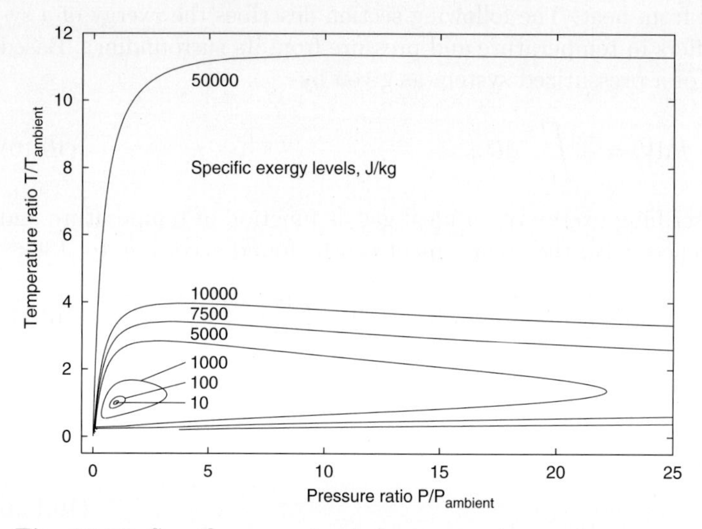

Fig. 10.12. Specific exergy levels for temperature and pressure ratios of the ideal gas. Note the increasing significance of the temperature contribution

A simulation yields Fig. 10.12. The temperature and pressure ratios are depicted for various exergy levels over four orders of magnitude. The graphs for low exergy levels are egg-shaped, whereas the graphs for relatively higher exergy values show the increasing significance of rising temperature over rising pressure. The pressure and temperature ratios contribute to the amount of exergy almost equally in the lower range, but for higher exergies the temperature is the dominant factor in determining the exergy content of an ideal gas.

All the graphs in Fig. 10.12 circle around the point (1,1). Temperature and pressure ratios smaller than one are possible, and the corresponding exergies do not become negative. This might sound contradictory at first, but can be understood with the previously mentioned idea of extractable energy when the pictured system is flooded by the environment. The ideal gas could be in a vacuum at low temperature.

A three-dimensional plot of the simulation data (Fig. 10.13) reveals the exergy valley of the ideal gas. The contour lines in the xy plane are similar in shape to those shown in Fig. 10.12.

Newton's method requires a guess of the first solution of the numerical problem. In a simulation, an initial guess for the pressure is entered and

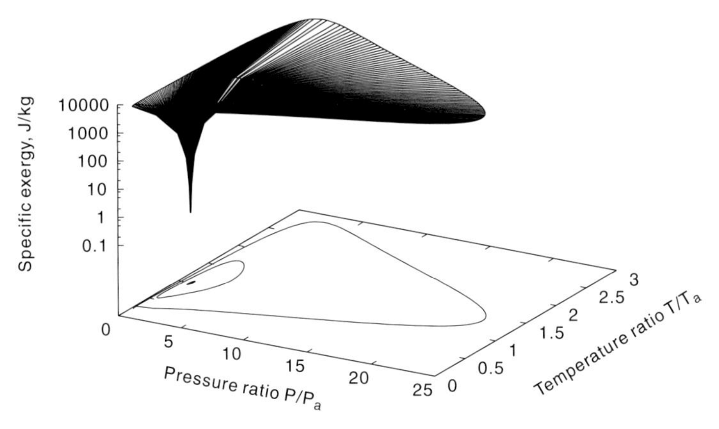

Fig. 10.13. The exergy valley. Specific exergy levels for temperature and pressure ratios of the ideal gas. The z axis is logarithmic

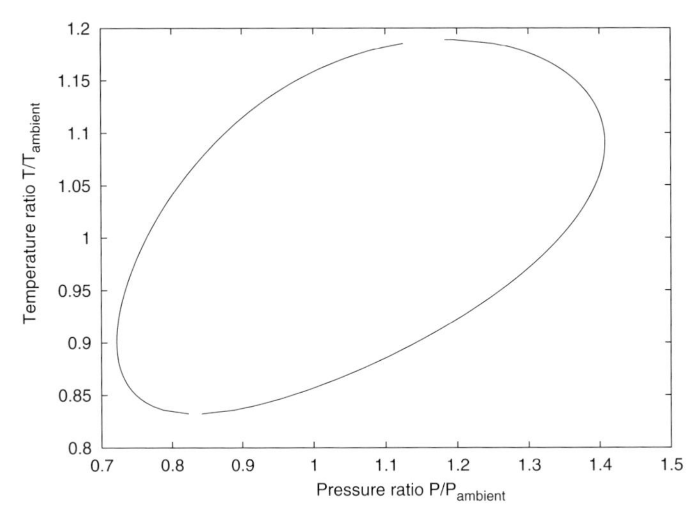

Fig. 10.14. Illustrating the sensitivity of finding solutions for the exergy of the ideal gas equation by Newton's method. Specific exergy level $E = 100\,\mathrm{J/kg}$. The left-hand side of the graph is created with normal ambient pressure and an initial guess of 2.0 for the resulting pressure, the right-hand side of the graph with very low ambient pressure and a 1.35 guess for the root

changed in a loop according to the results of (10.11) and (10.14). The result is evaluated along with its derivative (10.15). According to Newton's method a pressure is eventually found that satisfies the boundary conditions within the given tolerance.

Newton's method proves to be very powerful, but is also very sensitive to changes of the initial pressure value and the range of temperatures to be evaluated. The process of finding solutions requires good guessing and

Table 10.4. Values used in creating the exergy valley of the ideal gas by Newton's method, showing the sensitivity of the process. The values guessed for the initial pressure for the left-hand and the right-hand side of the graph correspond to ambient pressure P_a = 101 325 Pa and very low ambient pressure P_a = 1.01325 Pa, respectively

Specific exergy J/kg	T_{low} K	Step size -	P, left guess Pa	P, right guess Pa
$1.0 \cdot 10^{-1}$	293.15	1/100	2.0	1.1
$1.0 \cdot 10^{0}$	293.15	1/100	2.0	1.1
$1.0 \cdot 10^{1}$	283.15	1/50	2.0	1.2
$1.0 \cdot 10^{2}$	240.65	1/16	2.0	1.35
$1.0 \cdot 10^{3}$	160.15	1/5.5	2.0	1.715
$5.0 \cdot 10^{3}$	53.15	1/2.45	2.0	2.95
$7.5 \cdot 10^{3}$	23.15	1/1.95	2.0	5.425
$1.0 \cdot 10^{4}$	13.15	1/1.68	2.0	10.0
$5.0 \cdot 10^{4}$	0.00005	1.75/1	2.0	1000.0

patience. As seen in Fig. 10.12, the exergy equation for the ideal gas has two solutions for each temperature ratio within the boundaries of the constant exergy level. Newton's method approximates only one value at a time, and so the program has to be run twice in order to find the second root.

The second run requires another initial guess for the pressure, at a changed ambient pressure. This process is illustrated in Fig. 10.14 for the two sides of a graph resulting in the boundaries of a specific exergy of 100 J/kg of the ideal gas. The missing links are due to the step size for the temperature ratio (here 1/2000 over the assumed range of the graph) and to the tolerance, which was given as 10^{-6}.

Table 10.4 summarizes the values necessary for producing the graphs constituting the exergy valley. Note the sensitivity of the process, as the program has to deal with nonexisting solutions as well as small differences.

10.4 Exergy of a Concentrating Collector

Solar collector systems behave in a contradictory manner when evaluated by an exergy analysis, or a more conventional thermal analysis. A solar collector is exergetically best operated when its output temperature becomes closer to the reference, the radiance temperature of the sun at 5777 K. Conventionally, the collector reaches its highest thermal efficiency with a rather low output temperature, since thermal losses are minimized that way. Table 10.5 schematically expresses the two possible, yet contradictory, ways of optimizing a solar collector system. It appears that solar concentrators respond favorably to an exergy analysis due to their potential of delivering high output temperatures.

Our investigation reports on the exergetic principle of optimization for the case of a typical solar collector under test conditions.

Table 10.5. Energetical and exergetical optimization strategies for solar collectors

Energetically	Exergetically
Minimize losses	minimize losses
Minimize T_{output}	maximize T_{output}

Solar collectors for mid-range temperatures around 100–250℃ are generally built as evacuated tube-type collectors to minimize heat losses.

Incoming solar radiation is reflected or refracted with a small concentration factor of 1.1–1.5 onto an absorber. To avoid convective and conductive heat losses the concentrator is placed in an evacuated glass tube. Radiative losses are significant but can be restrained by choosing an appropriate selective surface with its temperature range of emittance exceeding the actual temperature of the absorber.

The most prominent example for an evacuated tube-type solar collector is the compound parabolic concentrator (CPC), where a parabolic shaped mirror reflects radiation within two acceptance half-angles onto an absorber in the center of the collector.

Exergy Balance of a Solar Collector

Following (10.9) one can define the exergy balance of a solar collector as the difference between exergy flow rates [169]:

$$\dot{E}_{\mathrm{o}} + \dot{E}_{\mathrm{i}} + \dot{E}_{\mathrm{l}} + \dot{E}_{\mathrm{a}} + \dot{E}_{\mathrm{s}} = 0\,. \tag{10.16}$$

For an explanation and illustration of these properties refer to Table 10.6 and Fig. 10.15.

The appropriate formulation of each exergy flow rate follows [169]. The exergy flow rate out of the system (unit collector, heat capacity c_{p} per collector area) is

$$\begin{aligned}\dot{E}_{\mathrm{o}} = & -c_{\mathrm{p}}\dot{m}\left(T_{\mathrm{o}} - T_{\mathrm{a}} - T_{\mathrm{a}}\ln\frac{T_{\mathrm{o}}}{T_{\mathrm{a}}}\right) \\ & + \frac{1}{A_{\mathrm{t}}}\left(\frac{1}{2}\frac{(A_{\mathrm{t}}\dot{m})^3}{(\rho A)^2} + \frac{\dot{m}\Delta P_{\mathrm{out}}}{\rho}\right)\,.\end{aligned} \tag{10.17}$$

Table 10.6. Exergy balance terms of a solar collector

$\dot{E}_\mathrm{o}$	exergy flow rate out of the system, with working fluid
$\dot{E}_\mathrm{i}$	exergy flow rate into the system, with working fluid, and with solar radiation
$\dot{E}_\mathrm{l}$	exergy leakage rate out of the system, losses due to heat transfer without substance due to evacuated tube-type collector
$\dot{E}_\mathrm{a}$	exergy annihilation rate, accounting for working fluid pressure drop, exergy losses during absorption of solar radiation, and exergy annihilation during heat conduction from absorber to working fluid
$\dot{E}_\mathrm{c}$	exergy storage rate, changes in the absorber without fluid flow

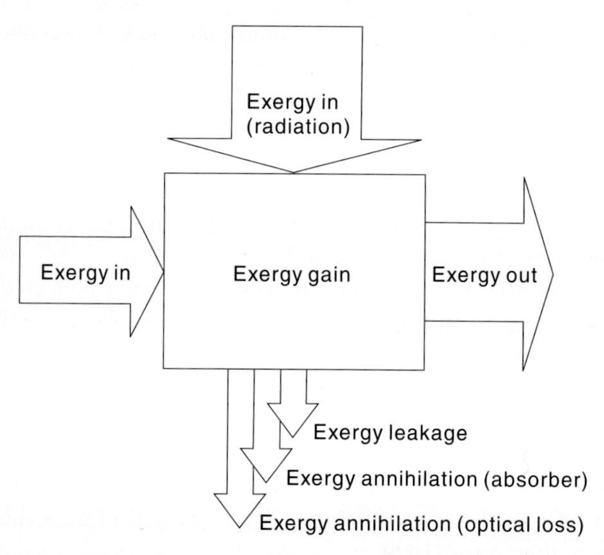

Fig. 10.15. Schematic of exergy balance applied to solar collector of evacuated tube-type

Exergy flows out of the system are as defined negative. The part on the second line in (10.17) refers to the static exergy due to the pressure of the fluid, which might have been provided by a pump. Solar collectors might use air as the working medium instead of the usual water or thermal oil. No longer the the medium can be assumed to be incompressible, and (10.11) describing the exergy of an ideal gas must be employed for $\dot{E}_\mathrm{o}$ and $\dot{E}_\mathrm{i}$. For our analysis we assume the working fluid to be water.

Exergy flowing into the collector system has two components. Exergy enters the absorber with the working fluid at a rate similar to $\dot{E}_\mathrm{o}$ except for the sign, and with $P_{\mathrm{in}} > P_{\mathrm{out}}$ due to a frictional pressure drop [44]:

$$\dot{E}_{\mathrm{i,f}} = c_{\mathrm{p}}\dot{m}\left(T_{\mathrm{o}} - T_{\mathrm{a}} - T_{\mathrm{a}}\ \ln\frac{T_{\mathrm{o}}}{T_{\mathrm{a}}}\right) + \frac{1}{A_{\mathrm{t}}}\left(\frac{1}{2}\frac{(A_{\mathrm{t}}\dot{m})^3}{(\rho A)^2} + \frac{\dot{m}\Delta P_{\mathrm{in}}}{\rho}\right) . \tag{10.18}$$

Solar radiation provides the second part of the exergy inflow. This exergy flow is defined by the solar radiance temperature $T_{\mathrm{s}} = 5777\,\mathrm{K}$, accounting for a large exergy input. One could criticize the exergy concept in this regard. It attributes high exergy input from the sun at a temperature that can be reached on earth only with the ideal collector concentration ratio of around 43 400. Solar radiation not intercepted by a collector eventually becomes completely destroyed, creating the reference temperature T_{a}. Thus, one could argue for comparing solar collector exergy efficiency with competing users, e.g. photosynthetic efficiency rather than theoretical concentration. This line of thought leads to the minimum size of the environment, to be defined as an exergy sink Exergy analysis only works if the assigned environment (the sink) is large enough to be a true reference point.

The collector's optical efficiency $(\tau\alpha)$ modifies the value of the unit collector's radiative intake:

$$\dot{E}_{\mathrm{i,r}} = (\tau\alpha)\,G_{\mathrm{T}}\left(1 - \frac{T_{\mathrm{a}}}{T_{\mathrm{s}}}\right) . \tag{10.19}$$

This equation closely resembles the efficiency of an ideal thermal machine creating mechanical energy (available work, exergy) from the heat transfer between two sources at temperatures $T_{\mathrm{upper}} > T_{\mathrm{lower}}$:

$$\dot{E} = \dot{Q}\left(1 - \frac{T_{\mathrm{lower}}}{T_{\mathrm{upper}}}\right) , \tag{10.20}$$

where $1 - T_{\mathrm{lower}}/T_{\mathrm{upper}}$ is again the Carnot factor [11, 57].

Undesired exergy losses from the collector are termed the exergy leakage rate $\dot{E}_{\mathrm{l}}$. The collector loss factor U_{L} for the unit collector, which depends on the mean absorber temperature T_{p}, is the decisive factor for heat transfer losses:

$$\dot{E}_{\mathrm{l}} = -U_{\mathrm{L}}\,(T_{\mathrm{p}} - T_{\mathrm{a}})\left(1 - \frac{T_{\mathrm{a}}}{T_{\mathrm{p}}}\right) . \tag{10.21}$$

In the case of an evacuated tube-type solar collector, U_{L} is almost exclusively defined by the emissivity of the absorber and the consecutive radiative losses. U_{L} can be considered constant for standard operation of most collectors. Values can be found in [37].

The exergy annihilation rate consists of three factors, namely the exergy annihilation rate due to the frictional pressure drop of the fluid $(\dot{E}_{\mathrm{a,p}})$, and the exergy annihilation rate attributed to absorbing solar radiation $(\dot{E}_{\mathrm{a,r}})$, and exergy annihilation during heat conduction from absorber to the medium ($\dot{E}_{\mathrm{a,c}}$). For the unit collector, the exergy destruction due to the pressure drop is

$$\dot{E}_{\mathrm{a,p}} = -\left(\frac{\dot{m}}{\rho A_{\mathrm{t}}}\right) \Delta P T_{\mathrm{a}} \frac{\ln\left(T_{\mathrm{o}}/T_{\mathrm{i}}\right)}{T_{\mathrm{o}} - T_{\mathrm{i}}} . \tag{10.22}$$

The most significant exergy annihilation process is the destruction of exergy while absorbing solar radiation per unit collector. At this point the large difference between the temperature of the sun T_{s} and the average absorber plate temperature T_{p} takes its toll, yielding the expression

$$\dot{E}_{\mathrm{a,r}} = (\tau\alpha)\, G_{\mathrm{T}} T_{\mathrm{a}} \left(\frac{1}{T_{\mathrm{p}}} - \frac{1}{T_{\mathrm{s}}}\right) , \tag{10.23}$$

where T_{p} stands for the mean absorber temperature. Exergy annihilation due to the heat conduction from the absorber to the fluid is, therefore, constant and can be described with the simplification of assuming a mean absorber temperature as

$$\dot{E}_{\mathrm{a,c}} = -c_{\mathrm{p}} \dot{m} T_{\mathrm{a}} \left(\ln \frac{T_{\mathrm{o}}}{T_{\mathrm{i}}} - \frac{T_{\mathrm{o}} - T_{\mathrm{i}}}{T_{\mathrm{p}}}\right) . \tag{10.24}$$

The last factor contributing to the exergy balance equation of a solar collector is the exergy change rate of the absorber, to be understood as a storage change, excluding the fluid, and is given by

$$\dot{E}_{\mathrm{c}} = -c_{\mathrm{abs}} m_{\mathrm{abs}} \left(1 - \frac{T_{\mathrm{a}}}{T_{\mathrm{p}}}\right) \frac{\mathrm{d}T_{\mathrm{p}}}{\mathrm{d}t} . \tag{10.25}$$

Combining (10.17)–(10.19) and (10.21)–(10.25) following (10.16) leads to

$$\begin{aligned} &\left(c_{\mathrm{p}} \dot{m} \left(T_{\mathrm{o}} - T_{\mathrm{i}}\right) - \frac{\dot{m}}{\rho A_{\mathrm{t}}} \Delta P\right) \left(1 - \frac{T_{\mathrm{a}} \ln\left(T_{\mathrm{o}}/T_{\mathrm{i}}\right)}{T_{\mathrm{o}} - T_{\mathrm{i}}}\right) \\ &\quad + c_{\mathrm{abs}} m_{\mathrm{abs}} \left(1 - \frac{T_{\mathrm{a}}}{T_{\mathrm{p}}}\right) \frac{\mathrm{d}T_{\mathrm{p}}}{\mathrm{d}t} \\ &= G_{\mathrm{T}} \left(1 - \frac{T_{\mathrm{a}}}{T_{\mathrm{s}}}\right) - \left(1 - (\tau\alpha)\right) G_{\mathrm{T}} \left(1 - \frac{T_{\mathrm{a}}}{T_{\mathrm{s}}}\right) + (\tau\alpha)\, G_{\mathrm{T}} T_{\mathrm{a}} \left(\frac{1}{T_{\mathrm{p}}} - \frac{1}{T_{\mathrm{s}}}\right) \\ &\quad + U_{\mathrm{L}} \left(T_{\mathrm{p}} - T_{\mathrm{a}}\right) \left(1 - \frac{T_{\mathrm{a}}}{T_{\mathrm{p}}}\right) + c_{\mathrm{p}} \dot{m} T_{\mathrm{a}} \left(\ln \frac{T_{\mathrm{o}}}{T_{\mathrm{i}}} - \frac{T_{\mathrm{o}} - T_{\mathrm{i}}}{T_{\mathrm{p}}}\right) . \end{aligned} \tag{10.26}$$

The efficiencies of solar collectors are often described in terms of their steady-state performance, i.e. the effects of starting up the system or of changing radiation or load are excluded. Thus, the term describing the changing storage of the collector itself becomes zero.

It can be shown that exergy destruction due to the frictional pressure drop is negligible [44]. Equation (10.26) for the steady-state operation of a unit collector and without pressure drop reduces to

$$\begin{aligned}
&c_{\mathrm{p}}\dot{m}\left(T_{\mathrm{o}}-T_{\mathrm{i}}\right)\left(1-\frac{T_{\mathrm{a}}\,\ln\left(T_{\mathrm{o}}/T_{\mathrm{i}}\right)}{T_{\mathrm{o}}-T_{\mathrm{i}}}\right)\\
&=G_{\mathrm{T}}\left(1-\frac{T_{\mathrm{a}}}{T_{\mathrm{s}}}\right)-\left(1-(\tau\alpha)\right)G_{\mathrm{T}}\left(1-\frac{T_{\mathrm{a}}}{T_{\mathrm{s}}}\right)+(\tau\alpha)\,G_{\mathrm{T}}T_{\mathrm{a}}\left(\frac{1}{T_{\mathrm{p}}}-\frac{1}{T_{\mathrm{s}}}\right)\\
&\quad+U_{\mathrm{L}}\left(T_{\mathrm{p}}-T_{\mathrm{a}}\right)\left(1-\frac{T_{\mathrm{a}}}{T_{\mathrm{p}}}\right)+c_{\mathrm{p}}\dot{m}T_{\mathrm{a}}\left(\ln\frac{T_{\mathrm{o}}}{T_{\mathrm{i}}}-\frac{T_{\mathrm{o}}-T_{\mathrm{i}}}{T_{\mathrm{p}}}\right)\,. \qquad (10.27)
\end{aligned}$$

The exergy balance (10.27) reminds us of (and can be transformed into) the most fundamental energy balance equation describing a solar unit collector, which is

$$c_{\mathrm{p}}\dot{m}\left(T_{\mathrm{o}}-T_{\mathrm{i}}\right)=(\tau\alpha)\,G_{\mathrm{T}}-U_{\mathrm{L}}\left(T_{\mathrm{p}}-T_{\mathrm{a}}\right)\,. \qquad (10.28)$$

Both the exergy balance equation (10.27) and the energy balance equation (10.28) can be transformed to express the efficiencies. The reference point for the efficiency is the incident solar radiation in its respective description. Using the basic definition

$$\varepsilon=\frac{\dot{E}_{\mathrm{out}}}{\dot{E}_{\mathrm{in}}}\,, \qquad (10.29)$$

the dimensionless exergetic efficiency ε of a solar unit collector can be expressed as

$$\begin{aligned}
\varepsilon&=\frac{c_{\mathrm{p}}\dot{m}\left(T_{\mathrm{o}}-T_{\mathrm{i}}-T_{\mathrm{a}}\,\ln\left(T_{\mathrm{o}}/T_{\mathrm{i}}\right)\right)}{G_{\mathrm{T}}\left(1-T_{\mathrm{a}}/T_{\mathrm{s}}\right)}\\
&=1-\left(1-(\tau\alpha)\right)+(\tau\alpha)\,\frac{1/T_{\mathrm{p}}-1/T_{\mathrm{s}}}{1/T_{\mathrm{a}}-1/T_{\mathrm{s}}}+\frac{U_{\mathrm{L}}\left(T_{\mathrm{p}}-T_{\mathrm{a}}\right)}{G_{\mathrm{T}}}\frac{1-T_{\mathrm{a}}/T_{\mathrm{p}}}{1-T_{\mathrm{a}}/T_{\mathrm{s}}}\\
&\quad+\frac{c_{\mathrm{p}}\dot{m}T_{\mathrm{a}}}{G_{\mathrm{T}}\left(1-T_{\mathrm{a}}/T_{\mathrm{s}}\right)}\left(\ln\frac{T_{\mathrm{o}}}{T_{\mathrm{i}}}-\frac{T_{\mathrm{o}}-T_{\mathrm{i}}}{T_{\mathrm{p}}}\right)\,. \qquad (10.30)
\end{aligned}$$

For the energetic efficiency η of a solar unit collector (using an average absorber temperature) we obtain

$$\begin{aligned}
\eta&=\frac{c_{\mathrm{p}}\dot{m}\left(T_{\mathrm{o}}-T_{\mathrm{i}}\right)}{G_{\mathrm{T}}}\\
&=F_{\mathrm{p}}\,(\tau\alpha)-\frac{F_{\mathrm{p}}U_{\mathrm{L}}\left(T_{\mathrm{p}}-T_{\mathrm{a}}\right)}{G_{\mathrm{T}}}\,. \qquad (10.31)
\end{aligned}$$

The variable F_{p} here in a simplification represents the collector efficiency factor F' for short collectors, where the temperature rise between T_{i} and T_{o} can be assumed to be linear. This step is taken in order to use the mean absorber temperature T_{p} in the ongoing calculations. The collector efficiency factor F' physically is the ratio of the heat transfer resistance from the fluid to the ambient air to the heat transfer resistance from the absorber plate to the ambient air [37]. The latter resistance has been called U_{L}. Typical values for these parameters are given in Table 10.7.

Table 10.7. Typical parameters of an evacuated tube-type solar collector; values for flat plate collector for comparison[a]

Solar collector	$(\tau\alpha)$	U_L	F'
	-	W/Km^2	-
Evacuated tube	0.47	1.1	0.99
Flat plate	0.82	5.0	0.97

[a] Source: [37].

Comparison of Energetic and Exergetic Efficiencies

Analysis of solar collectors should always start with the incident solar radiation. Solar insolation should be measured as the total irradiation G_T, in Watt, on a tilted plane. A particularly sunny day is illustrated in Fig. 10.16; the data is from Kyushu, Japan. On this date, 19 March 1990, the direct fraction of the total irradiation exceeded 70%, and there were hardly any clouds during the five minute intervals when data was taken.

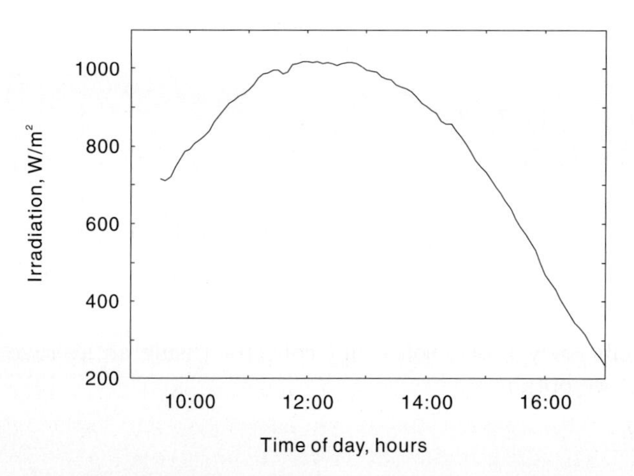

Fig. 10.16. Total solar irradiance on a plane tilted 33° towards south (latitude tilt) over time on 19 March 1990; five minute interval data on Kyushu Island, Japan

Data has been provided for a series of tests that were carried out with a prototype compound parabolic concentrator. Both exergetic and energetic efficiencies of the evacuated tube-type solar collector were calculated according to (10.30) and (10.31), and are plotted in Fig. 10.17. The data is for the irradiation in Fig. 10.16, after eliminating disturbing influences of changing irradiation due to cloud cover.

The efficiencies are plotted versus $(T_p - T_a)/G_T$ to account for ambient conditions, thus making collector testing under changing conditions possible.

The exergetic efficiency ε is plotted ten times its original size to make it easier to follow. The absolute values of both the energetic and exergetic efficiencies have no significance for our case.

The curves in Fig. 10.17 have been fitted by linear (η) and nonlinear (ε) curves using the Marquardt–Levenberg Algorithm for least-square fitting. The exergetic efficiency naturally runs through the point (0,0); the energetic efficiency reaches its highest point when there are no heat losses (T_p and T_a are equal). At this point, the energetic efficiency is equal to the optical efficiency of the collector.

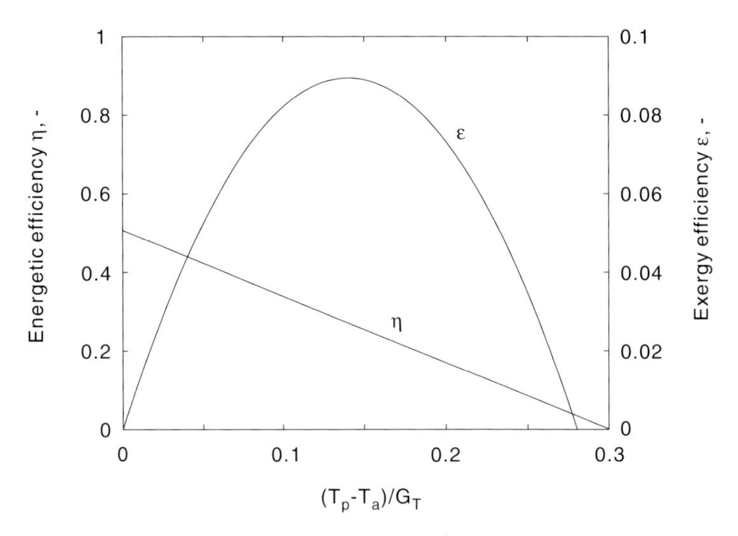

Fig. 10.17. Exergetic and energetic efficiencies for an evacuated tube-type solar collector; test data for a CPC on 19 March 1990 on Kyushu Island, Japan

As we can see from Fig. 10.17, the exergetic efficiency allows us to set a point of optimized operation, defined as the difference between the collector plate temperature T_p and the ambient temperature T_a, at a given irradiance G_T. The energetic efficiency is helpful for comparing different solar collectors and collector types, but it does not tell us anything about the optimal operational mode of a collector field made of one type of solar concentrator. This is the advantage and greatest merit of exergy analysis for a solar collector.

Further analysis of the exergetic properties of the solar concentrator reveals where exergy is lost and gained (following the concept of the black box in Fig. 10.15 and the terms in Table 10.6). Figure 10.18 shows the composition of the exergy losses, plotted as cumulative losses added to the exergetic gain of the system.

The data used is the same as that used in Fig. 10.17. Again, the curves have been fitted. The main losses are the exergy annihilation at the absorber and the optical losses. The annihilation of exergy indicated in this figure

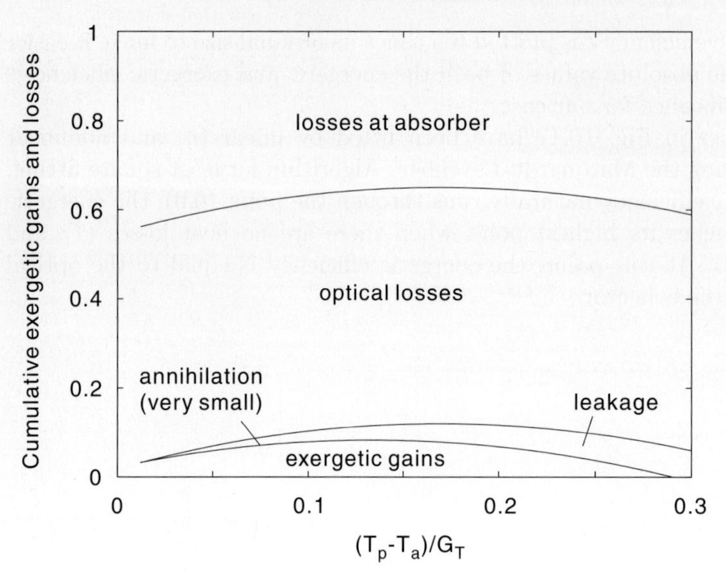

Fig. 10.18. Normalized cumulative exergetic gains and losses for an evacuated tube-type solar collector; test data for a CPC on 19 March 1990 on Kyushu Island, Japan

refers only to the exergy annihilation due to the heat transfer between the absorber and the working fluid. Note that the maximum for exergetic gains coincides with the maximum exergetic efficiency.

An energetic analysis cannot supply this wealth of information regarding the performance of a system. The exergetic analysis performed here has certainly proved itself to be worth the extra effort in computation, as it provided us with an insight into the performance parameters of a solar concentrator, and allowed us to set the optimum strategy for its operation.

Some care has to be taken when systems at different reference temperatures are to be compared. The exergy concept fails for these cases. One might consider an entropy analysis instead. Apart from this problem, a thorough exergy analysis should be added to studies analyzing a thermodynamic system, when otherwise the reasons for poor performance are unclear or an energetic analysis does not deliver strong clues regarding efficient operation. Exergy analysis is certainly a powerful tool.

11 Solar Concentration in Space

11.1 Space Concentrator Arrays

Solar concentrators go space. Lens and mirror-based solar concentrators have recently begun to boost photovoltaic power supplies for satellites in space. In 1998, the first mission carrying solar concentrators was launched. Deep Space 1 uses lenses of medium concentration at 7X (suns). The probe's main objective is the validation of three novel technologies. Besides the test of an autonomous navigation system, this means the successful demonstration of:

- solar electric propulsion (SEP), which is an ion-propulsion engine [26]. Xenon is ionized, then accelerated electrostatically, and exits at high speed. While its thrust (20–90 mN) is small in comparison to the impulse obtained with conventional chemical reactions, the engine is ten times more efficient: 81.5 kg of fuel are expected to last for 8000 hours of operation, allowing for a total change of speed of 3.6 km/s;
- solar concentrator arrays with refractive linear element technology (SCARLET), which provide the high voltage electric power for the ion-propulsion engine [114].

Lens Concentrators

In many ways, the Deep Space 1 mission is uniquely qualified to employ solar concentrators. The probe is on a roughly heliocentric orbit, and must be able to change its position in order to fulfill the goal of multiple fly-bys of interplanetary objects. The ion-propulsion engine used for these accelerations and orbital changes needs a strong electric power source.

The photovoltaic effect in a conventional semiconductor is observed only if the irradiation level exceeds a minimum value, which is approximately equal to the radiation flux at a distance of 1.5 AU from the sun. One astronomical unit (AU) is the mean distance between the earth and the sun. Radiation originating from the sun is diluted by the squared distance it travels. The solar radiation flux at 1.5 AU limits the efficient use of conventional flat plate photovoltaic systems to locations within the radius of the planet Mars.

Optical concentration can overcome the limit of flux dilution. Additionally, concentration increases the efficiency of the solar cells used (discussed in Sect. 9.4). Potentially, the concentrator is lighter than the photovoltaic cells, although the Deep Space 1 solar wings produce electricity at a power-to-weight ratio of only 60 W/kg, which is quite similar to the specific power delivered by the advanced flat plate system using the same triple-junction cells.

Weight reduction is a leading incentive for developing solar concentrators for space. The cost of launching one kilogram into space is about US$ 11 000–66 000 for low earth orbits (LEO) and geosynchronous orbits (GEO), respectively. In addition, limited stowage space demands tight packing of the solar concentrator panels, with strong implications for the design of the concentrator.

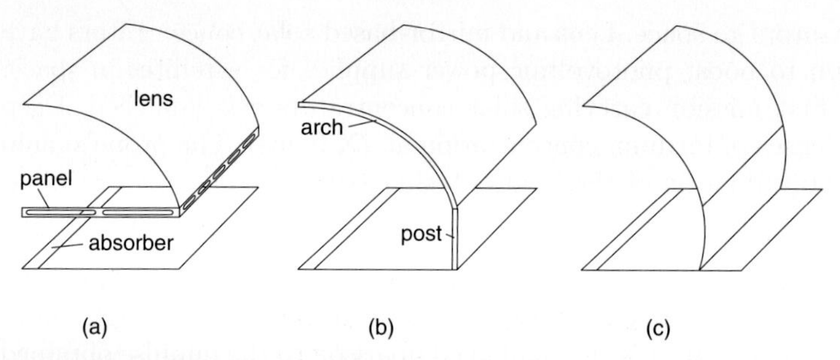

Fig. 11.1. Refractive concentrator concepts for solar photovoltaic power conversion in space, right-hand side of lenses shown only: (**a**) self-supporting Fresnel lens made of silicone laminated under glass (the panel and lens collapse for storage onto the absorber, similar to state-of-the-art SCARLET concept [40] launched in 1998 on Deep Space 1); (**b**) stretched Fresnel lens silicone film (the arch and post collapse with the lens onto absorber [130]); (**c**) inflatable Fresnel lens [132]. In contrast to the elliptical shape of lenses (a) and (b), the unsupported inflatable shape of lens (c) tends to be circular and not optimal. Note that all lenses have linear geometry. Schematic, not to scale

The SCARLET lens concept pictured in Fig. 11.1a, and described in Table 11.1 below, is the result of a long evolutionary process, which began with the invention of the arched bifocal Fresnel lens by O'Neill in 1978 [126]. Some five generations of terrestrial lenses preceded the space-qualified lens used for SCARLET [40]. As opposed to the schematic in Fig. 11.1a, the actual SCARLET lens is truncated, not at half-height over the absorber ($y = 0.5f$), but already approximately at $y = 0.8f$.

Due to this truncation, the geometrical concentration ratio of the linear lens is 8.0. Taking into account losses, the optical concentration ratio of the SCARLET lens is 7.14; the typical flux in the focal area is about 20X. Only part of the triple-junction cells constituting the absorber is brightly illuminated, a common feature of refractive concentrators (compare Fig. 9.4). The acceptance half-angle of the SCARLET lens is $\theta = \pm 2.25°$ to accommodate tracking errors. High tracking accuracy and reliability of the tracker is es-

Table 11.1. Solar photovoltaic concentrator technologies in space, aboard the satellites Deep Space 1 and Galaxy XI. Data for the nonconcentrating solar arrays of the International Space Station (ISS) are listed for comparison[a]

	Deep Space 1	Galaxy XI	ISS solar power module
Launch date	24 Oct 1998	14 Dec 1999	30 Nov 2000[b]
Mission	technical validation	communications satellite	manned space station
Orbit	heliocentric, 1–1.5 AU[c]	geostationary orbit (GEO), 36 000 km	low earth orbit (LEO), 400 km
Orbital period, min	–	1436	92.5
Mass, kg	489	< 5200	15 750
Number of solar wings	2	2	2[d]
Wing size (length × width), m	5.2 × 1.6	15.5	32.6 × 11.6
Power at 1 AU, kW	2.5	15	31
Power, end-of-life, kW	–	> 10	–
Solar concentrator type	linear Fresnel lenses	linear angled foil reflectors (boosters)	–[e]
Concentration ratio, optical (geometrical)	7.14 (8.0)	1.5 (1.8)	1.0
Acceptance half-angle[f], °	±2.25	±30	±90
Photovoltaic cell type	$GaInP_2$/GaAs/Ge	GaAs/Ge	Si
Cell efficiency, %	23	22	–
Maximum operating temperature, ℃	< 70	130–140[g]	–
Fill factor, %	81	–	–

[a] Sources: [17, 114, 130, 118, 167].
[b] Launch of Zarya module (the first for the International Space Station) on 20 November 1998. Solar power module delivered by Space Shuttle.
[c] Astronomical Units, mean distance earth–sun, 1 AU= $1.49597870 \cdot 10^{11}$ m.
[d] 8 wings total by 2006.
[e] No photovoltaic concentration planned, but solar thermal concentrators are under development (see text).
[f] Linear systems, only one acceptance half-angle pair considered.
[g] Design maximum 160℃.

sential for concentrating systems. For angles of incidence larger than the acceptance half-angles, the refracted rays miss the absorber, which leaves the photovoltaic cells, and hence the power supply of the satellite, literally in the dark.

The stretched-film Fresnel lens in Fig. 11.1b is intended to succeed the rigid SCARLET lens. The latter is made from 0.25 mm thick silicone, laminated under a 0.08 mm ceria-doped glass superstrate, to keep its shape. The foldable lens consists of a single layer of space-qualified silicone, namely Dow Corning DC93-500. The material favored for terrestrial applications, polymethylmethacrylate (PMMA), does not survive the strong ultraviolet radiation in space. The transmittance of the silicone material is good, and the $GaInP_2$/GaAs/Ge cells are illuminated over the whole range of usable wavelengths in the solar spectrum, 360–1800 nm.

The stretched-film lens is lighter than the rigid lens; a prototype presented in 2000 reportedly achieved a power-to-mass ratio of > 300 W/kg, for the panel only [133]. The stowage requirements are smaller due to the foldable nature of the lens, which collapses flat onto the absorber plate. The rigid lens and its supporting panel are lowered onto the absorber, too, but remain high in comparison with the folded lens, which represents the ultimate stowage solution.

An inflatable Fresnel lens (Fig. 11.1c) has been proposed by the same authors [132]. The inflated lens offers the advantages of less weight, reduced cost, and less fragility over its rigid brother. Possibly, deployment in space would be easier, as there are few moving parts, but this advantage may be offset by the novelty of having to deliver make-up gas to the solar wings. The inflatable lens was proposed in 1997, in response to the first successful deployment of an inflatable mirror in space. The lens in Fig. 11.1c requires some structural elements to achieve the same optimum elliptical shape of the lenses (a) and (b).

A problem concerning all structures in space in general, and inflated concentrators in particular, is micrometeroid puncturing, requiring small amounts of make-up gas.

Mirror Concentrators

Galaxy XI, a commercial telecommunications satellite launched in 1999, uses wing foils as reflective boosters to increase its photovoltaic electricity output by 50% to about 15 kW, making it the highest-powered satellite of its time. Like Deep Space 1, Galaxy XI carries a xenon ion-propulsion system; 5 kg of fuel annually are sufficient for stationkeeping.

Table 11.1 lists some of the power supply characteristics of the Deep Space 1 probe, the Galaxy XI satellite, and the International Space Station (ISS). The latter does not carry any photovoltaic concentrators, but an experiment is planned which will focus on the use of a mirror concentrator

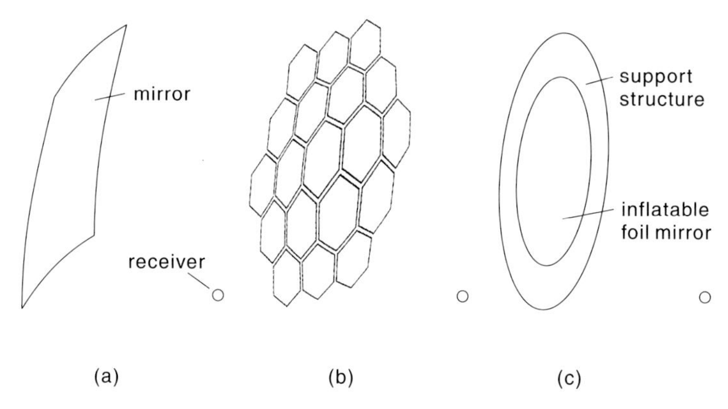

Fig. 11.2. Reflective concentrator concepts for solar thermal power conversion in space: (**a**) parabolic mirror intended to supply heat for Stirling engine in the Japan Experimental Module of the International Space Station [115]; (**b**) Fresnel mirror of solar dynamic power module operating a closed Brayton cycle, for the International Space Station [145]; (**c**) inflatable parabolic mirror for space power generation [174], deployed as state-of-the-art antenna in 1996. Note that all mirrors are three-dimensional point-focusing devices. Schematic, not to scale

intended for cascaded power conversion with a Stirling engine [115]. A schematic is given in Fig. 11.2a.

In the future, the growing power demand of the space station will be satisfied by so-called solar dynamic power modules, which are solar thermal power converters. Typically, Fresnel mirrors are configured as in a terrestrial dish-Stirling system. The concentrators are expected to deliver temperatures high enough to drive a closed Brayton cycle [145], i.e. a closed gas turbine for power generation. A concentrator similar to the one pictured in Fig. 11.2b has undergone extensive terrestrial tests and is deemed fit for space applications [95].

The inflatable mirror of Fig. 11.2c had been successfully deployed from the Space Shuttle in an antenna experiment in 1996. An extensive list of references to stretched membrane mirrors is given in [130]. Some of the characteristics of the systems pictured in Fig. 11.2(a)–(c) are collected in Table 11.2.

The inflatable mirror of Fig. 11.2c is an imaging off-axis parabolic concentrator with a half-angle $\theta = 30°$. The parabolic shape is created by inflating a balloon-like structure: the solar rays pass through a transparent foil which faces the sun. The rays are then reflected at a mirrored foil forming the back of the system before they pass the first foil once more on their way to the receiver. This basic geometry is shown in Fig. 11.3. The focal length of the parabola is [182]

$$f = a' (1 + \sin\theta) \ . \tag{11.1}$$

Table 11.2. Solar dynamic power systems for space applications[a]

	Parabolic mirror	Fresnel mirror	Inflatable mirror
Purpose	cascaded thermal power experiment on the ISS	closed Brayton gas turbine cycle for the ISS	antenna, thermal power
Status	conceptual	terrestrial test	experimental deployment 20 May 1996
Dimensions (height × width), m	1.5 × 0.86	2.5[b]	2.0 × 3.0
Secondary concentrator	CPC[c] cavity	possibly sapphire refractive secondary [63]	–
Cycle efficiency[d], %	–	29	–

[a] Sources and visualizations: see Fig. 11.2.
[b] Diameter of the assembly of seven Fresnel mirrors.
[c] Compound parabolic concentrator.
[d] Rated power is measured as electric output of the cycle; the collected heat can be calculated by multiplying the area of the mirror by the extraterrestrial irradiance constant at 1 AU, 1367 Wm^{-2}.

The notation follows that given for the edge ray principle in Sect. 2.1, with the entry aperture denoted by $2a$, the exit aperture by $2a'$, and the half-angle by θ. The parabola in Fig. 11.3 is the same as that which forms one side reflector of a compound parabolic concentrator (CPC). The figure can be understood as showing the right-hand side of a CPC in the cross-sectional plane, and tilted by 90°.

The procedure to determine the shape of the parabola segment is not trivial, and should be followed in [182]. Also, it should be noted that in contrast to the CPC, the off-axis parabolic concentrator is an imaging device. The overall length of the system is given by

$$l = a' (1 + \sin\theta) \frac{\cos\theta}{\sin^2\theta} . \tag{11.2}$$

We find that the distance a' between the mirror and the receiver F is reasonably small, allowing for storage and deployment. The off-axis concentrators discussed here allow the placement of the receiver on the body of the satellite. Neither mirror nor receiver obstruct the optical path.

The parabola segment is a truncated part of a full parabolic curve. This leads to an optical disadvantage. Since the mirror segment is not placed normal to the sun, it intercepts solar radiation only over an area filled by the

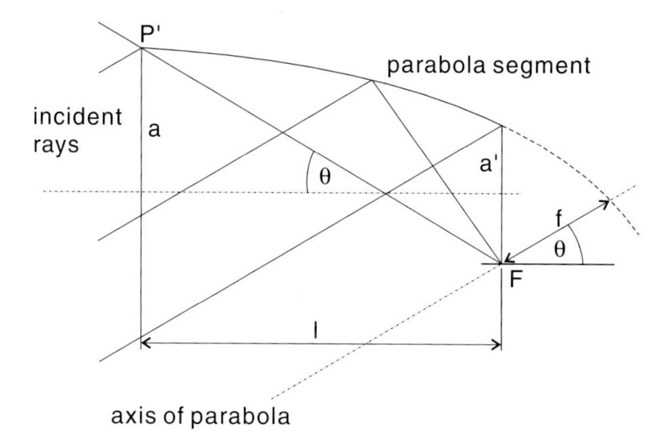

Fig. 11.3. Imaging off-axis parabolic concentrator of half-angle $\theta = 30°$. Light enters from the lower left side; the receiver is placed at point F. Two-dimensional view, following [182]

mirror's projection onto the plane normal to the incident rays. In Fig. 11.3, this area has a width equal to the normal distance between the two extreme rays. The phenomenon is somewhat equal to the cosine losses discussed in connection with radiation on a horizontal plate, and cannot entirely be avoided. In practice, the mirror material must be lighter than, say, photovoltaic panels covering the projected area, to result in an overall weight reduction.

The amount of radiation intercepted, and not simply mirror or lens area, determines the performance and the success of the concentrator system.

11.2 Design Challenges in Space

Solar concentrators for space applications face design challenges which are quite foreign to the designer of concentrators for the terrestrial collection of solar energy. Foremost, there is the fact that solar power conversion in space is vital to the survival of the satellite. A loss of power, even momentarily, can have catastrophic consequences. While the survival of humans on earth depends on solar power in the long term, space solar power is a vital non-hazardous source of energy available for future manned missions taking any length of time. Virtually all unmanned satellites in earth orbit depend on solar power.

Reliability affects the design of any power system. Flat plate photovoltaic systems, although proven for almost 50 years, do occasionally fail to deploy or do not operate properly. Concentrators are more fragile and complex, and require an eloquent system of cables and hinge lines for deployment of the assembly, which is folded and stowed during launch. This raises the potential for mechanical failures.

The design challenges for solar concentrator systems in space are the results of conditions which may be grouped into mechanical and environmental factors. Some of these will be discussed in more detail below; for more complete accounts refer to [168, 109], and other chapters in that handbook, as well as to [56]. The design challenges are:

- Mechanical challenges
  - accurate tracking
  - structural integrity
  - safe stowage
  - proper deployment
- Space environment
  - operating temperature and thermal cycling
  - ultraviolet radiation
  - charged particle radiation
  - solid particles and micrometeorites
  - electrostatic charge built-up
  - residual gas atmosphere (LEO) or high vacuum otherwise.

The space environment requires the use of space-qualified materials, and technical solutions for reliable operations. Note that cost is a factor usually addressed by means of the mass (weight) of the system, which is of equal importance to the mechanical durability, longevity, and operational reliability of the concentrator system.

Tracking

Once the concentrator is unfolded and successfully deployed, tracking of the sun becomes the main concern of the operator. Flat plate systems have an acceptance half-angle for solar angles of incidence of up to 90°, and will work quite well for angles of incidence below 60°. Concentrators, on the other hand, have acceptance half-angles of 2° or less. Imaging devices are very sensitive to incidence other than paraxial rays. While lenses are partially self-correcting for angular error, the misalignment of a ray reflected by a mirror is approximately twice as large as the error of the surface (Sect. 6.1).

Tracking errors may have two possible extreme effects: total loss of vital power, and destruction of the area close to the receiver due to the high flux and temperatures generated by the solar concentrator.

M. J. O'Neill has pointed out that the accuracy of the tracking mechanism for mirror concentrators must be one hundred times that for lenses. The underlying assumptions are that the mirror is used for thermal power conversion, while the lens concentrator works with photovoltaics. Although there is no technical reason why this distinction should be made, Sect. 11.1 has confirmed our introduction that in practice there is a division along these lines.

Now, if the goal is to achieve an energy conversion efficiency for sunlight into electricity of 30%, the photovoltaic concentration ratio with a nonimaging lens over advanced multijunction devices should be around 20X, while the thermal system must reach temperatures that require concentration ratios of close to 1000X (this is actually the case for the system in Fig. 11.2a). Accounting for the lens' advantage of angular self-correction with a factor of two, we find that the nominal concentration ratio of mirrors for thermal electricity generation should be one hundred times greater than the concentration ratio of lenses for photovoltaic conversion of sunlight.

The maximum concentration of refractive systems is limited by dispersion of the solar spectrum, but this effect is negligible for lenses of 20X, as shown earlier. Mirrors of high concentration have to track the sun in two axes, whereas one-axis tracking can be sufficient for linear lenses.

Splitting the required tracking accuracy into two axes turns out to be an advantage. For the nonimaging mirror concentrator, the allowed error for each axis must not exceed $\pm 1.8°$. The value is found by solving the ideal concentration equation $1/\left(\sin^2 \theta\right)$ for the acceptance half-angle θ. This angle must equal the maximum tracking error under the edge ray principle (Sect. 2.1). Including a margin of two for the beam spread error, the tracking accuracy is $\pm 0.9°$. Imaging optics is very sensitive to changes of incidence angles, and the effects are nonlinear. It is hardly possible to give anything more meaningful than the rules of thumb previously noted, e.g. in Table 8.1.

To give an example [118], the rotational pointing accuracy of the International Space Station's Bearing, Motor and Roll Ring Modules (which are responsible for rotating the solar wings of the ISS) is $\pm 1°$. With an orbital period of 92.5 minutes, the rotational tracking rate is 4°/min. The tracking system can become complex, if electricity or thermal energy have to be transported through roll rings or rotating joints, respectively.

Structural Design and Other Mechanical Considerations

The launch of the spacecraft is the most dangerous phase in the life of the concentrator, as it will be exposed to high acoustic and vibrational levels. Concentrators are tested on the ground to prove the technology and minimize in-flight risks. The ground-testing procedure for the solar wings of Deep Space 1 is described in [114]:

1. visual inspection test: workmanship, cleanliness, integrity;
2. physical test: weight, interfaces;
3. electrical functional test: power, isolation, capacitance;
4. deployment test: tiedowns, kinematics, alignment;
5. deployment stiffness and strength test: out-of-plane stiffness and strength, electrical functional test;
6. thermal cycle test: three cycles −123° to 113℃ (12 out of 360 SCARLET lens glass superstrates cracked);

7. visual inspection, passive electrical and deployment test: complete deployment and electrical functional test;
8. acoustic test: overall sound pressure level of 141.6 dB (31.5–10 000 Hz) for one minute (7 lenses out of 360 cracked);
9. visual inspection, passive electrical and deployment test: complete deployment and electrical functional test;
10. random vibration test (see also [161]): pre-test to match resonant frequency, maximum value $0.04\,g^2/\mathrm{Hz}$, at 20, 50, and 500 Hz, where $g = 9.80665\,\mathrm{m/s^2}$ is the standard acceleration due to gravity (3 out of 360 lenses cracked);
11. visual inspection and deployment test of the complete system;
12. electrical functional test.

The number of cracked lenses refers to cracked glass superstrates. The shape of the lens is kept by the silicone substrate, and the photovoltaic concentrator remains functional. These less-than-perfect results are cited as reason for the development of the stretched lens [130].

Space Environment

Another reason for using the stretched lens is the thermal expansion coefficient of silicone (0.3%/K), which is three times higher than that of glass. A solar panel in a geosynchronous orbit passes through the earth's shadow, or eclipse, several thousand times. The temperature of the array will vary between –180℃ and 80℃ or more. The contraction and expansion requires the concentrator to be given some freedom of movement within a permanent fixture.

The conditions of the space environment affect the design of the solar power system of the satellite in terms of lifetime durability. Space environment conditions are very demanding on all parts of the spacecraft, and some points have been identified to be significantly influenced by the presence of solar photovoltaic concentrators, i.e.

- the lens concentrator acts as a radiation shield for the photovoltaic panel, minimizing radiation damage to the cells, raising the typical end-of-life conversion efficiency;
- overheating may be destructive to cells or wing structures. Heat dissipation in space is only possible by heat conduction in the cells through an electric isolator to the wing structure, and by radiation into space. Thermal control is difficult, and occasionally, cell temperatures reach the limit of efficient cell operation (as with the 160℃ cell temperature in Galaxy XI). Tracking misalignment may damage structures near the cell or thermal receiver;
- high-powered satellites > 10 kW apparently are vulnerable to an electromagnetical interaction between the array and solar storms, leading to electrical malfunctions, shorting out circuits [173]. Solar concentrators increase the

electrical potentials locally and over whole circuits of powerful solar cells, and thus may contribute to the likelihood of such events;
- any object placed into the ionosphere and exposed to the thermal plasma there will develop an equilibrium charge different from zero. This effect is called charging. Experimentally, potentials of up to −10 kV have been measured, and current pulses have been observed [109]. Lower earth orbits are more prone to these effects due to the auroral activity there. The solar concentrator itself may be charged, and thus increase the electrical potentials between cell and lens, possibly leading to discharges and short-circuits.

There is no actual evidence for the contribution of concentrators towards the last two points. Galaxy XI, the only high powered satellite with solar concentrators, has been operating flawlessly in this respect, but its concentrators are booster wings, not lenses with much higher flux densities. Also, the craft is in a geostationary orbit, where charging is less problematic.

Increased radiation shielding is often seen as an advantageous side effect of the solar concentrator in space. The concentrator calls for the use of III–V solar cells, e.g. gallium arsenide-based multijunction cells, which offer higher radiation hardness than silicon cells in the first place. Second, the concentrator acts as a physical shield against damaging radiation, and may replace the currently used coverglass over the cell. This has the potential for weight reductions at increased performance.

11.3 Lenses and Mirrors!

Lenses or mirrors? Photovoltaic or thermal concentration? These were the questions of the introductory Sect. 1.1. A full book later, they have not been answered. Instead, the stereotypes of using lenses for photovoltaic concentration, on the one hand, and mirrors for thermal applications, on the other hand, have been reinforced in the previous sections on solar concentration in space.

What is the reasoning behind using lenses only for line-focusing photovoltaic concentration, and why have mirrors exclusively (Galaxy XI aside) been developed for point-focusing on thermal generators? History and tradition, if you would ask us.

The technical advantages of either option are not easy to find; if pressed we opt for slightly better performance under realistic conditions, and cheaper manufacturing of the nonimaging Fresnel lens, when compared to reflective concentrators. First, let us note that neither simple imaging Fresnel lenses, nor spherical or parabolic mirrors are optimum concentrators. The future of concentrators began with the development of nonimaging mirrors and nonimaging lenses. Right from the start, the design of nonimaging devices deals with the peculiarities of solar concentration, i.e. solar disk size and the earth's movements, which require tracking of the sun along one or two axes or large acceptance half-angles in stationary operation.

Regarding the differences between nonimaging lenses and nonimaging mirrors we found that

- lenses are more forgiving for angular incidence errors, which could be caused by tracking errors, or shape distortions;
- the steps of a Fresnel lens can be manufactured with higher accuracy than the large gradient surfaces of a mirror;
- Fresnel lenses may be designed as 'fast', or with an f/number or aspect ratio as low as 0.5. Mirrors require truncation for low aspect ratios (or an unrealistically high number of reflections);
- both lens and mirror materials are available that are high quality, UV-resistant, and qualified for space;
- both concepts can have approximately the same optical performance, as the transmittance losses of the lens are often quite similar to the reflection losses of a mirror. For shallow nonimaging mirrors with close to single reflections of the incident rays, the optical performance may be better than for lenses;
- plastic lens materials can be cheaper, and lenses can be manufactured at lower cost and higher volumes, than metal mirrors. Potentially, reflective foils can close this gap;
- mirrors are more attractive for the beginner in the field. While single reflectors may be produced from scratch, the diamond lathe and molding process of Fresnel lenses represent a high entry barrier. The solar performance of readily available imaging lenses tarnishes their reputation;
- metal mirrors are to be chosen for niche applications, e.g. inside evacuated tube-type concentrators (due to the high temperature present when the evacuation is carried out in manufacturing) or for some applications in optoelectronics, where extremely short wavelengths prohibit the use of lenses;
- dispersion is a limiting factor for refractive lenses of high concentration, but so are shape distortions for highly concentrating mirrors.

There may be other similarities and differences, but we would like to stress that the nonimaging lens performs better than most potential users used to think. Thus, we conclude that the question *lenses or mirrors?* is purely rhetorical. The answer, however, is not the common and correct *this depends*... Rather, for crying out loud, it is *solar concentrators!*

References

1. K. Akhmedov, R.A. Zakhidov, and S.I. Klychev. Optical and energy characteristics of round Fresnel lenses with flat bands. *Geliotekhnika*, 27(1):48–51, 1991. Translated into English in *Applied Solar Energy* 27(1):43–46, 1991.
2. G. Alefeld. Problems with the exergy concepts (or the missing second law). *Newsletter of the IEA Heat Pump Center*, 6(3):19–23, 1988.
3. V.M. Andreev. Concentrator solar cells. In V.M. Andreev, V.A. Grilikhes, and V.D. Rumyantsev, editors, *Photovoltaic Conversion of Concentrated Sunlight*. Wiley, Chichester, 1997.
4. V.M. Andreev, V.M. Lantratov, V.R. Larionov, V.D. Rumyantsev, and M.Z. Shvarts. Development of pv receivers for space line-focus concentrator modules. In *Proceedings of the 25th IEEE Photovoltaic Specialists Conference*, pages 341–344, Washington, D.C., May 1996.
5. B.A. Aničin, V.M. Babović, and D.M. Davidović. Fresnel lenses. *American Journal of Physics*, 57(4):312–316, 1989.
6. K. Araki and M. Yamaguchi. Silicon concentrator cells by low cost process. In *Proceedings of the 28th IEEE Photovoltaic Specialists Conference*, Anchorage, Alaska, September 2000.
7. K. Araki, M. Yamaguchi, M. Imaizumi, S. Matsuda, T. Takamoto, and H. Kurita. AM0 concentration operation of III–IV compound solar cells. In *Proceedings of the 28th IEEE Photovoltaic Specialists Conference*, Anchorage, Alaska, September 2000.
8. K. Araki and Y. Yamaguchi. Parameter analysis on solar concentrator cells under nonuniform illumination. In *16th European Photovoltaic Conference and Exhibition*, Glasgow, United Kingdom, May 2000.
9. Array Technologies, Inc., 3312 Stanford NE, Albuquerque, New Mexico 87107, http://www.wattsun.com/. *Wattsun Solar Trackers Installation Guide Version 3.5*, November 2000.
10. ASHRAE. *Psychrometrics: Theory and Practice*. American Society of Heating, Refrigerating and Air-Conditioning Engineers, Atlanta, Georgia, 1996.
11. V. Badescu. Letter to the editor. *Solar Energy*, 61(1):61–64, 1997.
12. I.M. Bassett and R. Winston. Limits to concentration in physical optics and wave mechanics. In *Proceedings of the SPIE International Conference on Nonimaging Concentrators*, volume 441, pages 2–6, San Diego, California, August 1983.
13. A. Bejan, G. Tsatsaronis, and M. Moran. *Thermal Design and Optimization*. Wiley, New York, 1996.
14. Solare Klimatisierung. Project Info Service, BINE Informationsdienst, Mechenstraße 57, 53129 Bonn, 1998. In German.
15. W. Bloss, H.P. Hertlein, W. Knaupp, S. Nann, and F. Pfisterer. Photovoltaic power stations. In C.-J. Winter, R.L. Sizmann, and L.L. Vant-Hull, editors, *Solar Power Plants*. Springer, Berlin, 1991.

16. W.H. Bloss, M. Griesinger, and E.R. Reinhardt. Dispersive concentrating (DISCO) systems based on transmission phase holograms for solar applications. In *Proceedings of the 16th IEEE Photovoltaic Specialists Conference*, pages 463–468, 1982.
17. Power to burn: Versatile new series answers customer needs. http://www.hughesspace.com/factsheets/702/702.html, 2000. Boeing Corp. website.
18. E.C. Boes and A. Luque. Photovoltaic concentrator technology. In T.B. Johansson, H. Kelly, A.K.N. Reddy, and R.H. Williams, editors, *Renewable Energy, Sources for Fuel and Electricity.* Island Press, Washington, D.C., 1992.
19. E.A. Boettner and N.E. Barnett. Design and construction of Fresnel optics for photoelectric receivers. *Journal of the Optical Society of America*, 41(11):849–857, 1951.
20. J. Boise Pearson and M.D. Watson. Analytical study of the relationship between absorber cavity and solar Fresnel concentrator. In *Proceedings of the ASME International Solar Energy Conference*, Albuquerque, New Mexico, June 1998.
21. M. Born and E. Wolf. *Principles of Optics.* Pergamon, Oxford, 6th edition, 1989.
22. F. Bosnjakovic, K.F. Knoche, and D. Stehmeier. Exergetic analysis of ammonia/water absorption heat pumps. In R.A. Gaggioli, editor, *Proceedings of the ASME Winter Annual Meeting*, volume 3, pages 93–104, Anaheim, California, 1986.
23. V.M. Brodyansky, M.V. Sorin, and P. Le Goff, editors. *The Efficiency of Industrial Processes: Exergy Analysis and Optimization.* Elseviers Science BV, Amsterdam, the Netherlands, 1994. Editors.
24. H. Broesamle, H. Mannstein, C. Schillings, and F. Trieb. Assessment of solar electricity potentials in North Africa based on satellite data and a geographic information system. *Solar Energy*, 70(1):1–12, 2001.
25. I.N. Bronštejn and K.A. Semendjajew. *Taschenbuch der Mathematik.* Verlag Harri Deutsch, Thun und Frankfurt am Main, Germany, 23rd edition, 1987. In German.
26. J. Brophy, R. Kakuda, J. Polk, J. Anderson, M. Marcucci, D. Brinza, M. Henry, K. Fujii, K. Mantha, J. Stocky, J. Sovey, M. Patterson, V. Rawlin, J. Hamley, T. Bond, J. Christensen, H. Cardwell, G. Benson, J. Gallagher, M. Matranga, and D. Bushway. Ion propulsion validation on DS1. In *Proceedings of the Deep Space 1 Technology Validation Symposium*, Pasadena, California, February 2000.
27. M.E. Bugulov, V.U. Butaev, A.B. Pinov, E.V. Tver'yanovich, and M.K. Khadikov. Optical-energetic characteristic of glass focons. *Geliotekhnika*, 23(4):18–20, 1987. Translated into English in *Applied Solar Energy* 23(4):19–22, 1987.
28. CADDET. South East Asia's largest solar heating system at Singapore's Changi Airport. http://194.178.172.86/, 1997.
29. D.E. Carlson and S. Wagner. Amorphous silicon photovoltaic systems. In T.B. Johansson, H. Kelly, A.K.N. Reddy, and R.H. Williams, editors, *Renewable Energy, Sources for Fuel and Electricity.* Island Press, Washington, D.C., 1992.
30. G.W. Chantry and J. Chamberlain. Far infrared spectra of polymers. In A.D. Jenkins, editor, *Polymer Science.* Amsterdam, 1972.
31. N.V. Chartčenko. *Thermische Solaranlagen.* Springer, Berlin, 1995. In German.
32. M.M. Chen, J.B. Berkowitz-Mattuck, and P.E. Glaser. The use of a kaleidoskope to obtain uniform flux over a large area in a solar or arc imaging furnace. *Applied Optics*, 2(3):265–271, 1963.

33. M. Collares-Pereira. High temperature solar collector with optimal concentration: Non-focusing Fresnel lens with secondary concentrator. *Solar Energy*, 23:409–420, 1979.
34. M. Collares-Pereira and A. Rabl. The average distribution of solar radiation – correlations between diffuse and hemispherical and between daily and hourly insolation values. *Solar Energy*, 22:155–164, 1979.
35. M. Collares-Pereira, A. Rabl, and R. Winston. Lens-mirror combinations with maximal concentration. *Applied Optics*, 16(10):2677–2683, 1977.
36. R.L. Donovan, H.C. Hunter, J.T. Smith, R.L. Jones, and S. Broadbent. Ten kilowatt photovoltaic concentrating array. In *Proceedings of the 13th IEEE Photovoltaic Specialists Conference*, pages 1125–1130, 1978.
37. J.A. Duffie and W.A. Beckman. *Solar Engineering of Thermal Processes*. Wiley, New York, 2nd edition, 1991.
38. Energy Information Administration EIA. Renewable energy annual, form EIA-63B, Annual Photovoltaic Module/Cell Manufacturer's Survey. http://www.eia.doe.gov/, 2000.
39. F. Erismann. Design of plastic aspheric Fresnel lens with a spherical shape. *Optical Engineering*, 36(4):988–991, 1997.
40. M.I. Eskenazi, D.M. Murphy, E.L. Ralph, and H.I. Yoo. Testing of dual-junction SCARLET modules and cells plus lessons learned. In *Proceedings of the 27th IEEE Photovoltaic Specialists Conference*, pages 831–834, Anaheim, California, September 1997.
41. F. Franc, V. Jirka, M. Malý, and B. Nábělek. Concentrating collectors with flat linear Fresnel lenses. *Solar & Wind Technology*, 3(2):77–84, 1986.
42. W. Fratzscher, V.M. Brodjanskij, and K. Michalek. *Exergie*. Leipzig, 1986. In German.
43. Fresnel Technologies, Inc., 101 West Morningside Drive, Fort Worth, Texas 76110, USA; http://www.fresneltech.com/. *Fresnel Lenses*, 1995.
44. M. Fujiwara. Exergy analysis for the performance of solar collectors. *Transactions of the ASME, Journal of Solar Energy Engineering*, 105:163–167, 1983.
45. D.Y. Goswami, F. Kreith, and J.F. Kreider. *Principles of Solar Engineering*. Taylor & Francis, Philadelphia, Pennsylvenia, 2nd edition, 1999.
46. W. Grasse, H.P. Hertlein, and C.-J. Winter. Thermal solar power plants experience. In C.-J. Winter, R.L. Sizmann, and L.L. Vant-Hull, editors, *Solar Power Plants*. Springer, Berlin, 1991.
47. M.A. Green. Crystalline- and polycrystalline-silicon solar cells. In T.B. Johansson, H. Kelly, A.K.N. Reddy, and R.H. Williams, editors, *Renewable Energy, Sources for Fuel and Electricity*. Island Press, Washington, D.C., 1992.
48. A.W. Greynolds. Ray tracing through non-imaging concentrators. In *Proceedings of the SPIE International Conference on Nonimaging Concentrators*, volume 441, pages 10–17, San Diego, California, August 1983.
49. V.A. Grilikhes. Transfer and distribution of radiant energy in concentration systems. In V.M. Andreev, V.A. Grilikhes, and V.D. Rumyantsev, editors, *Photovoltaic Conversion of Concentrated Sunlight*. Wiley, Chichester, 1997.
50. V.A. Grilikhes, V.D. Rumyantsev, and M.Z. Shvarts. Indoor and outdoor testing of space concentrator AlGaAs/GaAs photovoltaic modules with Fresnel lenses. In *Proceedings of the 25th IEEE Photovoltaic Specialists Conference*, pages 345–348, Washington, D.C., May 1996.
51. S. Harmon. Solar-optical analyses of mass-produced plastic circular Fresnel lens. *Solar Energy*, 19:105–108, 1977.
52. E. Hecht. *Optics*. Addison-Wesley, Reading, Massachusetts, 3rd edition, 1990.
53. M. Herzberger. Some remarks on ray tracing. *Journal of the Optical Society of America*, 41(11):805–807, 1951.

54. R.H. Hildebrand. Focal plane optics in far-infrared and submilimeter astronomy. In *Proceedings of the SPIE International Conference on Nonimaging Concentrators*, volume 441, pages 40–50, San Diego, California, August 1983.
55. T. Hirose, M. Goto, A. Kodama, and T. Kuma. Thermally activated honeycomb dehumidifier for adsorption cooling system. In *Proceedings of the 1995 ASME/JSME/JSES International Solar Energy Conference*, Maui, Hawaii, March 1995.
56. P.A. Iles and Y.C.M. Yeh. Silicon, gallium arsenide, and indium phosphide cells: Single junction, one sun space. In L.D. Partain, editor, *Solar Cells and Their Applications*. Wiley, New York, 1995.
57. M. Izquierdo Millán, F. Hernández, and E. Martín. Available solar exergy in an absorption cooling process. *Solar Energy*, 56(6):505–511, 1996.
58. C.P. Jacovides, L. Hadjioannou, S. Pashiardis, and L. Stefanou. On the diffuse fraction of daily and monthly global radiation for the island of Cyprus. *Solar Energy*, 56(6):565–572, 1996.
59. L.W. James. Effects of concentrator chromatic aberration on multi-junction cells. In *Proceedings of the 1994 IEEE First World Conference on Photovoltaic Energy Conversion, and the 24th IEEE Photovoltaic Specialists Conference*, pages 1799–1802, Waikoloa, Hawaii, December 1994.
60. L.W. James and J.K. Williams. Fresnel optics for solar concentration on photovoltaic cells. In *Proceedings of the 13th IEEE Photovoltaic Specialists Conference*, pages 673–679, 1978.
61. R.W. Jans. Acrylic polymers for optical applications. In *Proceedings of the SPIE, Physical Properties of Optical Materials*, volume 204, pages 1–8, San Diego, California, August 1979.
62. Japan Air Conditioning, Heating & Refrigeration News (JARN). *World Trends of Chillers and Large Air-Conditioning Equipment*, November 1997. Special edition.
63. D.A. Jaworske, W.A. Wong, and T.J. Skowronski. Optical evaluation of a refractive secondary concentrator. Technical Memorandum NASA/TM-1999-209375, National Aeronautics and Space Administration, NASA, Lewis Research Center, Cleveland, Ohio 44135-3191; ftp://ftp-letrs.lerc.nasa.gov/LeTRS/reports/1999/TM-1999-209375.pdf, July 1999.
64. R.W. Jebens. Fresnel lens concentrator. United States Patent 4799778, 1989.
65. D. Jenkins, J. O'Gallagher, and R. Winston. Attaining and using extremely high intensities of solar energy with non-imaging concentrators. In K.W. Böer, editor, *Advances in Solar Energy*, volume 11, chapter 2. American Solar Energy Society, Boulder, Colorado, 1997.
66. F.A. Jenkins and H.E. White. *Fundamentals of Optics*. McGraw-Hill, Singapore, 4th edition, 1981.
67. J. Kamoshida. Analysis of model Fresnel collector operating in the fixed-aperture mode with seasonally tracking absorber. In *Proceedings of the JSES/JWEA Joint Conference*, pages 415–418, November 2000. In Japanese.
68. S. Kaneff. Solar thermal power – a historical, technological, and economic overview. In *Proceedings of the 34th Annual Conference, Australia and New Zealand Solar Energy Society (ANZSES)*, pages 294–306, Darwin, Northern Territories, October 1996.
69. J.H. Keenan. A steam chart for second law analysis: A study of thermodynamic availability in the steam power plant. *Mechanical Engineering*, 54:195–204, 1932.
70. D.L. King, J.A. Kratochvil, and W.E. Boyson. Measuring solar spectral and angle-of-incidence effects on photovoltaic modules and solar irradiance sensors.

In *Proceedings of the 26th IEEE Photovoltaic Specialists Conference*, Anaheim, California, September 1997.
71. M.E. Klausmeier-Brown. Status, prospects, and economics of terrestrial, single junction GaAs concentrator cells. In L.D. Partain, editor, *Solar Cells and Their Applications*. Wiley, New York, 1995.
72. T. Kouchiwa. Design of a plastic lens for copiers. In *Proceedings of the SPIE International Lens Design Conference*, volume 554, pages 419–424, Cherry Hill, New Jersey, June 1985.
73. E.A. Krasina, E.V. Tver'yanovich, and A.V. Romankevich. Optical efficiency of solar-engineering Fresnel lenses. *Geliotekhnika*, 25(6):8–12, 1989. Translated into English in *Applied Solar Energy* 25(6):8–12, 1989.
74. A. Kribus, V. Krupin, A. Yogev, and W. Spirkl. Performance limit of heliostat fields. *Transactions of the ASME, Journal of Solar Energy Engineering*, 120:240–246, 1998.
75. E.M. Kritchman. Two stage linear Fresnel lenses. *Solar Energy*, 33(1):35–39, 1984.
76. E.M. Kritchman, A.A. Friesem, and G. Yekutieli. Efficient Fresnel lens for solar concentration. *Solar Energy*, 22:119–123, 1979.
77. E.M. Kritchman, A.A. Friesem, and G. Yekutieli. Highly concentrating Fresnel lenses. *Applied Optics*, 18(15):2688–2695, 1979.
78. S.R. Kurtz, D.J. Friedman, and J.M. Olson. The effect of chromatic aberrations on two-junction, two-terminal devices in a concentrator system. In *Proceedings of the 1994 IEEE First World Conference on Photovoltaic Energy Conversion, and the 24th IEEE Photovoltaic Specialists Conference*, Waikoloa, Hawaii, December 1994.
79. S.R. Kurtz, D. Myers, and J.M. Olson. Projected performance of three- and four-junction devices using GaAs and GaInP. In *Proceedings of the 26th IEEE Photovoltaic Specialists Conference*, Anaheim, California, September 1997.
80. S.R. Kurtz and M.J. O'Neill. Estimating and controlling chromatic aberration losses for two-junction, two-terminal devices in refractive concentrator systems. In *Proceedings of the 25th IEEE Photovoltaic Specialists Conference*, pages 361–364, Washington, D.C., May 1996.
81. F.M. Legge. Solar Track Pty Ltd, 26 Surrey Road, Wilson, Western Australia 6107; personal communication.
82. G. Létay. Radiation resistant space concentrator solar cells and modules. In *Proceedings of the 2nd Mini-Workshop on Space Solar Array Concentrators*, European Space Agency, Nordwijk, The Netherlands, 2000. Unpublished.
83. R. Leutz, A. Suzuki, A. Akisawa, and T. Kashiwagi. Design of a nonimaging Fresnel lens for solar concentrators. *Solar Energy*, 65(6):379–387, 1999.
84. R. Leutz, A. Suzuki, A. Akisawa, and T. Kashiwagi. Shaped nonimaging Fresnel lenses. *Journal of Optics A: Pure and Applied Optics*, 2:112–116, 2000.
85. J.L. Lindsey. *Applied Illumination Engineering*. The Fairmont Press, Lilburn, Georgia, 2nd edition, 1996.
86. G. Löf. *Active Solar Systems*, pages xi–xvi. Solar Heat Technologies: Fundamentals and Applications. MIT Press, Cambridge, Massachusetts, 1992.
87. G.O.G. Löf. Desiccant systems. In A.A.M. Sayigh and J.C. McVeigh, editors, *Solar Air Conditioning and Refrigeration*. Oxford, 1992.
88. W. Lorenz. Design guidelines for a glazing with seasonally dependent solar transmittance. *Solar Energy*, 63(2):79–96, 1998.
89. E. Lorenzo. Chromatic aberration effect on solar energy systems using Fresnel lenses. *Applied Optics*, 20(21):3729–3732, 1981.

90. E. Lorenzo and A. Luque. Fresnel lens analysis for solar energy applications. *Applied Optics*, 20(17):2941–2945, 1981.
91. E. Lorenzo and A. Luque. Comparison of Fresnel lenses and parabolic mirrors as solar energy concentrators. *Applied Optics*, 21(10):1851–1853, 1982.
92. W. Luft. Concentrator solar arrays for space. In *Proceedings of the 12th IEEE Photovoltaic Specialists Conference*, pages 456–461, 1976.
93. A. Luque. *Solar Cells and Optics for Photovoltaic Concentration.* Adam Hilger, Bristol, 1989.
94. A. Luque and E. Lorenzo. Conditions of achieving ideal and Lambertian symmetrical solar concentrators. *Applied Optics*, 21(20):3736–3738, 1982.
95. L.S. Mason. Technology projections for solar dynamic power. Technical Memorandum NASA/TM-1999-208851, National Aeronautics and Space Administration, NASA, Lewis Research Center, Cleveland, Ohio 44135-3191; ftp://ftp-letrs.lerc.nasa.gov/LeTRS/reports/1999/TM-1999-208851.pdf, January 1999.
96. E.L. Maxwell. A quasi-physical model for converting hourly global horizontal to direct normal insolation. Report SERI/TR-215-3087 DE, Solar Energy Research Institute, 1617 Cole Boulevard, Golden, Colorado 80401-3393, August 1987.
97. J.A. McGovern. Exergy analysis: A different perspective on energy, part i and ii. *Journal of Power and Energy*, 204:253–268, 1990.
98. J. McVeigh, D. Burtraw, J. Darmstadter, and K. Palmer. Winner, loser or innocent victim: Has renewable energy performed as expected? Research report 7, Renewable Energy Research Project (REPP), http://www.repp.org/, 1999.
99. A.B. Meinel and M.P. Meinel. *Applied Solar Energy – An Introduction.* Addison-Wesley, Reading, Massachusetts, 1976.
100. Global Meteorological Database Version 3.06. Meteotest, Fabrikstraße 14, 3012 Bern, Switzerland, http://www.meteotest.ch/, 1998.
101. D.R. Meyers, S.R. Kurtz, K. Emery, C. Whitaker, and T. Townsend. Outdoor meteorological broadband and spectral conditions for evaluating photovoltaic modules. In *Proceedings of the 28th IEEE Photovoltaic Specialists Conference*, Anchorage, Alaska, September 2000.
102. J.C. Miñano, P. Benítez, and J.C. González. RX: A nonimaging concentrator. *Applied Optics*, 34(13):2226–2235, 1995.
103. J.C. Miñano, P. Benítez, and M. Hernández. Space applications of the SMS concentrator design method. In *Proceedings of the 2nd Mini-Workshop on Space Solar Array Concentrators*, European Space Agency, Nordwijk, The Netherlands, 2000. Unpublished.
104. J.C. Miñano, J.C. González, and P. Benítez. A high-gain, compact, nonimaging concentrator: RXI. *Applied Optics*, 34(34):7850–7856, 1995.
105. M. Mijatović, D. Dimitrovski, and V. Veselinović. Fresnel lens-absorber system with uniform concentration and normal incoming rays to the absorber. *Journal of Optics (Paris)*, 18(5-6):261–264, 1987.
106. A.R. Miller. *Turbo Pascal Programs for Scientists and Engineers.* San Francisco, California, 1987.
107. O.E. Miller, J.H. McLeod, and W.T. Sherwood. Thin sheet plastic Fresnel lenses of high aperture. *Journal of the Optical Society of America*, 41(11):807–815, 1951.
108. D.R. Mills and G.L. Morrison. Compact linear Fresnel reflector solar thermal powerplants. *Solar Energy*, 68(3):263–283, 2000.
109. D.G. Mitchell. The space environment. In V.L. Pisacane and R.C. Moore, editors, *Fundamentals of Space Systems.* Oxford University Press, New York, 1994.

110. A.L. Moffat and R.S. Scharlack. The design and development of a high concentration and high efficiency photovoltaic concentrator utilizing a curved Fresnel lens. In *Proceedings of the 16th IEEE Photovoltaic Specialists Conference*, pages 601–606, 1982.
111. R.L. Moon, L.W. James, H.A. Vander Plas, T.O. Yep, G.A. Antypas, and Y. Chai. Multigap solar cell requirements and the performance of AlGaAs and Si cells in concentrated sunlight. In *Proceedings of the 13th IEEE Photovoltaic Specialists Conference*, pages 859–867, 1978.
112. K. Mori. Himawari (sunflower). La Foret Engineering Information Service Ltd, Himawari Building, 2-7-8 Toranomon, Tokyo, Japan, http://www.himawari-net.co.jp/, 1985. Personal Communication in Japanese.
113. A. Mouchot. *La Chaleur Solaire et ses Applications Industrielles.* Gauthier-Villars, Paris, 2nd edition, 1879. German translation F. Griese: Die Sonnenwärme und ihre industriellen Anwendungen, Olynthus Verlag Oberbözberg, Switzerland, 1987.
114. D.M. Murphy. Scarlet solar array: Deep Space 1 flight validation. In *Proceedings of the Deep Space 1 Technology Validation Symposium*, Pasadena, California, February 2000.
115. H. Naito, T. Hoshino, T. Fujihara, K. Eguchi, K. Tanaka, A. Yamada, and Y. Ohtani. Ground-engineering study on solar HP/TES receiver for future ISS-JEM experiment program. In *Proceedings of the 34th International Energy Conversion Engineering Conference*, 1999. Paper 1999-01-2587.
116. Y. Nakata. Fresnel-type aspheric prism lens. United States Patent 4904069, 1990.
117. Y. Nakata, N. Shibuya, T. Kobe, K. Okamoto, A. Suzuki, and T. Tsuji. Performance of circular Fresnel lens photovoltaic concentrator. *Japanese Journal of Applied Physics*, 19(Supplement 19-2):75–78, 1979. Proceedings of the 1st Photovoltaic Science and Engineering Conference in Japan.
118. Photovoltaic array assembly PVAA. http://www.shuttlepresskit.com/STS-97/payloads.htm, 2000. NASA-website.
119. Federal Climate Complex Global Surface Summary of Day Data Version 5. http://www.ncdc.noaa.gov/, 2000.
120. National Renewable Energy Laboratory (NREL). Resource assessment program. http://rredc.nrel.gov/solar/old_data/nsrdb/redbook/atlas/, 1994. In printed form available as Solar Radiation Data Manual for Flat-Plate and Concentrating Collectors, NREL, 1617 Cole Boulevard, Golden, Colorado 80401-3393.
121. D.T. Nelson, D.L. Evans, and R.K. Bansal. Linear Fresnel lens concentrators. *Solar Energy*, 17:285–289, 1975.
122. Nihon Fresnel Ltd., Tokyo, 1999. Personal communication.
123. Nineplanets. A multimedia tour of the solar system. http://seds.lpl.arizone.edu/nineplanets/nineplanets/, 1997.
124. J. Noring, D. Grether, and A. Hunt. Circumsolar radiation data: The Lawrence Berkeley Laboratory reduced data base. Report TP-262-44292, National Renewable Energy Laboratory (NREL), 1617 Cole Boulevard, Golden, Colorado 80401-3393, http://rredc.nrel.gov/solar/pubs/circumsolar/, 1991.
125. F.J. Olmo, F.J. Battles, and L. Alados-Arboledas. Performance of global to direct/diffuse decomposition models before and after the eruption of Mt. Pinatubo, June 1991. *Solar Energy*, 57(6):433–443, 1996.
126. M.J. O'Neill. Solar concentrator and energy collection system. United States Patent 4069812, 1978.
127. M.J. O'Neill. Bi-focussed solar energy concentrator. United States Patent 4545366, 1985.

128. M.J. O'Neill. Fourth-generation, line-focus, Fresnel lens photovoltaic concentrator. In *Proceedings of the Fourth Sunshine Workshop on Crystalline Silicon Solar Cells*, pages 153–160, Chiba, Japan, November 1992.
129. M.J. O'Neill. Color-mixing lens for solar concentrator system and methods of manufacture and operation thereof. United States Patent 6031179, 2000.
130. M.J. O'Neill. Stretched Fresnel lens solar concentrator for space power. United States Patent 6075200, 2000.
131. M.J. O'Neill and A.J. McDanal. Fourth-generation concentrator system: From the lab to the factory to the field. In *Proceedings of the 1994 IEEE First World Conference on Photovoltaic Energy Conversion, and the 24th IEEE Photovoltaic Specialists Conference*, volume 1, pages 816–819, Waikoloa, Hawaii, December 1994.
132. M.J. O'Neill and M.F. Piszczor. Inflatable lenses for space photovoltaic concentrator arrays. In *Proceedings of the 26th IEEE Photovoltaic Specialists Conference*, Anaheim, California, September 1997.
133. M.J. O'Neill, M.F. Piszczor, M.I. Eskenazi, C. Carrington, and H.W. Brandhorst. The stretched lens ultralight concentrator array. In *Proceedings of the 28th IEEE Photovoltaic Specialists Conference*, Anchorage, Alaska, September 2000.
134. I. Oshida. Elephant nose. Invention is manufactured by Sanyo Ltd., Japan. Personal communication.
135. I. Oshida. Step lenses and step prisms for utilization of solar energy. In *New Sources of Energy, Proceedings of the Conference, United Nations*, volume 4, pages S/22, 598–603, Rome, Italy, August 1961.
136. I. Oshida. *Lecture on Exergy.* Research Institute for Solar Energy, Tokyo, Japan, 1986.
137. L.D. Partain. Solar cell fundamentals. In L.D. Partain, editor, *Solar Cells and Their Applications.* Wiley, New York, 1995.
138. R.A. Pethrick. *Polymer Yearbook 8.* Chur, Switzerland, 1991.
139. R.B. Pettit. Characterization of the reflected beam profile of solar mirror materials. *Solar Energy*, 19:733–741, 1977.
140. V. Poulek, 1999. Poulek Solar Ltd., Kastanova, 25001 Brandys n.L., Prague East District, Czech Republic; Personal communication.
141. V. Poulek and M. Libra. New solar tracker. *Solar Energy Materials and Solar Cells*, 51:113–120, 1998.
142. A. Rabl. Optical and thermal properties of Compound Parabolic Concentrators. *Solar Energy*, 18:497–511, 1976.
143. A. Rabl. *Active Solar Collectors and Their Applications.* Oxford, New York, 1985.
144. D.S. Renné, R. Perez, A. Zelenka, C. Whitlock, and R. DiPasquale. Recent advances in assessing solar resources over large areas. In *Proceedings of the 28th IEEE Photovoltaic Specialists Conference*, Anchorage, Alaska, September 2000.
145. J.L. Rhatigan, E.L. Christiansen, and M.L. Fleming. On protection of Freedom's solar dynamic radiator from the orbital debris environment: Part I – preliminary analysis and testing. *Transactions of the ASME Journal of Solar Energy Engineering*, 114:135–141, 1992.
146. H. Ries, J.M. Gordon, and M. Lasken. High-flux photovoltaic solar concentrators with kaleidoskope-based optical designs. *Solar Energy*, 60(1):11–16, 1997.
147. R.J. Roman, J.E. Peterson, and D.Y. Goswami. An off-axis Cassegrain optimal design for short focal length parabolic solar concentrators. *Transactions of the ASME, Journal of Solar Energy Engineering*, 117:51–56, 1995.

148. M.C. Ruda. How and when to use a nonimaging concentrator. In *Proceedings of the SPIE International Conference on Nonimaging Concentrators*, volume 441, pages 51–58, San Diego, California, August 1983.
149. V.D. Rumyantsev. Fundamentals of photovoltaic conversion of concentrated sunlight. In V.M. Andreev, V.A. Grilikhes, and V.D. Rumyantsev, editors, *Photovoltaic Conversion of Concentrated Sunlight*. Wiley, Chichester, 1997.
150. V.D. Rumyantsev, V.M. Andreev, A.W. Bett, F. Dimroth, M. Hein, G. Lange, M.Z. Shvarts, and O.V. Sulima. Progress in development of all-glass terrestrial concentrator modules based on composite Fresnel lenses and III–IV solar cells. In *Proceedings of the 28th IEEE Photovoltaic Specialists Conference*, pages 1169–1172, Anchorage, Alaska, September 2001.
151. T. Saitō and K. Yoshioka. Preparation and properties of photovoltaic static concentrators. In A.A.M. Sayigh, editor, *Proceedings of the World Renewable Energy Congress V*, volume 1, pages 566–571, Florence, Italy, September 1998. Pergamon.
152. K. Sakuta, S. Sawata, and M. Tanimoto. Luminescent concentrator module: Improvement in cell coupling. In *Proceedings of the JSES/JWEA Joint Conference*, pages 85–88, November 1995. In Japanese.
153. G. Sala, J.C. Arboiro, A. Luque, J.C. Zamorano, J.C. Miñano, C. Dramsch, T. Bruton, and D. Cunningham. The EUCLIDES prototype: An efficient parabolic trough for pv concentration. http://www.users.globalnet.co.uk/~blootl/trackers/eucl.html, 1999.
154. A.A.M. Sayigh and J.C. McVeigh. *Solar Air Conditioning and Refrigeration*. Oxford, 1992.
155. Schott Computer Glaskatalog version 1.0. Schott Glaswerke, Mainz, 1992.
156. M. Schubnell. Sunshape and its influence on the flux distribution in imaging solar concentrators. *Transactions of the ASME, Journal of Solar Energy Engineering*, 114:260–266, 1992.
157. R.R. Shannon. *The Art and Science of Optical Design*. Cambridge University Press, Cambridge, 1997.
158. N.F. Shepard and T.S. Chan. The design and performance of a point-focus concentrator module. In *Proceedings of the 15th IEEE Photovoltaic Specialists Conference*, pages 336–341, 1981.
159. R.A. Sinton. Terrestrial silicon concentrator solar cells. In L.D. Partain, editor, *Solar Cells and Their Applications*. Wiley, New York, 1995.
160. R. Sizmann. Solar radiation conversion. In C.-J. Winter, R.L. Sizmann, and L.L. Vant-Hull, editors, *Solar Power Plants*. Springer, Berlin, 1991. With contributions of P. Köpke and R. Busen.
161. W.E. Skullney. Spacecraft configuration and structural design. In V.L. Pisacane and R.C. Moore, editors, *Fundamentals of Space Systems*. Oxford University Press, New York, 1994.
162. S.J. Slayzak, A.A. Pesaran, and C.E. Hancock. Experimental evaluation of commercial desiccant dehumidifier wheels. Technical report, National Renewable Energy Laboratory (NREL), 1617 Cole Boulevard, Golden, Colorado 80401-3393, http://www.nrel.gov/, 1998.
163. Solar-concentrator discussion group on the internet, 2000. Archives at ftp://ftp.scruz.net/users/cichlid/public/mailing-list-archives/.
164. A.A. Soluyanov and V.A. Grilikhes. A method for designing Fresnel lenses as solar radiation concentrators. *Geliotekhnika*, 29(5):48–53, 1993. Translated into English in *Applied Solar Energy* 29(5):57–62, 1993.
165. W. Spirkl, H. Ries, J. Muschawek, and R. Winston. Nontracking solar concentrators. *Solar Energy*, 62:113–120, 1998.

166. C.B. Stillwell and B.D. Shafer. Compatibility of Fresnel lenses and photovoltaic cells in concentrator modules. In *Proceedings of the 15th IEEE Photovoltaic Specialists Conference*, pages 160–164, 1981.
167. R. Stribling. The HS 702 concentrator solar array. In *Proceedings of the 28th IEEE Photovoltaic Specialists Conference*, Anchorage, Alaska, September 2000.
168. R.L. Sullivan. Space power systems. In V.L. Pisacane and R.C. Moore, editors, *Fundamentals of Space Systems*. Oxford University Press, New York, 1994.
169. A. Suzuki. General theory of exergy balance analysis and an application to solar collectors. *Energy*, 13:153–160, 1988.
170. A. Suzuki and S. Kobayashi. Yearly distributed insolation model and optimum design of a two dimensional compound parabolic concentrator. *Solar Energy*, 54(5):327–331, 1995.
171. W. Szulmayer. A solar strip concentrator. *Solar Energy*, 14:327–335, 1973.
172. A. Terao, W.P. Mulligan, S.G. Daroczi, O.C. Pujol, P.J. Verlinden, R.M. Swanson, J.C. Miñano, P. Benítez, and J.L. Alvarez. A mirror-lens design for micro-concentrator modules. In *Proceedings of the 28th IEEE Photovoltaic Specialists Conference*, Anchorage, Alaska, September 2000.
173. T.D. Thompson. Space log 1997–1998, volume 33–34. Published by TRW Space & Electronics Group, Editor, TRW Space Log, TRW Space & Electronics Group, One Space Park, Mail Station E2/3014, Redondo Beach, CA 90278, 1999.
174. C.M. Tolbert. Inflatable solar thermal concentrator. http://wwww.grc.nasa.gov/WWW/tmsb/concentrators/doc/inflatable.html, December 1999. NASA website.
175. E.V. Tver'yanovich. Profiles of solar-engineering Fresnel lenses. *Geliotekhnika*, 19(6):31–34, 1984. Translated into English in *Applied Solar Energy* 19(6):36–39, 1984.
176. Unknown Author. How to mass-produce inexpensive, high grade focusing lenses (continuation, UN memo). Two-page summary including 2 figures and 1 equation, 1999. Posted until early January 1999 at http://www.xs4all.nl/~solomon/05sam.html; other sources unknown.
177. L.L. Vant-Hull. Concentrator optics. In C.-J. Winter, R.L. Sizmann, and L.L. Vant-Hull, editors, *Solar Power Plants*. Springer, Berlin, 1991.
178. J. Walker. Home Planet. Public domain program for Windows, http://www.fourmilab.ch/, 1997.
179. G. Wall. *Exergy: A Useful Concept.* Ph.D. thesis, Chalmers University of Technology, Göteborg, Sweden, 1986. http://exergy.se/goran/thesis/.
180. W.T. Welford. Connections and transitions between imaging and nonimaging optics. In *Proceedings of the SPIE International Conference on Nonimaging Concentrators*, volume 441, pages 7–9, San Diego, California, August 1983.
181. W.T. Welford and R. Winston. *The Optics of Nonimaging Concentrators.* Academic Press, New York, 1978.
182. W.T. Welford and R. Winston. *High Collection Nonimaging Optics.* Academic Press, San Diego, California, 1989.
183. G.R. Whitfield, R.W. Bentley, C.K. Weatherby, A. Hunt, H.-D. Mohring, F.H. Klotz, P. Keuber, J.C. Miñano, and E. Alarte-Garvi. The development and testing of small concentrating pv systems. In *Proceedings of the ISES Solar World Congress*, Jerusalem, Israel, July 1999.
184. J.A. Wiebelt and J.B. Henderson. Selected ordinates for total solar radiation property evaluation from spectral data. *Transactions of the ASME, Journal of Heat Transfer*, 101:101–107, 1979.

185. S. Wieder. *An Introduction to Solar Energy for Scientists and Engineers.* New York, 1982.
186. R. Winston. Principles of solar collectors of a novel design. *Solar Energy*, 16:89–95, 1974.
187. V. Wittwer, K. Heidler, A. Zastrow, and A. Goetzberger. Efficiency and stability of experimental fluorescent planar concentrators (FPC). In *Proceedings of the 14th IEEE Photovoltaic Specialists Conference*, pages 760–764, 1980.
188. E. Wolf, editor. *Progress in Optics, Volume XXVII.* North-Holland, Amsterdam, The Netherland, 1989.
189. World Radiation Data Center (WRDC), 2000. Maintained by the World Meteorological Organisation (WMO), http://wrdc-mgo.nrel.gov/.
190. Y. Yatabe. On Fresnel lenses. *Taiyō Enerugie (Solar Energy)*, 4(3):19–24, 1978. In Japanese.
191. K. Yoshioka, K. Endō, M. Kobayashi, A. Suzuki, and T. Saitō. Design and properties of a refractive static concentrator module. *Solar Energy Materials and Solar Cells*, 34:125–131, 1994.
192. K. Zweibel and A.M. Barnett. Polycrystalline thin-film photovoltaics. In H. Kelly, A.K.N. Reddy, and R.H. Williams, editors, *Renewable Energy, Sources for Fuel and Electricity.* Island Press, Washington, D.C., 1992.

Index

Springer Series in

OPTICAL SCIENCES

New editions of volumes prior to volume 60

1 **Solid-State Laser Engineering**
By W. Koechner, 5th revised and updated ed. 1999, 472 figs., 55 tabs., XII, 746 pages

14 **Laser Crystals**
Their Physics and Properties
By A. A. Kaminskii, 2nd ed. 1990, 89 figs., 56 tabs., XVI, 456 pages

15 **X-Ray Spectroscopy**
An Introduction
By B. K. Agarwal, 2nd ed. 1991, 239 figs., XV, 419 pages

36 **Transmission Electron Microscopy**
Physics of Image Formation and Microanalysis
By L. Reimer, 4th ed. 1997, 273 figs. XVI, 584 pages

45 **Scanning Electron Microscopy**
Physics of Image Formation and Microanalysis
By L. Reimer, 2nd completely revised and updated ed. 1998,
260 figs., XIV, 527 pages

Published titles since volume 60

60 **Holographic Interferometry in Experimental Mechanics**
By Yu. I. Ostrovsky, V. P. Shchepinov, V. V. Yakovlev, 1991, 167 figs., IX, 248 pages

61 **Millimetre and Submillimetre Wavelength Lasers**
A Handbook of cw Measurements
By N. G. Douglas, 1989, 15 figs., IX, 278 pages

62 **Photoacoustic and Photothermal Phenomena II**
Proceedings of the 6th International Topical Meeting, Baltimore, Maryland,
July 31 - August 3, 1989
By J. C. Murphy, J. W. Maclachlan Spicer, L. C. Aamodt, B. S. H. Royce (Eds.),
1990, 389 figs., 23 tabs., XXI, 545 pages

63 **Electron Energy Loss Spectrometers**
The Technology of High Performance
By H. Ibach, 1991, 103 figs., VIII, 178 pages

64 **Handbook of Nonlinear Optical Crystals**
By V. G. Dmitriev, G. G. Gurzadyan, D. N. Nikogosyan,
3rd revised ed. 1999, 39 figs., XVIII, 413 pages

65 **High-Power Dye Lasers**
By F. J. Duarte (Ed.), 1991, 93 figs., XIII, 252 pages

66 **Silver-Halide Recording Materials**
for Holography and Their Processing
By H. I. Bjelkhagen, 2nd ed. 1995, 64 figs., XX, 440 pages

67 **X-Ray Microscopy III**
Proceedings of the Third International Conference, London, September 3-7, 1990
By A. G. Michette, G. R. Morrison, C. J. Buckley (Eds.), 1992, 359 figs., XVI, 491 pages

68 **Holographic Interferometry**
Principles and Methods
By P. K. Rastogi (Ed.), 1994, 178 figs., 3 in color, XIII, 328 pages

69 **Photoacoustic and Photothermal Phenomena III**
Proceedings of the 7th International Topical Meeting, Doorwerth, The Netherlands,
August 26-30, 1991
By D. Bicanic (Ed.), 1992, 501 figs., XXVIII, 731 pages

Springer Series in

OPTICAL SCIENCES

70 **Electron Holography**
By A. Tonomura, 2nd, enlarged ed. 1999, 127 figs., XII, 162 pages

71 **Energy-Filtering Transmission Electron Microscopy**
By L. Reimer (Ed.), 1995, 199 figs., XIV, 424 pages

72 **Nonlinear Optical Effects and Materials**
By P. Günter (Ed.), 2000, 174 figs., 43 tabs., XIV, 540 pages

73 **Evanescent Waves**
From Newtonian Optics to Atomic Optics
By F. de Fornel, 2001, 277 figs., XVIII, 268 pages

74 **International Trends in Optics and Photonics**
ICO IV
By T. Asakura (Ed.), 1999, 190 figs., 14 tabs., XX, 426 pages

75 **Advanced Optical Imaging Theory**
By M. Gu, 2000, 93 figs., XII, 214 pages

76 **Holographic Data Storage**
By H.J. Coufal, D. Psaltis, G.T. Sincerbox (Eds.), 2000
228 figs., 64 in color, 12 tabs., XXVI, 486 pages

77 **Solid-State Lasers for Materials Processing**
Fundamental Relations and Technical Realizations
By R. Iffländer, 2001, 230 figs., 73 tabs., XVIII, 350 pages

78 **Holography**
The First 50 Years
By J.-M. Fournier (Ed.), 2001, 266 figs., XII, 460 pages

79 **Mathematical Methods of Quantum Optics**
By R.R. Puri, 2001, 13 figs., XIV, 285 pages

80 **Optical Properties of Photonic Crystals**
By K. Sakoda, 2001, 95 figs., 28 tabs., XII, 223 pages

81 **Photonic Analog-to-Digital Conversion**
By B.L. Shoop, 2001, 259 figs., 11 tabs., XIV, 330 pages

82 **Spatial Solitons**
By S. Trillo, W.E. Torruellas (Eds), 2001, 194 figs., 7 tabs., XVIII, 445 pages

83 **Nonimaging Fresnel Lenses**
Design and Performance of Solar Concentrators
By R. Leutz, A. Suzuki, 2001, 139 figs., 44 tabs., XII, 272 pages

Printing (Computer to Film): Saladruck Berlin
Binding: Stürtz AG, Würzburg